[redacted]

[redacted]

[redacted]

This book is concerned with the theory of unbounded derivations in C*-algebras, a subject whose study was motivated by questions in quantum physics and statistical mechanics, and to which the author has made a considerable contribution. This is an active area of research, and one of the most ambitious aims of the theory is to develop quantum statistical mechanics within the framework of the C*-theory. The presentation, which is based on lectures given by the author in Newcastle upon Tyne and Copenhagen, concentrates on topics from quantum statistical mechanics and differentiations on manifolds. One of the goals is to formulate the absence theorem of phase transitions in its most general form within the C* setting. For the first time, the author constructs, within that global setting, derivations for a fairly wide class of interacting models, and presents a new axiomatic treatment for the construction of time evolutions and KMS states.

The wealth of new insight offered here will make the book essential reading for graduate students and professionals working in operator algebras, mathematical physics and functional analysis.

ENCYCLOPEDIA OF MATHEMATICS AND ITS APPLICATIONS

EDITED BY G.-C. ROTA

Volume 41

Operator algebras in dynamical systems

ENCYCLOPEDIA OF MATHEMATICS AND ITS APPLICATIONS

1 Luis A. Santalo *Integral geometry and geometric probability*
2 George E. Andrews *The theory of partitions*
3 Robert J. McEliece *The theory of information and coding: a mathematical framework for communication*
4 Wallard Miller, Jr *Symmetry and separation of variables*
5 David Ruelle *Thermodynamic formalism: the mathematical structures of classical equilibrium statistical mechanics*
6 Henryk Minc *Permanents*
7 Fred S. Roberts *Measurement theory with applications to decisionmaking, utility, and the social sciences*
8 L.C. Biedenham and J.D. Louck *Angular momentum in quantum physics: theory and application*
9 L.C. Biedenham and J.D. Louck *The Racah–Wigner algebra in quantum theory*
10 W. Dollard and Charles N. Friedman *Product integration with application to differential equations*
11 William B. Jones and W.J. Thron *Continued fractions: analytic theory and applications*
12 Nathaniel F.G. Martin and James W. England *Mathematical theory of entropy*
13 George A. Baker, Jr and Peter R. Graves-Morris *Padé approximants, Part I: Basic theory*
14 George A. Baker, Jr and Peter R. Graves-Morris *Padé approximants, Part II: Extensions and applications*
15 E.C. Beltrametti and G. Cassinelli *The logic of quantum mechanics*
16 G.D. James and A. Kerber *The representation theory of the symmetric group*
17 M. Lothaire *Combinatorics on words*
18 H.O. Fattorini *The Cauchy problem*
19 G.G. Lorentz, K. Jetter, and S.D. Riemenschneider *Birkhoff interpolation*
20 Rudolf Lidl and Harald Niederreiter *Finite fields*
21 William T. Tutte *Graph theory*
22 Julio R. Bastida *Field extensions and Galois theory*
23 John R. Cannon *The one-dimensional heat equation*
24 Stan Wagon *The Banach–Tarski paradox*
25 Arto Salomaa *Computation and automata*
26 Neil White (ed) *Theory of matroids*
27 N.H. Bingham, C.M. Goldie & J.L. Teugels *Regular variation*
28 P.P. Petrushev & V.A. Popov *Rational approximation of real functions*
29 Neil White (ed) *Combinatorial geometries*
30 M. Pohst and H. Zassenhaus *Algorithmic algebraic number theory*
31 J. Aczel & J. Dhombres *Functional equations containing several variables*
32 Marek Kuczma, Bogden Chozewski & Roman Ger *Iterative functional equations*
33 R.V. Ambartzumian *Factorization calculus and geometric probability*
34 G. Gripenberg, S.-O. Londen and O. Staffans *Volterra integral and functional equations*
35 George Gasper & Mizan Rahman *Basic hypergeometric series*
36 Erik Torgersen *Comparison of statistical experiments*
37 Arnold Neumaier *Interval methods for systems of equations*
38 N. Korneichuk *Exact constants in approximation theory*
41 Shôichirô Sakai *Operator algebras in dynamical systems: the theory of unbounded derivations in C*-algebras*

ENCYCLOPEDIA OF MATHEMATICS AND ITS APPLICATIONS

Operator algebras in dynamical systems

The theory of unbounded derivations in C*-algebras

SHÔICHIRÔ SAKAI
Department of Mathematics
College of Humanities and Sciences
Nihon University, Tokyo, Japan

The right of the University of Cambridge to print and sell all manner of books was granted by Henry VIII in 1534. The University has printed and published continuously since 1584.

CAMBRIDGE UNIVERSITY PRESS
Cambridge
New York Port Chester
Melbourne Sydney

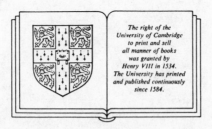

Published by the Press Syndicate of the University of Cambridge
The Pitt Building, Trumpington Street, Cambridge CB2 1RP
40 West 20th Street, New York, NY 10011-4211, USA
10 Stamford Road, Oakleigh, Melbourne 3166, Australia

First published 1991

Printed in Great Britain at the University Press, Cambridge

British Library cataloguing in publication data

Sakai, Shôichirô
Operator algebras in dynamical systems.
1. Operator algebras
I. Title
512.55

Library of Congress cataloguing in publication data

Sakai, Shôichirô
Operator algebras in dynamical systems: the theory of unbounded derivations in C*-algebras/Shôichirô Sakai.
p. cm. – (Encyclopedia of mathematics and its applications; v. 41)
Includes bibliographical references.
ISBN 0 521 40096 1
1. C*-algebras. 2. Differentiable dynamical systems. 3. Harmonic analysis. 4. Operator theory. I. Title. II. Series: Encyclopedia of mathematics and its applications; v. 39.
QA326.S26 1991
512′.55—dc20 90-48013 CIP

ISBN 0 521 40096 1 hardback

CONTENTS

PREFACE

Derivations appeared for the first time at a fairly early stage in the young field of C*-algebras, and their study continues to be one of the central branches in the field. During the past four decades, the study of derivations has made great strides. Their theory divides naturally into two major parts: bounded derivations and unbounded derivations. About thirty years ago, Kaplansky (in an excellent survey [97] on derivations) brought together two apparently unrelated results which stimulated research on continuous derivations. The first, related to quantum mechanics and due to Wielandt [190] in 1949, proved that the commutation relation $ab - ba = 1$ cannot be realized by bounded operators. The second, involving differentiation and due to Šilov [180] in 1947, proved that if a Banach algebra A of continuous functions on the unit interval contains all infinitely differentiable functions, then A contains all n-times differentiable functions for some n.

It is noteworthy that Kaplansky's observations three decades ago are still applicable to recent developments in the theory of unbounded derivations, although one has to replace quantum mechanics by all of quantum physics. Furthermore, the work of Šilov continues to have a strong influence on the study of unbounded derivations in commutative C*-algebras.

At an early stage, mathematicians devoted most of their efforts to the study of bounded derivations, rather than unbounded ones, even though the work of Šilov and Wielandt had already suggested the importance of unbounded derivations. This, of course, is understandable because bounded derivations are more easily handled than unbounded ones. The study of bounded derivations has led to a beautiful mathematical theory that provides (among other things) essential tools for the study of unbounded derivations. In contrast to the study of bounded derivations, which is now approaching completion, the study of unbounded derivations is still in an early stage. Their study began much later and initially was motivated by the problem of constructing the dynamics in statistical mechanics. It soon became apparent that the work of Šilov was also important in the study of unbounded drivations in commutative C*-algebras.

The contents of the present book are based in large part on the author's lecture notes on the theory of unbounded derivations in C*-algebras given at the Universities of Copenhagen and Newcastle upon Tyne in 1977 [170]; also included are some of the extensive new developments discovered since 1977. Unbounded derivations is a fast growing field and the subjects it involves are quite diverse. Therefore I have made no attempt to give complete coverage of the theory. Rather, I have taken a somewhat personal stand on the selection of material. This selection is concentrated, for the most part, on the topics involving quantum statistical mechanics and differentiations on manifolds.

There is a good possibility that the theory of quantum lattice systems in statistical mechanics may also be developed naturally within the context of unbounded derivations in C*-algebras (although its phase transition has not yet been established even for the three-dimensional Heisenberg ferromagnet). In fact, many theorems in the theory of quantum lattice systems have already been formulated for normal *-derivations in UHF algebras. One of the most ambitious programs in the theory of unbounded derivations is to develop quantum statistical mechanics within the C*-theory. Of particular importance is the generalization of phase transition theory to quantum lattice systems. The prospect of success is not necessarily gloomy, because the absence theorem for the phase transition theory (in one-dimensional quantum lattice systems with bounded surface energy) has already been successfully formulated in the quite general setting of UHF algebras ([5], [104], [168], [169]). This 'absence theorem' refers to a result where, in a restricted context, the temperature states can be shown to be unique. Hence the 'absence' of phase transitions. One of the goals of the present book will be to formulate the absence theorem for phase transitions in its most general form within the C*-setting. Compared with the theory of quantum lattice systems, the theory of continuous quantum systems is somewhat incomplete. Except for a few examples, time evolutions have never been constructed for interesting interaction models. As a matter of fact, even derivations have never been constructed globally.

In the present book, we shall construct derivations globally for a fairly wide class of interaction models in the C*-setting. Also, an axiomatic treatment of the construction of time evolutions and KMS states will be attempted. However, it is not thoroughly discussed, and much is left to future research.

A few words concerning the theory of unbounded derivations in commutative C*-algebras: I have not attempted to give complete coverage of recent important developments. (The reason for this is that the results are so numerous that it would be more suitable to publish another book concentrated on the subject.) Instead, I have included comments on matters related to transformation groups on locally euclidean spaces because I believe it may turn out to be one of the most important problems in the subject for future research.

With respect to non-commutative manifolds, there is an excellent book by Bratteli [19], so I will not discuss them in the present work.

References to theorems refer to the section number for example Theorem 5.2.2 refers to the theorem in Section 5.2.2.

The author expresses his appreciation of the friendly care with which Professor P.E.T. Jorgensen has read the manuscript, of many valuable suggestions he made, and of his help in the proof reading.

The author acknowledges also the invaluable assistance of Dr David Tranah and Ms Frances Fawkes, Cambridge University Press.

Shôichirô Sakai

Sendai, Japan

1

Preliminaries

In this chapter we shall state some fundamental facts used in this book. We shall also use elementary facts on C*-algebra and W*-algebras freely. The reader can find the proofs of them in the author's book [165] or other standard text books [41], [42], [139], [193], [194], [195], [196]. In addition we shall use some common notation without definition. The reader can find the appropriate definitions in [165].

1.1 Banach algebras, C*-algebras and W*-algebras

Let $\mathscr{A}$ be a linear associative algebra over the complex numbers. The algebra $\mathscr{A}$ is called a normed algebra if there is associated to each element x a real number $\|x\|$, called the norm of x, with the properties:

(1) $\|x\| \geqslant 0$, and $\|x\| = 0$ if and only if $x = 0$;
(2) $\|x + y\| \leqslant \|x\| + \|y\|$;
(3) $\|\lambda x\| = |\lambda| \|x\|$, where λ is a complex number;
(4) $\|xy\| \leqslant \|x\| \|y\|$.

If $\mathscr{A}$ is complete with respect to the norm (i.e., if $\mathscr{A}$ is also a Banach space), then it is called a Banach algebra. A mapping $x \mapsto x^*$ of $\mathscr{A}$ onto $\mathscr{A}$ itself is called an involution if it satisfies the following conditions:

(1) $(x^*)^* = x$;
(2) $(x + y)^* = x^* + y^*$;
(3) $(xy)^* = y^* x^*$;
(4) $(\lambda x)^* = \bar{\lambda} x^*$, λ a complex number.

An algebra with an involution is called a *-algebra. A Banach *-algebra is called a C*-algebra if it satisfies $\|x^* x\| = \|x\|^2$ for $x \in \mathscr{A}$. It is easily seen that $\|x^* x\| = \|x\|^2$ for $x \in \mathscr{A}$ implies $\|x^*\| = \|x\|$ $(x \in \mathscr{A})$.

A C*-algebra $\mathcal{M}$ is called a W*-algebra if it is a dual space as a Banach space (i.e. there is a Banach space $\mathcal{M}_*$ such that the dual $(\mathcal{M}_*)^*$of $\mathcal{M}_*$ is $\mathcal{M}$). We call $\mathcal{M}_*$ the *pre-dual* of $\mathcal{M}$. In general, a dual Banach space is not necessarily the dual space of a unique Banach space. However the W*-algebra has a unique pre-dual space. The W*-algebras are also called von Neumann algebras and the present definition (W*-algebras) agrees with the familiar alternative definitions of von Neumann algebras. This Theorem is due to the author.

1.2 Topologies on C*-algebras and W*-algebras

The topology defined by the norm $\|\cdot\|$ on a C*-algebra $\mathcal{A}$ is called the uniform topology (or the norm topology). The weak *-topology $\sigma(\mathcal{M},\mathcal{M}_*)$ on $\mathcal{M}$ is called the weak topology or the σ-weak topology.

1.3 *-Homomorphisms, *-isomorphisms and *-automorphisms

Let $\mathcal{A}$, $\mathcal{B}$ be C*-algebras. A linear mapping Φ of $\mathcal{A}$ into $\mathcal{B}$ is said to be a *-homomorphism if it satisfies:

(1) $\Phi(xy)=\Phi(x)\Phi(y)$;
(2) $\Phi(x^*)=\Phi(x)^*$.

A *-homomorphism Φ of a C*-algebra $\mathcal{A}$ into $\mathcal{B}$ is always norm-decreasing (i.e. $\|\Phi(x)\| \leqslant \|x\|$). Moreover the image $\Phi(\mathcal{A})$ is automatically uniformly closed in $\mathcal{B}$.

A one-to-one *-homomorphism of $\mathcal{A}$ into $\mathcal{B}$ is called a *-isomorphism. A *-isomorphism Φ is always isometric (i.e. $\|\Phi(x)\| = \|x\|$).

A *-isomorphism ρ of $\mathcal{A}$ onto $\mathcal{A}$ itself is called a *-automorphism of $\mathcal{A}$. Let $\mathcal{M}$, $\mathcal{N}$ be W*-algebras and let Φ be a *-isomorphism of $\mathcal{M}$ onto $\mathcal{N}$; then Φ is bicontinuous with respect to $\sigma(\mathcal{M},\mathcal{M}_*)$ and $\sigma(\mathcal{N},\mathcal{N}_*)$.

1.4 Self-adjointness and positivity

An element a in a C*-algebra $\mathcal{A}$ is said to be self-adjoint if $a^* = a$. An element h in a C*-algebra $\mathcal{A}$ is said to be positive if it can be expressed as $h=a^*a$ for some $a\in\mathcal{A}$. It can be shown that the set of all positive elements in $\mathcal{A}$ forms a convex cone.

1.5 Positive linear functionals and states

A linear functional f on a C*-algebra $\mathcal{A}$ is said to be positive if $f(x^*x)\geqslant 0$ for $x\in\mathcal{A}$. The reader can readily verify the following facts. If $\mathcal{A}$ has an identity, then a positive linear functional is bounded and $\|f\| = f(1)$. A bounded linear functional f on $\mathscr{A}$ is positive if there exists a positive element h (>0) in $\mathscr{A}$ such that $f(h) = \|f\|\,\|h\|$.

A positive linear functional ϕ on $\mathscr{A}$ is said to be a *state* if $\|\phi\| = 1$.

A state ϕ on $\mathscr{A}$ is said to be *tracial* if $\phi(xy) = \phi(yx)$ for $x, y \in \mathscr{A}$.

1.6 Commutative C*- and W*-algebras

This section gives the spectral representation for commutative algebras (both C* and W*). Let Ω be a locally compact Hausdorff space, and let $C_0(\Omega)$ be the algebra of all complex-valued continuous functions on Ω vanishing at infinity. Define $\|a\| = \sup_{t\in\Omega}|a(t)|$ and $a^*(t) = \overline{a(t)}$ for $a \in C_0(\Omega)$. Then $C_0(\Omega)$ is a commutative C*-algebra. If K is a compact Hausdorff space, $C(K)$ (the algebra of all complex-valued continuous functions on K) is a commutative C*-algebra with identity. The converse statement is also true: If $\mathscr{A}$ is a given commutative C*-algebra, then it is *-isomorphic to $C_0(\Omega)$, where Ω is some locally compact Hausdorff space; if $\mathscr{A}$ is assumed to be commutative with identity, then $\mathscr{A}$ is *-isomorphic to $C(K)$, where K is a compact Hausdorff space.

Let (Ω, μ) be a given measure space with $\mu(\Omega) < +\infty$, let $L^\infty(\Omega, \mu)$ be the C*-algebra of all essentially bounded μ-measurable functions on Ω and let $L^1(\Omega, \mu)$ be the Banach space of all μ-integrable functions on Ω. Then by the Random–Nikodym theorem, $L^1(\Omega, \mu)^* = L^\infty(\Omega, \mu)$. Hence $L^\infty(\Omega, \mu)$ is a commutative W*-algebra. More generally, let (Γ, ν) be a localizable measure space (i.e. a direct sum of finite measure spaces) and let $L^\infty(\Gamma, \nu)$ be the C*-algebra of all ν-essentially bounded measurable functions on Γ, and let $L^1(\Gamma, \nu)$ be the Banach space of all ν-integrable functions on Γ. Then $(L^1(\Gamma, \nu))^* = L^\infty(\Gamma, \nu)$; hence $L^\infty(\Gamma, \nu)$ is a commutative W*-algebra.

Conversely, if $\mathscr{M}$ is a commutative W*-algebra, then it is *-isomorphic to $L^\infty(\Gamma, \nu)$, where (Γ, ν) is a localizable measure space. Hence we also have a spectral representation for the commutative W*-algebras.

1.7 Concrete C*- and W*-algebras

Let $\mathscr{H}$ be a complex Hilbert space and let $B(\mathscr{H})$ be the algebra of all bounded linear operators on $\mathscr{H}$. We can define various topologies on $B(\mathscr{H})$. Although well known, they will be reviewed below.

(1) The uniform topology This is given by the operator norm $\|a\|$ ($a \in B(\mathscr{H})$), where $\|a\| = \sup_{\substack{\|\xi\|\leq 1\\ \xi\in\mathscr{H}}} \|c\xi\|$. Take the adjoint operation $a \mapsto a^*$ as the involution* (i.e. $(a\xi, \eta) = (\xi, a^*\eta)$ $(\xi, \eta \in \mathscr{H})$), where (,) is the scalar product (inner product) of $\mathscr{H}$). Then, $B(\mathscr{H})$ is a C*-algebra and moreover *any*

uniformly closed self-adjoint subalgebra $\mathscr{A}$ (i.e. $a \in \mathscr{A}$ implies $a^* \in \mathscr{A}$) of $B(\mathscr{H})$) is itself a C*-algebra.

(2) The strong operator topology Let $\xi \in \mathscr{H}$. The function $a \mapsto \|a\xi\|$ is then a semi-norm on $B(\mathscr{H})$. The set of all such semi-norms $\{\|a\xi\| \mid \xi \in \mathscr{H}\}$ defines a Hausdorff locally convex topology on $B(\mathscr{H})$, called the strong operator topology.

(3) The weak operator topology For each pair $\xi, \eta \in \mathscr{H}$, the function $a \to |(a\xi, \eta)|$ defines a semi-norm on $B(\mathscr{H})$. The set of all such semi-norms $\{|(a\xi, \eta)| \mid \xi, \eta \in \mathscr{H}\}$ defines a Hausdorff locally convex topology on $B(\mathscr{H})$, called the weak operator topology.

(4) The σ-weak topology Let Tr be the trace function in $B(\mathscr{H})$ and let $T(\mathscr{H})$ be the set of all trace class operators on $\mathscr{H}$. For $a \in T(\mathscr{H})$, define $\|a\|_1 = \mathrm{Tr}(|a|)$, where $|a| = (a^*a)^{1/2}$. Then $T(\mathscr{H})$ becomes a Banach space with the norm $\|\cdot\|_1$. Let f be a bounded linear functional on $B(\mathscr{H})$ which is assumed to be continuous on bounded spheres of $B(\mathscr{H})$ with respect to the weak operator topology. Then there is a unique trace class operator a on $\mathscr{H}$ such that $f(x) = \mathrm{Tr}(xa)$ for $x \in B(\mathscr{H})$ and $\|f\| = \mathrm{Tr}(|a|)$. Therefore $T(\mathscr{H})$ can be identified with the Banach space of all bounded linear functionals of $B(\mathscr{H})$ which are continuous with respect to the weak operator topology on bounded spheres of $B(\mathscr{H})$. Then it is known that $T(\mathscr{H})^* = B(\mathscr{H})$. Hence $B(\mathscr{H})$ is a W*-algebra and $\sigma(B(\mathscr{H}), B(\mathscr{H})_*) = \sigma(B(\mathscr{H}), T(\mathscr{H}))$ is the σ-weak topology of $B(\mathscr{H})$.

Let $\mathscr{M}$ be a self-adjoint subalgebra of $B(\mathscr{H})$ which is closed with respect to the weak operator topology; then it is $\sigma(B(\mathscr{H}), B(\mathscr{H})_*)$-closed. Let $\mathscr{M}^0$ be the polar of $\mathscr{M}$ in $T(\mathscr{H})$; then $(T(\mathscr{H})/\mathscr{M}^0)^* = \mathscr{M}$. Hence $\mathscr{M}$ is a W*-algebra and the restriction of $\sigma(B(\mathscr{H}), B(\mathscr{H})_*)$ to $\mathscr{M}$ is the σ-weak topology on $\mathscr{M}$.

Let $\mathscr{M}$ be a self-adjoint subalgebra of $B(\mathscr{H})$; then the following six properties are equivalent:

(1) $\mathscr{M} \cap S$ is closed with respect to the weak operator topology, where S is the unit sphere of $B(\mathscr{H})$;
(2) $\mathscr{M}$ is closed with respect to the weak operator topology;
(3) $\mathscr{M} \cap S$ is closed with respect to the strong operator topology;
(4) $\mathscr{M}$ is closed with respect to the strong operator topology;
(5) $\mathscr{M} \cap S$ is $\sigma(B(\mathscr{H}), B(\mathscr{H})_*)$-closed;
(6) $\mathscr{M}$ is $\sigma(B(\mathscr{H}), B(\mathscr{H})_*)$-closed.

The weak operator topology and $\sigma(\mathscr{M}, \mathscr{M}_*)$ are equivalent on $\mathscr{M} \cap S$. Therefore without confusion, we can call a self-adjoint subalgebra of $B(\mathscr{H})$, which is closed with respect to the weak operator topology or $\sigma(B(\mathscr{H}), B(\mathscr{H})_*)$, a weakly closed self-adjoint subalgebra of $B(\mathscr{H})$.

1.8 Representation theorems for C*- and W*-algebras

(1) Let $\mathscr{A}$ be a C*-algebra; then $\mathscr{A}$ is *-isomorphic to a uniformly closed self-adjoint subalgebra of $B(\mathscr{H})$, where $\mathscr{H}$ is a Hilbert space. (the Gelfand–Naimark theorem).

(2) Let $\mathscr{M}$ be a W*-algebra; then $\mathscr{M}$ is *-isomorphic to a weakly closed self-adjoint subalgebra of $B(\mathscr{H})$, where $\mathscr{H}$ is a Hilbert space. Moreover $\sigma(\mathscr{M}, \mathscr{M}_*)$ coincides with the restriction of $\sigma(B(\mathscr{H}), B(\mathscr{H})_*)$ under the *-isomorphism.

The next two theorems are fundamental to operator algebra theory. They will be stated without proof.

1.9 Commutation theorem (von Neumann's double commutant theorem)

Let $\mathscr{M}$ be a self-adjoint subalgebra of $B(\mathscr{H})$ containing the identity operator on a Hilbert space $\mathscr{H}$. Then, $\mathscr{M}$ is weakly closed if and only if $\mathscr{M}'' = \mathscr{M}$, where $\mathscr{M}' = \{x \in B(\mathscr{H}) \mid ax = xa \text{ for } a \in \mathscr{M}\}$ and $\mathscr{M}'' = (\mathscr{M}')'$.

1.10 Kaplansky's density theorem (Bounded approximation)

Let $\mathscr{B}$ be a self-adjoint subalgebra of $B(\mathscr{H})$ and let $\bar{\mathscr{B}}$ be the closure of $\mathscr{B}$ in $B(\mathscr{H})$ with respect to the weak operator topology; then $\bar{\mathscr{B}}$ is a weakly closed self-adjoint subalgebra of $B(\mathscr{H})$ and moreover $\bar{\mathscr{B}} \cap S$ is contained in the closure of $\mathscr{B} \cap S$ with respect to the strong operator topology, where S is the unit sphere of $B(\mathscr{H})$.

1.11 GNS (Gelfand–Naimark–Segal) representations

(1) From state to representation Let ϕ be a bounded positive linear functional on a C*-algebra $\mathscr{A}$. Introduce a conjugate bilinear functional $(x, y) = \phi(y^*x)$ in $\mathscr{A}$. Let $\mathscr{I} = \{x \mid \phi(x^*x) = 0, x \in \mathscr{A}\}$. Then $\mathscr{I}$ is a closed left ideal of $\mathscr{A}$. Define a conjugate bilinear functional on the quotient linear space $\mathscr{A}/\mathscr{I}$ such that if $x \in x_\phi$, $y \in y_\phi$, then $(x_\phi, y_\phi) = \phi(y^*x)$ (here x_ϕ(resp. y_ϕ) is the class containing x(resp. y)). The expression (x_ϕ, y_ϕ) does not then depend on a special choice of the representatives x, y. It will define a scalar product on $\mathscr{A}/\mathscr{I}$ under which $\mathscr{A}/\mathscr{I}$ will become a pre-Hilbert space. Let $\mathscr{H}_\phi$ be the completion with respect to this scalar product. Then $\mathscr{H}_\phi$ is a Hilbert space. Now we shall construct a *-representation of $\mathscr{A}$ on $\mathscr{H}_\phi$ (i.e. a *-homomorphism of $\mathscr{A}$ into $B(\mathscr{H}_\phi)$) via ϕ, denoted by $\{\pi_\phi, \mathscr{H}_\phi\}$. Put $\pi_\phi(a)x_\phi = (ax)_\phi$. Then $\pi_\phi(a)$ $(a \in \mathscr{A})$ is a linear operator on $\mathscr{A}/\mathscr{I}$. Moreover it is a bounded linear operator on the pre-Hilbert space $\mathscr{A}/\mathscr{I}$;

hence it can be uniquely extended to a bounded linear operator on $\mathscr{H}_\phi$, denoted by $\pi_\phi(a)$ again. Then, $a \mapsto \pi_\phi(a)$ $(a \in \mathscr{A})$ is a *-homomorphism of $\mathscr{A}$ into $B(\mathscr{H}_\phi)$. The *-representation $\{\pi_\phi, \mathscr{H}_\phi\}$ is called the GNS representation constructed via ϕ. It is known that there is a vector ξ_0 in $\mathscr{H}_\phi$ such that $\phi(x) = (x\xi_0, \xi_0)$ for $x \in \mathscr{A}$, and moreover $[\pi_\phi(\mathscr{A})\xi_0] = \mathscr{H}_\phi$ (i.e. $\{\pi_\phi, \mathscr{H}_\phi\}$ is cyclic), where $[(\cdot)]$ is the closure of $(\cdot)$ in $\mathscr{H}_\phi$.

If $\mathscr{A}$ has an identity 1 and ϕ is a state on $\mathscr{A}$, then $\xi_0 = 1_\phi$.

(2) From representation to state Conversely, if π is a *-representation of $\mathscr{A}$ into $B(\mathscr{H})$ such that $[\pi(\mathscr{A})\xi_0] = \mathscr{H}$, then $\{\pi, \mathscr{H}\}$ is unitarily equivalent to $\{\pi_\phi, \mathscr{H}_\phi\}$, where $\varphi(x) = (x\xi_0, \xi_0)$ $(x \in \mathscr{A})$.

(3) Normality Let ψ be a positive linear functional on a W*-algebra $\mathscr{M}$. ψ is then said to be *normal* if it is $\sigma(\mathscr{M}, \mathscr{M}_*)$-continuous. Let ψ be a normal positive linear functional on $\mathscr{M}$; then the GNS representation $\{\pi_\psi, \mathscr{H}_\psi\}$ satisfies the following conditions:

(1) $\pi_\psi(\mathscr{M})$ is weakly closed in $B(\mathscr{H}_\psi)$;
(2) π_ψ is a continuous mapping of $\mathscr{M}$ with $\sigma(\mathscr{M}, \mathscr{M}_*)$ into $B(\mathscr{H}_\psi)$ with $\sigma(B(\mathscr{H}_\psi), B(\mathscr{H}_\psi)_*)$.

1.12 Factorial and pure states

A W*-algebra $\mathscr{M}$ is called a factor if its center consists of scalar multiples of the identity. Let $\mathscr{A}$ be a C*-algebra and let ϕ be a state on $\mathscr{A}$. ϕ is said to be *factorial if* $\pi_\phi(\mathscr{A})''$ is a factor; ϕ is said to be pure if $\pi_\phi(\mathscr{A})'' = B(\mathscr{H}_\phi)$.

Let $\mathscr{S}_\mathscr{A}$ be the set of all states on $\mathscr{A}$; then a state ϕ on $\mathscr{A}$ is pure if and only if ϕ is an extreme point in the convex set $\mathscr{S}_\mathscr{A}$. If $\mathscr{A}$ has an identity, then $\mathscr{S}_\mathscr{A}$ is $\sigma(\mathscr{A}^*, \mathscr{A})$-compact. The space $\mathscr{S}_\mathscr{A}$ with $\sigma(\mathscr{A}^*, \mathscr{A})$ is called the state space of $\mathscr{A}$.

In Chapter 4, we shall often use harmonic functions, and so we shall now state the fundamental facts on harmonic function, which we shall need in Chapter 4. Let $\mathbb{C}$ be the complex plane, and let D be the closed unit disk in $\mathbb{C}$ such that $D = \{z \in \mathbb{C} \mid |z| \leqslant 1\}$. Let β be a non-zero real number and let S_β be the closed strip of the complex plane such that

$$S_\beta = \{z \in \mathbb{C} \mid 0 \leqslant \mathrm{Im}(z) \leqslant \beta \text{ if } \beta > 0 \text{ or } \beta \leqslant \mathrm{Im}(z) \leqslant 0 \text{ if } \beta < 0\}.$$

Let D^0 (resp. $S_\beta{}^0$) be the interior of D(resp. S_β) and let $D_1 = \{z \in D \mid z \neq -1, 1\}$. Consider a mapping $\omega = (\beta/\pi)\log \mathrm{i}(1-z)/(1+z)$ of D_1 in the z-complex plane onto S_β in the ω-complex plane. Then it is a one-to-one bicontinuous mapping of D_1 onto S_β and moreover it is a conformal mapping of D^0 onto $S_\beta{}^0$. Therefore, by Poisson's formula and the result from the Dirichlet problem, we have the following theorem (cf. [77]).

1.13 Theorem (Poisson Kernel for the strip)

There exist two positive continuous functions $K_1(z,t), K_2(z,t)$ on $S_\beta{}^0 \times \mathbb{R}$ such that

$$\int_{-\infty}^{\infty} K_1(z,t)\,\mathrm{d}t + \int_{-\infty}^{\infty} K_2(z,t)\,\mathrm{d}t = 1 \qquad \text{for } z\in S_\beta{}^0$$

and for any two real-valued, bounded continuous functions f_1, f_2 on $\mathbb{R}$,

$$u(z) = \int_{-\infty}^{\infty} f_1(t)K_1(z,t)\,\mathrm{d}t + \int_{-\infty}^{\infty} f_2(t)K_2(z,t)\,\mathrm{d}t \qquad \text{for } z\in S_\beta{}^0$$

is harmonic on $S_\beta{}^0$ and moreover

$$\lim_{\substack{z\to t\\ z\in S_\beta{}^0}} u(z) = f_1(t) \qquad \textit{and} \qquad \lim_{\substack{z\to t+i\beta\\ z\in S_\beta{}^0}} u(z) = f_2(t)\,(t\in\mathbb{R}).$$

Furthermore if u_1 is a bounded continuous function on S_β which is harmonic on $S_\beta{}^0$, and $u_1(t) = f_1(t)$, $u_1(t+i\beta) = f_2(t)$ $(t\in R)$, then $u_1(z) = u(z)$ for $z\in S_\beta$.

We shall not use the explicit form for the functions K_1, and K_2 but only the representation theorems stated here and their corollary:

1.14 Corollary (analytic version)

Let $f(z)$ be a bounded continuous function on S_β which is analytic on $S_\beta{}^0$; then

$$f(z) = \int_{-\infty}^{\infty} f(t)K_1(z,t)\,\mathrm{d}t + \int_{-\infty}^{\infty} f(t+\mathrm{i}\beta)K_2(z,t)\,\mathrm{d}t \qquad \textit{for } z\in S_\beta{}^0,$$

Perturbation Theory

We shall also need some fundamental results on bounded perturbations in Chapter 4. Here we mention some of them.

Let A, B be bounded linear operators on a Banach space E. Let $u_t = \exp(tA)$ and $T_t = \exp t(A+B)(t\in R)$ be exponential one-parameter groups.

1.15 Perturbation expansion theorem ($t\in\mathbb{R}$)

We have

$$T_t = u_t + \sum_{n=1}^{\infty} \int_{0\leqslant t_1\leqslant t_2\leqslant\cdots\leqslant t_n\leqslant t} u_{t_1}Bu_{t_2-t_1}Bu_{t_3-t_2}\cdots u_{t_n-t_{n-1}}Bu_{t-t_n}\,\mathrm{d}t_1\,\mathrm{d}t_2\cdots\mathrm{d}t_n.$$

where the series is norm-convergent for all t.

Proof Let $T_t^{(0)} = u_t$ and

$$T_t^{(n)} = \int_{0 \leqslant t_1 \leqslant t_2 \leqslant \cdots \leqslant t_n \leqslant t} u_{t_1} B u_{t_2 - t_1} B_{t_3 - t_2} \cdots u_{t_n - t_{n-1}} B u_{t - t_n} \, dt_1 \, dt_2 \cdots dt_n.$$

Then

$$\begin{aligned} T_t^{(n)} &= \int_0^t u_{t_1} B \, dt_1 \int_{t_1 \leqslant t_2 \leqslant \cdots \leqslant t_n \leqslant t} u_{t_2 - t_1} B u_{t_3 - t_2} \cdots u_{t_n - t_{n-1}} B u_{t - t_n} \, dt_2 \, dt_3 \cdots dt_n \\ &= \int_0^t u_{t_1} B \, dt_1 \int_{0 \leqslant t_2 \leqslant t_3 \leqslant \cdots \leqslant t_n \leqslant t - t_1} u_{t_2} B_{t_3 - t_2} \cdots u_{t_n - t_{n-1}} B u_{t - t_1 - t_n} \, dt_2 \, dt_3 \cdots dt_n \\ &= \int_0^t u_{t_1} B T_{t - t_1}{}^{(n-1)} \, dt_1. \end{aligned}$$

On the other hand,

$$\left\| \sum_{n=0}^{\infty} T_t^{(n)} \right\| \leqslant \sum_{n=0}^{\infty} \left\| T_t^{(n)} \right\| \leqslant \sum_{n=0}^{\infty} \frac{M^{n+1} \| B \|^n}{n!} t^n,$$

where $M = \sup_{|s| \leqslant t} \| u_s \|$.

Therefore,

$$\sum_{n=1}^{\infty} T_t^{(n)} = \int_0^t u_{t_1} B \sum_{n=0}^{\infty} T_{t - t_1}{}^{(n)} \, dt_1.$$

Put $w_t = \sum_{n=0}^{\infty} T_t^{(n)}$; then by the previous equality, we have

$$w_t = u_t + \int_0^t u_s B w_{t-s} \, ds.$$

Hence,

$$\begin{aligned} w_{t_1} w_{t_2} &= \left(u_{t_1} + \int_0^{t_1} u_s B w_{t_1 - s} \, ds \right) w_{t_2} \\ &= u_{t_1} w_{t_2} + \int_0^{t_1} u_s B w_{t_1 - s} w_{t_2} \, ds \\ &= u_{t_1} \left(u_{t_2} + \int_0^{t_2} u_s B w_{t_2 - s} \, ds \right) + \int_0^{t_1} u_s B w_{t_1 - s} w_{t_2} \, ds \\ &= u_{t_1 + t_2} + \int_0^{t_2} u_{s + t_1} B w_{t_2 - s} \, ds + \int_0^{t_1} u_s B w_{t_1 - s} w_{t_2} \, ds \\ &= u_{t_1 + t_2} + \int_{t_1}^{t_1 + t_2} u_s B w_{t_1 + t_2 - s} \, ds + \int_0^{t_1} u_s B w_{t_1 - s} w_{t_2} \, ds \end{aligned}$$

$$= u_{t_1+t_2} + \int_0^{t_1+t_2} u_s B w_{t_1+t_2-s}\,\mathrm{d}s + \int_0^{t_1} u_s B(w_{t_1-s}w_{t_2} - w_{t_1+t_2-s})\,\mathrm{d}s$$

$$= w_{t_1+t_2} + \int_0^{t_1} u_s B(w_{t_1-s}w_{t_2} - w_{t_1+t_2-s})\,\mathrm{d}s.$$

If we replace B by $zB(z\in\mathbb{C})$, denote the corresponding w by w^z and put $G_{t_1}(z) = w_{t_1}{}^z w_{t_2}{}^z - w_{t_1+t_2}{}^z\ (z\in\mathbb{C})$, then $G_{t_1}(z)$ is entirely analytic with respect to z, and

$$G_{t_1}(z) = z\int_0^{t_1} u_s BG_{t_1-s}(z)\,\mathrm{d}s = z\int_0^{t_1} u_{t_1-s}BG_s(z)\,\mathrm{d}s.$$

Since $G_t(z) = \sum_{n=0}^{\infty}(G_t{}^{(n)}(0)/n!)z^n$,

$$\sum_{n=0}^{\infty}\frac{G_t{}^{(n)}(0)}{n!}z^n = z\int_0^t u_{t-s}B\sum_{n=0}^{\infty}\frac{G_s{}^{(n)}(0)}{n!}z^n\,\mathrm{d}s$$

$$= \sum_{n=0}^{\infty} z^{n+1}\int_0^t u_{t-s}B\frac{G_s{}^{(n)}(0)}{n!}\,\mathrm{d}s \qquad \text{for } z\in\mathbb{C}.$$

Hence $G_t{}^{(0)}(0) = G_t{}^{(0)}(0) = \cdots = G_t{}^{(n)}(0) = \cdots = 0$ and so $G_t(z) = 0$. Therefore, $w_{t_1+t_2} = w_{t_1}w_{t_2}$.

On the other hand,

$$w_t = u_t + \int_0^t u_s B w_{t-s}\,\mathrm{d}s = u_t + \int_0^t u_{t-s}Bw_s\,\mathrm{d}s$$

$$= u_t + u_t\int_0^t u_{-s}Bw_s\,\mathrm{d}s.$$

Hence

$$\frac{\mathrm{d}}{\mathrm{d}t}w_t|_{t=0} = \frac{\mathrm{d}}{\mathrm{d}t}u_t|_{t=0} + u_t u_{-t}Bw_t|_{t=0}$$

$$= A + B.$$

Therefore $w_t = \exp t(A+B)$. This completes the proof. □

1.16 Corollary (complex version)

We have

$$\exp z(A+B) = \exp(zA) + \sum_{n=1}^{\infty} z^n \int_{0\leqslant t_1\leqslant t_2\leqslant\cdots\leqslant t_n\leqslant 1} u_{t_1 z}Bu_{(t_2-t_1)z}B\cdots$$

$$\times\, u_{(t_n-t_{n-1})z}Bu_{(1-t_n)z}\,\mathrm{d}t_1\,\mathrm{d}t_2\cdots\mathrm{d}t_n. \qquad (z\in\mathbb{C})$$

where the series is norm convergent for all z.

Proof

$$\begin{aligned}
\exp z(A+B) &= \exp 1(zA+zB) \\
&= \exp zA + \sum_{n=1}^{\infty} \int_{0\leqslant t_1 \leqslant t_2 \leqslant \cdots \leqslant t_n \leqslant 1} u_{t_1 z}(zB)u_{(t_2-t_1)z}(zB)u_{(t_3-t_2)z}\cdots \\
&\quad \times u_{(t_n-t_{n-1})z}(zB)u_{(1-t_{n-1})z}\,\mathrm{d}t_1\,\mathrm{d}t_2\cdots \mathrm{d}t_n \\
&= \exp zA + \sum_{n=1}^{\infty} z^n \int_{0\leqslant t_1 \leqslant t_2 \leqslant \cdots \leqslant t_n \leqslant 1} u_{t_1 z}Bu_{(t_2-t_1)z}Bu_{(t_3-t_2)z}\cdots \\
&\quad \times u_{(t_n-t_{n-1})z}Bu_{(1-t_n)z}\,\mathrm{d}t_1\,\mathrm{d}t_2\cdots \mathrm{d}t_n.
\end{aligned}$$

This completes the proof. □

Next we shall consider a generalization of Theorem 1.15. Let A be a self-adjoint linear operator in a Hilbert space $\mathscr{H}$ and let B be a bounded self-adjoint linear operator on $\mathscr{H}$. The two operators A and B will be specified this way throughout the discussion to follow.

Geometric vectors

A vector ξ in $\mathscr{H}$ is said to be geometric with respect to A if there is a positive number M_ξ such that $\|A^n\xi\| \leqslant M_\xi{}^n\|\xi\|$ for $n=1,2,3,\ldots$. Let $G(A)$ be the set of all geometric elements in $\mathscr{H}$ with respect to A. Let $A=\int_{-\infty}^{\infty}\lambda\,\mathrm{d}E_\lambda$ be the spectral decomposition of A and let $E_n=\int_{-n}^{n}\lambda\,\mathrm{d}E_\lambda$; then $\bigcup_{n=1}^{\infty}E_n\mathscr{H}\subset G(A)$ and so $G(A)$ is dense in $\mathscr{H}$.

Now assume that (1) $B\cdot G(A)\subset G(A)$; then for $\xi\in G(A)$,

$$(\exp t_1A)B(\exp(t_2-t_1)A)B\cdots(\exp(t_n-t_{n-1})A)B(\exp(t-t_n)A)\xi$$

is well-defined and it belongs to $G(A)$. In fact, for $t\in\mathbb{R}$, $\exp(tA)\xi=\sum_{n=0}^{\infty}(t^n/n!)A^n\xi$, and $\sum_{n=0}^{\infty}\|(t^n/n!)A^n\xi\|\leqslant\sum_{n=0}^{\infty}(M_\xi{}^n|t|^n/n!)\|\xi\|=\exp(M_\xi|t|)\|\xi\|$; hence $\exp(tA)\xi\in\mathscr{H}$. Moreover,

$$A^m\exp(tA)\xi=\sum_{n=0}^{\infty}\frac{t^n}{n!}A^{m+n}\xi$$

and

$$\begin{aligned}
\sum_{n=0}^{\infty}\frac{|t|^n}{n!}\|A^{m+n}\xi\| &\leqslant M_\xi{}^m\sum_{n=0}^{\infty}\frac{M_\xi{}^n|t|^n}{n!}\|\xi\| \\
&= M_\xi{}^m\exp(M_\xi|t|)\|\xi\|.
\end{aligned}$$

Now put $\|\xi\|=V_t\|(\exp tA)\xi\|$; then $\|A^m\exp(tA)\xi\|\leqslant M_\xi{}^m\exp(M_\xi|t|)$

$V_t \| \exp(tA)\xi \|$. Let $N_t = \max\{\exp(M_\xi |t|)V_t, 1\}$; then

$$\| A^m \exp(tA)\xi \| \leqslant (M_\xi N_t)^m \| \exp(tA)\xi \| \qquad (m = 1, 2, \ldots)$$

and so $\exp(tA)G(A) \subset G(A)$ for $t \in \mathbb{R}$.

Next, assume that (2) $\rho_t(B) = \exp(tA)B\exp(-tA)$ is a bounded linear operator in $\mathscr{H}$ for each $t \in \mathbb{R}$, and that $\| \overline{\rho_t(B)} \| \leqslant \exp(k|t|) \| B \|$ for $t \in \mathbb{R}$, where k is a positive number and $\overline{\rho_t(B)}$ is the closure of $\rho_t(B)$. Then

$$\begin{aligned} &\exp(t_1 A)B(\exp(t_2 - t_1)A)B \cdots (\exp(t_n - t_{n-1})A)B(\exp(t - t_n)A)\xi \\ &\quad = \rho_{t_1}(B)\rho_{t_2}(B) \cdots \rho_{t_n}(B)(\exp tA)\xi \end{aligned}$$

and

$$\begin{aligned} &\| \exp(t_1 A)B(\exp(t_2 - t_1)A)B \cdots (\exp(t_n - t_{n-1})A)B(\exp(t - t_n)A)\xi \| \\ &\quad \leqslant \| \rho_{t_1}(B) \| \, \| \rho_{t_2}(B) \| \cdots \| \rho_{t_n}(B) \| \, \| \exp(tA)\xi \| \\ &\quad \leqslant \exp(k(|t_1| + |t_2| + \cdots + |t_n|)) \| B \|^n \| \exp(tA)\xi \|. \end{aligned}$$

Therefore, for $\xi \in G(A)$, the vector given by the series

$$\exp(tA)\xi + \sum_{n=1}^{\infty} \int_{0 \leqslant t_1 \leqslant t_2 \leqslant \cdots \leqslant t_n \leqslant t} \rho_{t_1}(B)\rho_{t_2}(B) \cdots \rho_{t_n}(B) \exp(tA) \, \mathrm{d}t_1 \, \mathrm{d}t_2 \cdots \mathrm{d}t_n \xi$$

is well-defined and it belongs to $\mathscr{H}$. Then by a method similar to that used in the proof of Theorem 1.15, we have the following theorem.

1.17 Theorem (convergence on geometric vectors)

Suppose the above Conditions (1) *and* (2); *then*

$$\begin{aligned} \exp t(A + B)\xi = \exp tA\xi + \sum_{m=0}^{\infty} \int_{0 \leqslant t_1 \leqslant t_2 \leqslant \cdots \leqslant t_n \leqslant t} &\exp(t_1 A)B(\exp(t_2 - t_1)A)B \cdots \\ &\times (\exp(t_n - t_{n-1})A)B \exp(t - t_n)A\xi \, \mathrm{d}t_1 \, \mathrm{d}t_2 \cdots \mathrm{d}t_n \end{aligned}$$

for all $\xi \in G(A)$.

Automorphism groups

Let $\mathscr{A}$ be a C*-algebra, and let ρ be a *-automorphism on $\mathscr{A}$ – i.e. ρ is a one-to-one linear mapping of $\mathscr{A}$ onto $\mathscr{A}$ satisfying $\rho(ab) = \rho(a)\rho(b)$ and $\rho(a^*) = \rho(a)^*$ $(a, b \in \mathscr{A})$; then ρ is an isometry – i.e. $\| \rho(a) \| = \| a \|$ $(a \in \mathscr{A})$. Let $\mathrm{Aut}(\mathscr{A})$ be the group of all *-automorphisms on $\mathscr{A}$. The norm topology (or the uniform topology) on $\mathrm{Aut}(\mathscr{A})$ is defined by the norm $\| \rho_1 - \rho_2 \| = \sup_{\substack{\|a\| \leqslant 1 \\ a \in \mathscr{A}}} \| \rho_1(a) - \rho_2(a) \|$. Then, $\mathrm{Aut}(\mathscr{A})$ is a topological group under this topology.

The strong topology on $\mathrm{Aut}(\mathscr{A})$ is defined as follows: for $\varepsilon > 0$, $a_1, a_2, \ldots, a_n \in \mathscr{A}$,

$$U(\rho_1; a_1, a_2, \ldots, a_n; \varepsilon) = \{\rho \in \mathrm{Aut}(\mathscr{A}) \mid \|\rho(a_i) - \rho_1(a_i)\| < \varepsilon,\ i = 1, 2, \ldots, n\} \qquad (n = 1, 2, 3, \ldots)$$

is a neighborhood of $\rho_1 (\in \mathrm{Aut}(\mathscr{A}))$. Moreover, $\mathrm{Aut}(\mathscr{A})$ is a topological group under the strong topology.

Let G be a locally compact group. Let $t \mapsto \alpha_t$ $(t \in G)$ be a homomorphism of G into $\mathrm{Aut}(\mathscr{A})$. Then α is said to be norm-continuous (or uniformly continuous) if the homomorphism $t \mapsto \alpha_t$ is a continuous mapping of G into $\mathrm{Aut}(\mathscr{A})$ with the norm topology. The triple $\{\mathscr{A}, G, \alpha\}$ is said to be a norm-continuous (or uniformly continuous) C*-dynamical system. α is said to be strongly continuous if the homomorphism $t \mapsto \alpha_t$ is a continuous mapping of G into $\mathrm{Aut}(\mathscr{A})$ with the strong topology. In this case the triple $\{\mathscr{A}, G, \alpha\}$ is simply said to be a C*-dynamical system.

Now let $\mathscr{M}$ be a W*-algebra – i.e., $\mathscr{M}$ is a C*-algebra and is a dual Banach space. Then (as we have noted) there is a unique Banach space $\mathscr{M}_*$ (called the pre-dual of $\mathscr{M}$) such that $\mathscr{M} = (\mathscr{M}_*)^*$. The topology $\sigma(\mathscr{M}, \mathscr{M}_*)$ on $\mathscr{M}$ is said to be the σ-weak topology or the σ-topology on $\mathscr{M}$. Since every *-automorphism ρ on $\mathscr{M}$ is automatically σ-weakly continuous, there is an isometry ρ_* on $\mathscr{M}_*$ such that $(\rho_*)^* = \rho$, where $(\rho_*)^*$ is the dual mapping of ρ_*. Let $\mathrm{Aut}(\mathscr{M})_* = \{\rho_* \mid \rho \in \mathrm{Aut}(\mathscr{M})\}$; then for $\rho_1, \rho_2 \in \mathrm{Aut}(\mathscr{M})$, $(\rho_1\rho_2)_* = \rho_{2*}\rho_{1*}$ so that $\mathrm{Aut}(\mathscr{M})_*$ is a group. By using the strong topology on $\mathrm{Aut}(\mathscr{M})_*$, we obtain a topology on $\mathrm{Aut}(\mathscr{M})$. This topology is called the σ-weak topology on $\mathrm{Aut}(\mathscr{M})$. Clearly $\mathrm{Aut}(\mathscr{M})$ is then a topological group. Let $t \mapsto \alpha_t$ $(t \in G)$ be a homomorphism of G into $\mathrm{Aut}(\mathscr{M})$. The triple $\{\mathscr{M}, G, \alpha\}$ is said to be a W*-dynamical system if the homomorphism $t \mapsto \alpha_t$ is a continuous mapping of G into $\mathrm{Aut}(\mathscr{M})$ with the σ-weak topology.

Calculus of representations

Now let G be a locally compact group and let $t \mapsto \alpha_t$ be a homomorphism of G into $\mathrm{Aut}(\mathscr{M})$. Suppose that $t \mapsto \alpha_t(a)$ is σ-weakly continuous for each $a \in \mathscr{M}$ – i.e. for $f \in \mathscr{M}_*$, $\langle \alpha_t(a), f \rangle = f(\alpha_t(a))$ is a continuous function on G. Then,

$$\langle \alpha_t(a), f \rangle = \langle a, (\alpha_t)_* f \rangle \qquad (a \in \mathscr{M}, f \in \mathscr{M}_*).$$

Therefore the mapping $f \mapsto (\alpha_t)_* f$ of G into $\mathscr{M}_*$ is $\sigma(\mathscr{M}_*, \mathscr{M})$-continuous; hence the mapping $t \mapsto (\alpha_t)_*$ is weakly continuous. Let μ be a right invariant Haar measure on G and let $L^1(G, \mu)$ be the group algebra of G consisting of all μ-integrable functions on G. For $\xi \in L^1(G, \mu)$, define $T_\xi = \int (\alpha_t)_* \xi(t)\, \mathrm{d}\mu(t)$, where the integration is defined by using the weak topology on $\mathcal{M}_*$ – i.e.

$$\langle a, T_\xi f\rangle = \int \langle a, (\alpha_t)_* f\rangle \xi(t)\,\mathrm{d}\mu(t) \qquad \text{for } a\in\mathcal{M}, f\in\mathcal{M}_*.$$

Then,

$$\begin{aligned}\|(\alpha_t)_* T_\xi f - T_\xi f\| &= \left\| \int (\alpha_t)_*(\alpha_s)_* f\xi(s)\,\mathrm{d}\mu(s) - \int (\alpha_s)_* f\xi(s)\,\mathrm{d}\mu(s)\right\| \\ &= \left\| \int (\alpha_s\alpha_t)_* f\xi(s)\,\mathrm{d}\mu(s) - \int (\alpha_s)_* f\xi(s)\,\mathrm{d}\mu(s)\right\| \\ &= \left\| \int (\alpha_s)_* f\xi(st^{-1})\,\mathrm{d}\mu(s) - \int (\alpha_s)_* f\xi(s)\,\mathrm{d}\mu(s)\right\| \\ &\leqslant \|(\alpha_s)_* f\| \int |\xi(st^{-1}) - \xi(s)|\,\mathrm{d}\mu(s) \to 0 \qquad (t\to e),\end{aligned}$$

where e is the unit element of G. (Note that $\|(\alpha_s)_* f\| = \|f\|$.)

Since $L^1(G,\mu)$ has an approximate identity constructed in small neighborhoods of e with integral one, the linear subspace of $\mathcal{M}_*$ spanned by $\{T_\xi f \mid f\in\mathcal{M}_*, \xi\in L^1(G,\mu)\}$ is $\sigma(\mathcal{M}_*,\mathcal{M})$-dense in $\mathcal{M}_*$; hence it is norm-dense in $\mathcal{M}_*$. Since $\|(\alpha_t)_*\| = \|\alpha_t\| = 1$ for $t\in G$, $\|(\alpha_t)_* g - g\| \to 0$, $(t\to e)$ for $g\in\mathcal{M}_*$. Hence the mapping $t\mapsto\alpha_t$ of G into $\mathrm{Aut}(\mathcal{M})$ with the σ-weak topology is continuous, so that the triple $\{\mathcal{M}, G, \alpha\}$ is a W*-dynamical system.

Conversely it is clear that the mapping $t\mapsto\alpha_t(a)$ is σ-weakly continuous for each $a\in\mathcal{M}$ if $\{\mathcal{M}, G, \alpha\}$ is a W*-dynamical system. Therefore without confusion, we can say that the mapping $t\mapsto\alpha_t$ is σ-weakly continuous if $\{\mathcal{M}, G, \alpha\}$ is a W*-dynamical system with a locally compact group G.

Actions by Lie groups

Let G be a connected finite-dimensional real Lie-group and let $\{\mathcal{A}, G, \alpha\}$ be a C*-dynamical system. Let $\mathfrak{g}$ be the Lie algebra of G. Let $C_c^\infty(G)$ be the space of all infinitely differentiable functions on G with compact support, and let $L^1(G,\nu)$ be the group algebra of all integrable functions on G with respect to a left invariant Haar measure ν of G. For $f\in C_c^\infty(G)$ and $a\in\mathcal{A}$, let $T_f(a) = \int_G \alpha_t(a) f(t)\,\mathrm{d}\nu(t)$ and let $V(\alpha)$ be the linear subspace of $\mathcal{A}$ spanned by $\{T_f(a) \mid f\in C_c^\infty(G) \text{ and } a\in\mathcal{A}\}$. Then for $x\in\mathfrak{g}$, $\lambda\mapsto\alpha_{\exp(\lambda x)}$ $(\lambda\in\mathbb{R}^1)$ is a strongly continuous one-parameter group of *-automorphisms on $\mathcal{A}$; therefore there exists the infinitesimal generator $\mathrm{d}\tilde{\alpha}(x)$ such that $\alpha_{\exp(\lambda x)} = \exp(\lambda\,\mathrm{d}\tilde{\alpha}(x))$ and $\mathrm{d}\tilde{\alpha}(x)$ is a closed *-derivation in $\mathcal{A}$. Let $\{f_n\}$ $(\subset C_c^\infty(G))$ be an approximate identity of $L^1(G,\nu)$ constructed in small neighborhoods of e with integral one; then for $a\in\mathcal{D}(\mathrm{d}\tilde{\alpha}(x))$, where $\mathcal{D}(\mathrm{d}\tilde{\alpha}(x))$ is the domain of $\mathrm{d}\tilde{\alpha}(x)$, $\|T_{f_n}(a) - a\| \to 0$ and $\mathrm{d}\tilde{\alpha}(x) T_{f_n}(a) = T_{f_n}(\mathrm{d}\tilde{\alpha}(x)) \to \mathrm{d}\tilde{\alpha}(x)(a)$; hence $\overline{\mathrm{d}\tilde{\alpha}(x)|V(\alpha)} = \mathrm{d}\tilde{\alpha}(x)$, where

$d\tilde{\alpha}(x)|V(\alpha)$ is the restriction of $d\tilde{\alpha}(x)$ to $V(\alpha)$ and $\overline{d\tilde{\alpha}(x)|V(\alpha)}$ is the closure of $d\tilde{\alpha}(x)|V(\alpha)$. Now let $D(\alpha)$ be the *subalgebra of $\mathscr{A}$ generated by $V(\alpha)$ (actually $D(\alpha) = V(\alpha)$, see Dixmier–Malliavin [54]) and let $C^\infty(\alpha)$ be the *-subalgebra of all infinitely differentiable elements in $\mathscr{A}$ with respect to α – i.e., $t \mapsto \alpha_t(a)$ $(a \in C^\infty(\alpha))$ is infinitely differentiable with respect to G.

Then $D(\alpha) \subset C^\infty(\alpha)$. It is clear that $C^\infty(\alpha) \subset \bigcap_{x \in g} \mathscr{D}(d\tilde{\alpha}(x))$. We shall write $d\alpha(x) = d\tilde{\alpha}(x)|C^\infty(\alpha)$; then one can easily see that $d\alpha([x, y]) = [d\alpha(x), d\alpha(y)]$ for $x, y \in g$ and so the mapping $x \mapsto d\alpha(x)$ $(x \in g)$ is a Lie homomorphism and $\overline{d\alpha(x)} = d\tilde{\alpha}(x)$ $(x \in g)$.

Let $\mathscr{H}$ be a Hilbert space. The W*-algebra of all bounded operators on $\mathscr{H}$ is denoted by $B(\mathscr{H})$. A W*-algebra $\mathscr{M}$ in a Hilbert space $\mathscr{H}$ means a weakly closed *-subalgebra of $B(\mathscr{H})$.

A factor is a W*-algebra with its center consisting of scalar multiples of the identity. By $\mathscr{M}'$, we shall denote the commutant of $\mathscr{M}$ – i.e.

$$\mathscr{M}' = \{x' \in B(\mathscr{H}) | x'a = ax' \text{ for all } a \in \mathscr{M}\}.$$

If $\mathscr{M}$ is a W*-algebra containing the identity operator in a Hilbert space $\mathscr{H}$, then $(\mathscr{M}')'$ (denoted by $\mathscr{M}''$) $= \mathscr{M}$. In this case, $\mathscr{M}$ is also called a von Neumann algebra. Let $\mathscr{M}$ be a W*-algebra containing the identity operator $1_{\mathscr{H}}$ in a Hilbert space $\mathscr{H}$, and let $\mathscr{M}'$ be the commutant of $\mathscr{M}$. A linear operator T in $\mathscr{H}$ is said to be *affiliated* to $\mathscr{M}$ if for each unitary $u' \in \mathscr{M}'$, $u'\mathscr{D}(T) = \mathscr{D}(T)$ and $u'^*Tu' = T$, where $\mathscr{D}(T)$ is the domain of T. If T is affiliated to $\mathscr{M}$, we shall denote this by $T\eta\mathscr{M}$.

Next we shall discuss the relationship between C*- and W*-dynamical systems.

1.18 Proposition (the restricted C*-system)

Let $\{\mathscr{M}, G, \alpha\}$ be a W-dynamical system with a locally compact group G; then there exists a C*-dynamical system $\{\mathscr{A}, G, \hat{\alpha}\}$ such that $\mathscr{A}$ is a σ-weakly dense C*-subalgebra of $\mathscr{M}$ with an identity and $\hat{\alpha}$ is the restriction of α to $\mathscr{A}$.*

Proof Let ν be a left invariant Haar measure on G and let $L^1(G, \nu)$ be the group algebra of G. For $a \in \mathscr{M}$, $f \in L^1(G, \nu)$, put $T_f(a) = \int \alpha_t(a) f(t)\, d\nu(t)$, where the integral is defined by using the σ-weak topology on $\mathscr{M}$; then one can easily see that $T_f(a) \in \mathscr{M}$. Moreover,

$$\begin{aligned} \| \alpha_t(T_f(a)) - T_f(a) \| &= \left\| \int \alpha_{ts}(a) f(s)\, d\nu(s) - \int \alpha_s(a) f(s)\, d\nu(s) \right\| \\ &= \left\| \int \alpha_s(a) f(t^{-1}s)\, d\nu(s) - \int \alpha_s(a) f(s)\, d\nu(s) \right\| \\ &\leqslant \|a\| \int |f(t^{-1}s) - f(s)|\, d\nu(s). \end{aligned}$$

Hence $\lim_{t\to e}\|\alpha_t(T_f(a))-T_f(a)\|=0$. For $f_1, f_2 \in L^1(G,\nu)$, $a_1, a_2 \in \mathcal{M}$,

$$\begin{aligned}
&\|\alpha_t(T_{f_1}(a_1)T_{f_2}(a_2))-T_{f_1}(a_1)T_{f_2}(a_2)\|\\
&\quad\leqslant \|\alpha_t(T_{f_1}(a_1)T_{f_2}(a_2))-\alpha_t(T_{f_1}(a_1))T_{f_2}(a_2)\\
&\qquad+\alpha_t(T_{f_1}(a_1))T_{f_2}(a_2)-T_{f_1}(a)T_{f_2}(a_2)\|\\
&\quad\leqslant \|\alpha_t(T_{f_1}(a_1))\|\,\|\alpha_t(T_{f_2}(a_2))-T_{f_2}(a_2)\|\\
&\qquad+\|\alpha_t(T_{f_1}(a_1))-T_{f_1}(a_1)\|\,\|T_{f_2}(a_2)\|\to 0 \qquad (t\to e).
\end{aligned}$$

Let $\mathcal{A}_0$ be the *-subalgebra of $\mathcal{M}$ generated by $\{T_f(a) \mid f\in L^1(G,\nu),\ a\in\mathcal{M}\}$; then by the above consideration, $\|\alpha_t(x)-x\|\to 0$ $(t\to e)$ for $x\in\mathcal{A}_0$. Let $\mathcal{A}$ be the uniform closure of $\mathcal{A}_0$. For $y\in\mathcal{A}$, $\varepsilon>0$, let $x\in\mathcal{A}_0$ be an element such that $\|y-x\|<\varepsilon$; then

$$\|\alpha_t(y)-y\|\leqslant\|\alpha_t(y)-\alpha_t(x)\|+\|\alpha_t(x)-x\|+\|x-y\|.$$

Hence $\overline{\lim}_{t\to e}\|\alpha_t(y)-y\|\leqslant 2\varepsilon$. Since ε is arbitrary, $\lim_{t\to e}\|\alpha_t(y)-y\|=0$. Therefore $\{\mathcal{A},G,\hat{\alpha}\}$ is a C*-dynamical system, where $\hat{\alpha}$ is the restriction of α to $\mathcal{A}$. It is easily seen that $\mathcal{A}$ is σ-weakly dense in $\mathcal{M}$. □

Unitary representations

Now let $\mathcal{M}$ be a W*-algebra containing the identity operator $1_{\mathcal{H}}$ on a Hilbert space $\mathcal{H}$, and let $t\mapsto u_t$ be a strongly continuous unitary representation of a topological group G on $\mathcal{H}$ such that $u_t\mathcal{M}u_t^*=\mathcal{M}$ $(t\in G)$. Then, put $\alpha_t(a)=u_tau_t^*$ $(t\in G, a\in\mathcal{M})$. It follows that each α_t is a *-automorphism of $\mathcal{M}$ and that the mapping $t\mapsto\alpha_t$ is a homomorphism of G into $\mathrm{Aut}(\mathcal{M})$. For $\xi\in\mathcal{H}$, let $\phi_\xi(a)=(a\xi,\xi)$ $(a\in\mathcal{M})$, where (,) is the inner product of $\mathcal{H}$. Then

$$\begin{aligned}
&\sup_{\|a\|\leqslant 1}|(u_tau_t^*\xi,\xi)-(a\xi,\xi)|\\
&\quad=\sup_{\|a\|\leqslant 1}|(au_t^*\xi,u_t^*\xi)-(a\xi,\xi)|\\
&\quad\leqslant\sup_{\|a\|\leqslant 1}|(au_t^*\xi,u_t^*\xi)-(au_t^*\xi,\xi)|+\sup_{\|a\|\leqslant 1}|(au_t^*\xi,\xi)-(a\xi,\xi)|\\
&\quad\leqslant\|\xi\|\,\|u_t^*\xi-\xi\|+\|u_t^*\xi-\xi\|\,\|\xi\|\to 0 \qquad (t\to e).
\end{aligned}$$

Hence $\|(\alpha_t)_*\phi_\xi-\phi_\xi\|\to 0$ $(t\mapsto e)$. Since all finite linear combinations of all vector states on $\mathcal{M}$ are dense in $\mathcal{M}_*$, $\|(\alpha_t)_*f-f\|\to 0$ $(t\to e)$ for $f\in\mathcal{M}_*$; hence $\{\mathcal{M},G,\alpha\}$ is a W*-dynamical system.

2

Bounded derivations

Introduction

Derivations are defined by the familiar Leibnitz formula. As operators they may be *bounded* or *unbounded*. The bounded case will be discussed first.

Derivations appear in various branches of mathematics and physics. They clearly have their origin in the concept of differentiation developed by Newton and Leibnitz. Lie algebras and their theory have a long history in mathematics. They may also be considered to be part of the theory of pure algebras. However, the study of derivations in operator algebras originated in quantum mechanics rather than in Newtonian mechanics.

In late 1924, and early 1925, Heisenberg and Schrödinger both proposed explanations for the empirical quantization rules of Bohr and Sommerfeld. These explanations, which were originally known as matrix mechanics and wave mechanics respectively, are now known as quantum mechanics in the present theory of atomic structure. In 1931 Stone and von Neumann showed that these two formulations were equivalent. Heisenberg's formalism identified the coordinates of particle momentum and position with operators p_i and q_j satisfying the canonical commutation relations:

$$p_i p_j - p_j p_i = q_i q_j - q_j q_i = 0; \qquad p_i q_j - q_j p_i = -ih\delta_{ij}1,$$

where h is Planck's constant.

Heisenberg's formalism was tentatively proposed in terms of matrix operators. However, a simple calculation with the commutation relations shows that they cannot be matrix operators. In the 1940s, it was of central interest for mathematicians as to whether or not the commutation relations could be realized by bounded linear operators on Banach spaces. Subsequently the studies of bounded linear operators on a Banach space have been extensively exercised on the commutators of bounded operators.

Let E be a Banach space and $B(E)$ the algebra of all bounded linear operators on E. For $X, Y \in B(E)$, put $\delta_X(Y) = [X, Y] = XY - YX$; then δ_X is a bounded derivation on $B(E)$. Therefore, the theory of commutators stimulates the study of bounded derivations on Banach algebras.

In this chapter we shall mainly discuss bounded derivations and, to some extent, unbounded derivations which arise from quantum field-theoretic observations.

2.1 Introduction to derivations

Let $\mathscr{A}$ be a Banach algebra, δ a linear mapping in $\mathscr{A}$. Then δ is said to be a derivation in $\mathscr{A}$ if it satisfies the following conditions:

(1) the domain $\mathscr{D}(\delta)$ of δ is a dense subalgebra in $\mathscr{A}$;
(2) $\delta(ab) = \delta(a)b + a\delta(b)$ $(a, b \in \mathscr{D}(\delta))$.

If $\mathscr{D}(\delta) = \mathscr{A}$, then δ is said to be a derivation on $\mathscr{A}$. If δ is bounded, then δ can be uniquely extended to a bounded derivation on $\mathscr{A}$ and so it may be considered to be defined on $\mathscr{A}$. Let $\mathscr{A}$ be a *-Banach algebra, δ a derivation in $\mathscr{A}$. Then δ is said to be a *-derivation if it satisfies:

(3) $a \in \mathscr{D}(\delta)$ implies $a^* \in \mathscr{D}(\delta)$ and $\delta(a^*) = \delta(a)^*$.

A derivation δ in a Banach algebra is said to be *closed* if $x_n \in \mathscr{D}(\delta)$, $x_n \to x$ and $\delta(x_n) \to y$ implies $x \in \mathscr{D}(\delta)$ and $\delta(x) = y$; δ is said to be *closable* if $x_n \in \mathscr{D}(\delta)$, $x_n \to 0$ and $\delta(x_n) \to y$ implies $y = 0$. If δ is closable, then δ can be extended uniquely to the least closed derivation $\bar{\delta}$, called the *closure* of δ.

2.2 The commutation relation $ab - ba = 1$

Let $L^2(\mathbb{R})$ be the Hilbert space of all square integrable, complex-valued functions on the real line $\mathbb{R}$ with respect to the Lebesgue measure, f a complex-valued measurable function on $\mathbb{R}$. For $g \in L^2(\mathbb{R})$, define $T_f g = f \cdot g$; then T_f will define a closed linear operator in $L^2(\mathbb{R})$ with a dense domain. Let $a = \mathrm{d}/\mathrm{d}t$, $b = T_{f_0}$, where $f_0(t) = t$. Then for $g \in L^2(\mathbb{R})$ with $g' \in L^2(\mathbb{R})$,

$$\begin{aligned}(ab - ba)g(t) &= \frac{\mathrm{d}}{\mathrm{d}t}(tg(t)) - t\frac{\mathrm{d}}{\mathrm{d}t}g(t) \\ &= g(t) + tg'(t) - tg'(t) = g(t).\end{aligned}$$

Hence $ab - ba = 1$ on a dense subset of $L^2(\mathbb{R})$; therefore the commutation relation is realized by linear operators in $L^2(\mathbb{R})$; however a and b are unbounded. If we take $L^2([0, 1])$ as a Hilbert space, then b is bounded, but a is unbounded. However they cannot both be bounded; see later.

Wintner [191] (1947) for Hilbert spaces and Wielandt [190] (1949) for general Banach spaces proved that this is impossible for bounded operators.

After these results, the theory of bounded linear operators on a Banach space has been extensively developed around the commutation relation. In the following, we shall discuss parts of these developments.

2.2.1 Theorem (Kleinecke [110], Sirokov [182]) *Let $\mathscr{A}$ be a Banach algebra, δ a bounded derivation on $\mathscr{A}$. Suppose that $\delta^2(a) = 0$; then $\delta(a)$ is a generalized nilpotent – i.e., $(\|\delta(a)^n\|)^{1/n} \to 0$ $(n \to \infty)$.*

Proof $\delta(a) = 1!\delta(a)^1$. Suppose that $\delta^n(a^n) = n!\delta(a)^n$; then by Leibnitz's formula,

$$\begin{aligned}\delta^{n+1}(a^{n+1}) = \delta^{n+1}(a^n a) &= \delta^{n+1}(a^n)a + \binom{n+1}{1}\delta^n(a^n)\delta(a)\\ &= \delta(\delta^n(a^n))a + (n+1)n!\delta(a)^{n+1}\\ &= \delta(n!\delta(a)\cdots\delta(a))a + (n+1)!\delta(a)^{n+1}\\ &= (n+1)!\delta(a)^{n+1}.\end{aligned}$$

Hence

$$\delta^n(a^n) = n!\delta(a)^n \qquad (n = 1, 2, 3, \ldots).$$

$$(\|\delta(a)^n\|)^{1/n} = (\|\delta^n(a^n)\|/n!)^{1/n} \leqslant (\|\delta\|\|a\|/(n!)^{1/n}) \to 0 \ (n \to \infty). \qquad \square$$

2.2.2 Corollary (Wielandt [190], Wintner [191]) *Let $\mathscr{A}$ be a Banach algebra. Then there are no two elements a, b in $\mathscr{A}$ such that $ab - ba = 1$.*

Proof For $x \in \mathscr{A}$, let

$$\delta_a(x) = ax - xa = [a, x] \qquad (x \in \mathscr{A});$$

then δ_a is a bounded derivation on $\mathscr{A}$. $\delta_a^2(b) = \delta_a(1) = 0$ if $ab - ba = 1$. Hence by Theorem 2.2.1, $\delta_a(b)$ is a generalized nilpotent, a contradiction. $\square$

2.2.3 Corollary (Singer–Wermer [181]) *Let $\mathscr{A}$ be a commutative Banach algebra, δ a bounded derivation on $\mathscr{A}$; then $\delta(\mathscr{A})$ is contained in the radical of $\mathscr{A}$. In particular, if $\mathscr{A}$ is semi-simple, then $\delta \equiv 0$.*

Proof Let $B(\mathscr{A})$ be the algebra of all bounded linear operators on $\mathscr{A}$. For $a, x \in \mathscr{A}$, let $L_a x = ax$; then $L_a \in B(\mathscr{A})$. $[\delta, L_a](x) = (\delta L_a - L_a\delta)(x) = \delta(ax) - a\delta(x) = \delta(a)x = L_{\delta(a)}x$. Hence $[\delta, L_a] = L_{\delta(a)}$ and so $[L_a, [L_a, \delta]] = -[L_a, L_{\delta(a)}] = 0$, for a commutes with $\delta(a)$. Hence by Theorem 2.2.1, $L_{\delta(a)}$ is a generalized nilpotent, and so $(\|L_{\delta(a)^n}\|)^{1/n} = (\|\delta(a)^n\|)^{1/n} \to 0$. $\square$

2.2.4 Corollary (Šilov [180]) *Let $C^\infty([0, 1])$ be the algebra of all infinitely differentiable functions on the unit interval $[0, 1]$. Then there is no norm on $C^\infty([0, 1])$ under which $C^\infty([0, 1])$ becomes a Banach algebra.*

Proof Suppose that $C^\infty([0,1])$ is a Banach algebra under a norm $|||\cdot|||$. Consider the differential operator $\mathrm{d}/\mathrm{d}t$ on $C^\infty([0,1])$. For each $t \in [0,1]$, put $X_t(f) = f(t)$ $(f \in C^\infty([0,1]))$; then X_t is a character and so it is continuous under the norm $|||\cdot|||$. Hence $\|f\| \leqslant |||f|||$ for $f \in C^\infty([0,1])$, where $\|\cdot\|$ is the uniform norm. Suppose that $|||f_n||| \to 0$ and $|||f'_n - g||| \to 0$, where g is an element of $C^\infty([0,1])$. Then $\|f'_n - g\| \to 0$. Hence $\int_0^x f'_n(t)\,\mathrm{d}t = f_n(x) - f_n(0) \to \int_0^x g(t)\,\mathrm{d}t$. Hence $\int_0^x g(t)\,\mathrm{d}t = 0$ for $x \in [0,1]$ and so $g = 0$. By the closed graph theorem, the linear operator $\mathrm{d}/\mathrm{d}t$ on $C^\infty([0,1])$ is bounded with respect to the norm $|||\cdot|||$. By Corollary 2.2.3, $\mathrm{d}/\mathrm{d}t\,(C^\infty([0,1]))$ is then contained in the radical of $C^\infty([0,1])$. Since $\bigcap_{t\in[0,1]} \mathscr{M}_t = 0$, where $\mathscr{M}_t = \{f \in C^\infty([0,1]) \mid f(t) = 0\}$, $C^\infty([0,1])$ is semi-simple and so $\mathrm{d}/\mathrm{d}t = 0$, a contradiction. □

2.2.5 Theorem (cf. Rosenblum [157]) *Let $\mathscr{A}$ be a C*-algebra, δ a bounded derivation on $\mathscr{A}$. Suppose that $\delta(x) = 0$ for a normal element x (i.e. $x^*x = xx^*$) of $\mathscr{A}$; then $\delta(x^*) = 0$.*

Proof

$$\begin{aligned}\delta(\exp(\mathrm{i}\lambda x^*)) &= \delta(\exp(\mathrm{i}\lambda x^*)\exp(\mathrm{i}\bar{\lambda}x)\exp(-\mathrm{i}\bar{\lambda}x)) \\ &= \delta(\exp(\mathrm{i}\lambda x^*)\exp(\mathrm{i}\bar{\lambda}x))\exp(-\mathrm{i}\bar{\lambda}x) \\ &\quad + \exp(\mathrm{i}\lambda x^*)\exp(\mathrm{i}\bar{\lambda}x)\delta(\exp(-\mathrm{i}\bar{\lambda}x)), \qquad \lambda \in \mathbb{C}\end{aligned}$$

Suppose that $\delta(x^n) = 0$; then $\delta(x^{n+1}) = \delta(x^n)x + x^n\delta(x) = 0$. By induction $\delta(x^n) = 0$ $(n = 1, 2, 3, \ldots)$. Since $\delta(1) = 0$ and δ is bounded, $\delta(\exp(-\mathrm{i}\bar{\lambda}x)) = 0$. Therefore,

$$\delta(\exp(\mathrm{i}\lambda x^*)) = \delta(\exp(\mathrm{i}\lambda x^*)\exp \mathrm{i}\bar{\lambda}x))\exp(-\mathrm{i}\bar{\lambda}x).$$

Moreover

$$\begin{aligned}\delta(\exp(\mathrm{i}\lambda x^*))\exp(-\mathrm{i}\lambda x^*) &= \delta(\exp \mathrm{i}\lambda x^* \exp \mathrm{i}\bar{\lambda}x)\exp(-\mathrm{i}\bar{\lambda}x)\exp(-\mathrm{i}\lambda x^*) \\ &= \delta(\exp \mathrm{i}(\lambda x^* + \bar{\lambda}x))\exp - \mathrm{i}(\bar{\lambda}x + \lambda x^*).\end{aligned}$$

Put

$$f(\lambda) = \delta(\exp \mathrm{i}(\lambda x^* + \bar{\lambda}x))\exp - \mathrm{i}(\bar{\lambda}x + \lambda x^*) = \delta(\exp(\mathrm{i}\lambda x^*))\exp(-\mathrm{i}\lambda x^*);$$

then $f(\lambda)$ is complex-analytic on the whole plane of the complex field. Moreover $\|f(\lambda)\| \leqslant \|\delta\|$, for $\exp \mathrm{i}(\lambda x^* + \bar{\lambda}x)$ and $\exp - \mathrm{i}(\lambda x^* + \bar{\lambda}x)$ are unitaries. By Liouville's theorem, $f(\lambda) =$ a constant, and so

$$f(\lambda) = \delta(\exp(\mathrm{i}\lambda x^*))\exp(-\mathrm{i}\lambda x^*) = 0 \qquad \text{for } \lambda \in \mathbb{C};$$

hence $\delta(\exp(\mathrm{i}\lambda x^*)) = 0$ and so $(\mathrm{d}/\mathrm{d}\lambda)\delta(\exp(\mathrm{i}\lambda x^*))|_{\lambda=0} = \delta(\mathrm{i}x^*) = 0$. □

Fuglede's theorem follows as a corollary.

2.2.6 Corollary (Fuglede [62]) *Let T be a bounded normal operator on a Hilbert space $\mathscr{H}$ and S a bounded operator on $\mathscr{H}$. If $[S, T] = 0$, then $[S, T^*] = 0$.*

2.2.7 Theorem *Let $\mathscr{A}$ be a C*-algebra, δ a derivation on $\mathscr{A}$. If $[\delta(x), x] = 0$ for a normal element x of $\mathscr{A}$, then $\delta(x) = 0$.*

Proof By Fuglede's theorem, $[\delta(x), x] = 0$ implies $[\delta(x), x^*] = 0$.

$$\begin{array}{c}\delta(x^*x) = \delta(x^*)x + x^*\delta(x)\\ \|\\ \delta(xx^*) = \delta(x)x^* + x\delta(x^*).\end{array}$$

Hence $[\delta(x^*), x] = 0$ and so $[\delta(x^*), x^*] = 0$.

Let $\mathscr{B}$ be a C*-subalgebra of $\mathscr{A}$ generated by $\{1, x, \delta(x), \delta(x^*)\}$; then x belongs to the center of $\mathscr{B}$. Let $x = x_1 + ix_2$ ($x_1{}^* = x_1, x_2{}^* = x_2$) and let P be any closed primitive ideal of $\mathscr{B}$; then there is a real number λ such that $x_1 - \lambda 1 = a_1{}^2 - a_2{}^2$, where a_1, $a_2 \in P \cap \mathscr{B}^+$.

$$\delta(x_1) = \delta(x_1 - \lambda 1) = \delta(a_1)a_1 + a_1\delta(a_1) - \delta(a_2)a_2 - a_2\delta(a_2).$$

Clearly $\delta(x_1) \in \mathscr{B}$. Let ϕ be any state on $\mathscr{A}$ such that $\phi(P) = 0$; then

$$\begin{aligned}|\phi(\delta(x_1))| &\leqslant |\phi(\delta(a_1)a_1)| + |\phi(a_1\delta(a_1))| + |\phi(\delta(a_2)a_2)| + |\phi(a_2\delta(a_2))|\\ &\leqslant \phi(\delta(a_1)^*\delta(a_1))^{1/2}\phi(a_1{}^2)^{1/2} + \phi(\delta(a_2)^*\delta(a_2))^{1/2}\cdot\phi(a_2{}^2)^{1/2}\\ &\quad + \phi(\delta(a_1)\delta(a_1)^*)^{1/2}\phi(a_1{}^2)^{1/2} + \phi(\delta(a_2)\delta(a_2)^*)^{1/2}\cdot\phi(a_2{}^2)^{1/2} = 0;\end{aligned}$$

hence $\delta(x_1) \in P$, and so $\delta(x_1) \in \bigcap_{\forall P} P = (0)$ for C*-algebra are semi-simple.

Similarly $\delta(x_2) = 0$. Hence $\delta(x) = 0$. □

The five consequences following from this are now recorded.

2.2.8 Corollary (Singer [96]) *Let $\mathscr{A}$ be a commutative C*-algebra and let δ be a derivation on $\mathscr{A}$; then $\delta = 0$.*

2.2.9 Corollary (Putnam [149]) *Let T be a bounded normal operator on a Hilbert space $\mathscr{H}$ and let S be a bounded operator on $\mathscr{H}$. If $[T, S]$ commutes with T, then $[T, S] = 0$.*

Proof Let $B(\mathscr{H})$ be the C*-algebra of all bounded operators on $\mathscr{H}$. Put $\delta_S(x) = [S, X]$ ($X \in B(\mathscr{H})$); then $[\delta_S(T), T] = 0$. By Theorem 2.2.7, $\delta_S(T) = 0$. □

2.2.10 Corollary (Putnam [149]) *Let $A \in B(\mathscr{H})$ and suppose that $[A, [A^*, A]] = 0$; then $[A^*, A] = 0$ – i.e. A is normal.*

Proof Put $\delta_A(X) = [A, X]$ ($X \in B(\mathscr{H})$). Then $\delta_A{}^2(A^*) = 0$. Hence by Theorem 2.2.1, $\delta_A(A^*)$ is a generalized nilpotent. Since $AA^* - A^*A$ is self-adjoint, $AA^* - A^*A = 0$. □

2.2.11 Corollary (Kato–Taussky [99]) *Suppose that* $[[A^*, A], [A, [A^*, A]]] = 0$ *for some* $A \in B(\mathscr{H})$; *then* $[A, A^*] = 0$.

Proof $[A^*, A]$ is self-adjoint, and $[A^*, A]$ commutes with $[[A^*, A], A]$; hence by Theorem 2.2.7, $[[A^*, A], A] = 0$ and so by Corollary 2.2.10, $[A, A^*] = 0$. □

2.2.12 Corollary *Let* $C(E)$ *be the algebra of all compact operators on an infinite-dimensional Banach space* E. *For* $c \in C(E)$, *there are no two elements* a, b *in* $B(E)$ *such that* $ab - ba = \lambda 1 + c$ $(\lambda \neq 0)$, *where* $B(E)$ *is the Banach algebra of all bounded operators on* E.

Proof Suppose that $ab - ba = \lambda 1 + c$ $(\lambda \neq 0)$. Consider the Banach algebra $B(E)/C(E)$; then $\tilde{a}\tilde{b} - \tilde{b}\tilde{a} = \lambda \tilde{1}$ in $B(E)/C(E)$, where $\tilde{a}$, $\tilde{b}$, $\tilde{1}$ are the canonical images of a, b, 1 in $B(E)/C(E)$ respectively, a contradiction. □

For separable infinite-dimensional Hilbert spaces $\mathscr{H}$, the converse of this corollary is also true.

2.2.13 Theorem (Brown–Pearcy [45]) *Suppose that* $x \neq \lambda 1 + c$ *where* $\lambda \neq 0$, $c \in C(\mathscr{H})$; *then there are two elements* a, b *in* $B(\mathscr{H})$ *such that* $ab - ba = x$.

2.2.14 Notes and remarks The theorem of Kleinecke [110] and Sirokov (Theorem 2.2.1) is the affirmative solution of a conjecture of Kaplansky (cf. [213]) which originated from a theorem of Jacobson [214] developed in 1935 on an algebra of characteristic zero and probably from the Wintner and Wielandt theorem (Corollary 2.2.2). Many mathematicians tried to solve the conjecture without success before the solution of Kleinecke and Sirokov. The proof, presented in 2.2.1, is due to Kleinecke and is surprisingly simple and elementary. It is remarkable to realize that mathematicians took many years to find the affirmative solution of the conjecture which requires such an elementary proof. However such a situation often occurs in mathematics. 'Ingenious' does not necessarily mean 'difficult' or 'complex'. The theorem of Fuglede (Corollary 2.2.6) is the solution of an outstanding problem in functional analysis, proposed by von Neumann. Fuglede was then a young mathematician who had just started research and was visiting the Institute for Advanced Studies, Princeton, when he solved it. The beautiful proof, presented here, is essentially due to Rosenblum [157]. Putnam's theorem (2.2.9) has an interesting application which we shall describe in §2.6.

In general, commutators $[X, Y]$ appear in various branches of functional analysis. We will mention some of them in the following. In the lecture notes *Integral Operators, Commutators, Traces, Index and Homology* by Helton and

Howe [75], the commutators $[X, Y]$ are assumed to belong to the trace class. Commutators also appear in the invariant subspace problem of bounded operators on a Hilbert space. Let A be a bounded operator on an infinite-dimensional Hilbert space $\mathscr{H}$ and suppose that Rank $[A, K] \leqslant 1$ for some non-zero compact operator K on $\mathscr{H}$; then A has a proper invariant closed subspace ([121], [51]). If Rank $[T, S] = 1$ (S, $T \in B(\mathscr{H})$), and S and S^* have non-zero finite-dimensional kernels, then T has a proper invariant closed subspace ([101], [102]). An interesting problem for bounded derivations is whether the invariant subspace problem of general bounded operators on a Hilbert space could be treated within the theory. Finally, the author would like to recommend the text book for a more detailed study of commutators by Putnam [150].

2.3 Continuity of everywhere-defined derivations

Let δ be a bounded derivation on a Banach algebra $\mathscr{B}$; then $\{\exp(t\delta) | t \in \mathbb{R}\}$ is a norm-continuous one-parameter group of automorphisms on $\mathscr{B}$. One of the main problems in the theory of Banach algebras is the uniqueness of norms – i.e. if $\|\cdot\|$ is the original norm of $\mathscr{B}$ and $\|\cdot\|'$ another norm on $\mathscr{B}$ under which $\mathscr{B}$ is again a Banach algebra, can we conclude that $\|\cdot\|'$ is equivalent to $\|\cdot\|$? (i.e. there are two positive numbers γ_1, γ_2 such that $\gamma_1 \|x\| \leqslant \|x\|' \leqslant \gamma_2 \|x\|$ for all $x \in \mathscr{B}$.) This problem was initiated by Gelfand [197] in normed ring theory. Since then, many mathematicians have extended it to various Banach algebras. The most general theorem on the uniqueness of norms is due to Johnson [211].

The uniqueness of norms implies the automatic continuity of automorphisms. Therefore, as its infinitesimal form, the automatic continuity of everywhere-defined derivations on Banach algebras was naturally conjectured. This will be discussed in this section. The first result is due to the author.

2.3.1 Theorem *Let $\mathscr{A}$ be a C*-algebra and let δ be a derivation on $\mathscr{A}$ (i.e., $\mathscr{D}(\delta) = \mathscr{A}$); then δ is bounded.*

Proof It is enough to assume that $\mathscr{A}$ has an identity. In fact, if $\mathscr{A}$ has no identity, we shall consider a C*-algebra $\mathscr{A}_1$ obtained by adjoining an identity to $\mathscr{A}$, and define $\delta(1) = 0$; then δ can be uniquely extended to a derivation on $\mathscr{A}_1$. Let $\delta^*(x) = \delta(x^*)^*$ ($x \in \mathscr{A}$); then

$$\delta = \frac{\delta + \delta^*}{2} + i\frac{i\delta^* - i\delta}{2}$$

and so one may assume that δ is a *-derivation.

For $x (= x^*) \in \mathscr{A}$, let ϕ be a state on $\mathscr{A}$ such that $|\phi(x)| = \|x\|$ then we shall show that $\phi(\delta(x))=0$. We may assume that $\phi(x)=\|x\|$; otherwise, consider $-x$ instead of x.

Put $\|x\|1-x=h^2$ ($h\geqslant 0$, $h\in\mathscr{A}$); then

$$\begin{aligned}|-\phi(\delta(x))| &= |\phi(\delta(\|x\|1-x))| = |\phi(\delta(h^2))| \\ &= |\phi(h\delta(h))+\phi(\delta(h)h)| \\ &\leqslant \phi(h^2)^{1/2}\phi(\delta(h)^2)^{1/2}+\phi(\delta(h)^2)^{1/2}\phi(h^2)^{1/2}=0.\end{aligned}$$

Hence $\phi(\delta(x))=0$. Suppose that $x_n(=x_n^*)\to 0$ and $\delta(x_n)\to y(\neq 0)$. Let ϕ_n be a state on $\mathscr{A}$ such that $|\phi_n(y+x_n)|=\|y+x_n\|$, and let ϕ_0 be an accumulation point of $\{\phi_n\}$ in the state space of $\mathscr{A}$; then

$$\begin{aligned}|\phi_{n_j}(y+x_{n_j})-\phi_0(y)| &\leqslant |\phi_{n_j}(y+x_{n_j})-\phi_{n_j}(y)|+|\phi_{n_j}(y)-\phi_0(y)| \\ &\leqslant \|y+x_{n_j}-y\|+|\phi_{n_j}(y)-\phi_0(y)|\to 0\end{aligned}$$

for some subsequence (n_j) of (n) depending on y. Hence $|\phi_0(y)|=\|y\|$ and so $\phi_0(\delta(y))=0$. On the other hand,

$$0=\phi_{n_j}(\delta(y+x_{n_j}))=\phi_{n_j}(\delta(y)+\delta(x_{n_j}))\to\phi_0(\delta(y)+y).$$

Hence $\phi_0(y)=0$, a contradiction. By the closed graph theorem, δ is bounded. □

Theorem 2.3.1 can be extended to general semi-simple Banach algebras, though the proof is much more complicated and will be omitted here.

2.3.2 Theorem (Johnson–Sinclair [84]) *Let $\mathscr{A}$ be a semi-simple Banach algebra and let δ be a derivation on $\mathscr{A}$; then δ is continuous.*

2.3.3 Notes and remarks The notion of derivations can be extended to a linear mapping of a subalgebra into a larger algebra as follows. Let $\mathscr{B}$ be a Banach algebra and let $\mathscr{D}$ be a closed subalgebra of $\mathscr{B}$. δ is said to be a derivation of $\mathscr{B}$ into $\mathscr{D}$ if it satisfies the following conditions:

$$\delta(ax)=\delta(a)x+a\delta(x) \qquad \text{for } a, x\in\mathscr{D}.$$

Ringrose [208] proved that a derivation of a C*-subalgebra into a larger C*-algebra is automatically continuous. Various results on derivations have been extended to higher dimensional cohomology in studies by Kadison–Ringrose [209], Johnson [210] and others.

Theorem 2.3.1 was first proved by the author [161]. It is the affirmative solution to a conjecture by Kaplansky [96]. A simple proof of the theorem was found by Kishimoto [103].

References [152], [161].

2.4 Quantum field-theoretic observations (some unbounded derivations)

When a local quantum field theory is formulated in C*- or W*-dynamical systems, the positivity of energy and momenta corresponds to the positivity of the Hamiltonians defined by the corresponding dynamical systems. A vacuum state is naturally defined as the state with the lowest energy in a dynamical system. Physical notions like local commutativity, and asymptotic abelianness are analogously introduced into dynamical systems.

In this section, we shall mainly discuss the observability of those positive hamiltonians, and we shall develop a beautiful theory concerning the so-called spectrum condition. We shall also discuss the problem of extending this spectrum condition in field theory to a more general mathematical setting.

Case 1 The case of positive energy with a vacuum state

2.4.1 Theorem (Araki [2]) *Let $t \mapsto \alpha_t$ be a σ-weakly continuous one-parameter group of *-automorphisms of a W*-algebra $\mathcal{M}$ on a Hilbert space $\mathcal{H}$. Suppose that there exists a strongly continuous one-parameter unitary group $t \mapsto U_t \in B(\mathcal{H})$ with non-negative spectrum (i.e. in the Stone representation $U_t = \exp(\mathrm{i}tH)$, H is a positive self-adjoint operator in $\mathcal{H}$) such that $\alpha_t(a) = U_t a U_t^*$ ($a \in \mathcal{M}$) and H has a zero-eigenvalue vector ξ_0 with $\|\xi_0\| = 1$ and ξ_0 is a cyclic vector for $\mathcal{M}$. Then $U_t \in \mathcal{M}$ ($t \in R$) – namely, $\{U_t\}$ is observable.*

Proof Since $U_t \mathcal{M} U_t^* = \mathcal{M}$, it follows that $U_t \mathcal{M}' U_t^* = \mathcal{M}'$. Let $x' \in \mathcal{M}'$ and $a, b \in \mathcal{M}$. Then

$$\begin{aligned} f(t) &= (U_t x' U_t^* a\xi_0, b\xi_0) = (aU_t x' U_t^* \xi_0, b\xi_0) \\ &= (aU_t x'\xi_0, b\xi_0) = (a\exp(\mathrm{i}tH)x'\xi_0, b\xi_0). \end{aligned}$$

Since $H \geqslant 0$, $f(t)$ can be extended to a bounded analytic function $\tilde{f}$ on the upper half-plane of the complex field as follows;

$$\begin{aligned} \tilde{f}(t+\mathrm{i}s) &= (a\exp \mathrm{i}(t+\mathrm{i}s)Hx'\xi_0, b\xi_0) \\ &= (a\exp(\mathrm{i}tH)\exp(-sH)x'\xi_0, b\xi_0) \qquad (s \geqslant 0). \end{aligned}$$

On the other hand,

$$\begin{aligned} f(t) &= (U_t x' U_t^* a\xi_0, b\xi_0) = (a\xi_0, bU_t x'^* U_t^* \xi_0) \\ &= (a\xi_0, bU_t x'^* \xi_0) = (U_t^* b^* a\xi_0, x'^* \xi_0) \\ &= (\exp(-\mathrm{i}tH)b^* a\xi_0, x'^* \xi_0). \end{aligned}$$

Therefore $f(t)$ can be extended to a bounded analytic function $\tilde{\tilde{f}}$ on the lower half-plane of the complex field. Since $\tilde{f}(t) = \tilde{\tilde{f}}(t) = f(t)$ ($t \in R$), $f(t)$ can be extended to a bounded analytic function on the whole complex plane

(Painleve's theorem) (cf. [77]). Hence by Liouville's theorem, $f(t) = \text{constant}$ and so $(U_t x' U_t^* a\xi_0, b\xi_0) = (x' a\xi_0, b\xi_0)$. Since $[\mathscr{M}\xi_0] = \mathscr{H}$, $U_t x' U_t^* = x'$ and so $U_t \in \mathscr{M}'' = \mathscr{M}$. □

Case 2 **The general case** (the case of positive energy without the assumption of the existence of a vacuum state)

In this case, the theory becomes more complicated. We will use the idea of Arveson's spectral theory of one-parameter automorphism groups [6]. Let $t \mapsto U_t$ be a strongly continuous one-parameter group of unitary operators on a Hilbert space, with the Stone representation $U_t = \exp(itH) = \int_{-\infty}^{\infty} \exp(it\gamma)\,\mathrm{d}P(\gamma)$. Let $\beta_t(a) = U_t a U_t^*$ $(a \in B(\mathscr{H}))$; then $t \mapsto \beta_t$ is a σ-weakly continuous one-parameter group of *-automorphisms on $B(\mathscr{H})$. Let $L^1(\mathbb{R})$ be the group algebra of the real line $\mathbb{R}$ consisting of all Lebesque integrable functions on $\mathbb{R}$. For $f \in L^1(\mathbb{R})$, let $\hat{f}(\gamma) = \int_{-\infty}^{\infty} f(t) \exp(it\gamma)\,\mathrm{d}t$, $\beta(f) = \int_{-\infty}^{\infty} f(t)\beta_t\,\mathrm{d}t$ (the integral is defined by using the $\sigma(B(\mathscr{H}), B(\mathscr{H})_*)$-topology) and $U(f) = \int_{-\infty}^{\infty} f(t) U_t\,\mathrm{d}t$. $\mathrm{Supp}(\hat{f})$ is the support of $\hat{f}$ in the dual group $\hat{\mathbb{R}}$ of $\mathbb{R}$. The following fact is known: given a compact $K \subset \hat{\mathbb{R}}$ and an open set $W \supset K$, there is an $h \in L^1(\mathbb{R})$ such that $\hat{h}(\gamma) = 1$ for $\gamma \in K$ and $\hat{h}(\gamma) = 0$ for $\gamma \in \hat{\mathbb{R}} \setminus W$.

2.4.2 Lemma *If* $\mathrm{supp}(\hat{f}) \subset [\mu, \infty)$, *then* $\beta(f)(a)P([\lambda, \infty))\mathscr{H} \subset P([\lambda + \mu, \infty))\mathscr{H}$ *for* $\lambda, \mu \in \mathbb{R}$, $a \in B(\mathscr{H})$, *where* $P([l, \infty)) = \int_l^{\infty} \mathrm{d}P(\gamma)$ $(l \in \mathbb{R})$.

Proof Suppose $g \in L^1(\mathbb{R})$ and $\mathrm{supp}(\hat{g}) \subset (\lambda - \varepsilon, \infty)$ with $\varepsilon > 0$; then it is enough to show that $U(h)\beta(f)(a)U(g) = 0$ for $h \in L^1(\mathbb{R})$ with $\mathrm{supp}(\hat{h}) \subset (-\infty, \lambda + \mu - \varepsilon)$. We may also assume that $\hat{f}$, $\hat{g}$, $\hat{h}$ have compact supports respectively.

$$U(h)\beta(f)(a)U(g) = \left(\int h(t)U_t\,\mathrm{d}t\right)\left(\int f(s)U_s a U_s^*\,\mathrm{d}s\right)\left(\int g(r)U_r\,\mathrm{d}r\right)$$
$$= \iiint \mathrm{d}t\,\mathrm{d}s\,\mathrm{d}r\,h(t)f(s)g(r)U_{s+t}aU_{-s+r}.$$

After the change of variables $x = t$, $y = s + t$, $z = r - s$ (note the Jacobian $(J(x, y, z)/J(t, s, r)) = 1$), and an application of Fubini's theorem, one finds

$$\iiint \mathrm{d}t\,\mathrm{d}s\,\mathrm{d}r\,h(t)f(s)g(r)U_{t+s}U_{-s+r}$$
$$= \iiint \mathrm{d}x\,\mathrm{d}y\,\mathrm{d}z\,h(x)f(y-x)g(z+y-x)U_y a U_z$$
$$= \iint h*(f\cdot g_z)(y)U_y a U_z\,\mathrm{d}y\,\mathrm{d}z,$$

where $g_z(x) = g(z+x)$.

$$h*(f\cdot g_z) = \hat{h}\cdot(\hat{f}*\hat{g}_z)$$
$$\operatorname{supp}(\hat{f}*\hat{g}_z) \subset \operatorname{supp}(\hat{f}) + \operatorname{supp}(\hat{g}_z)$$
$$= \operatorname{supp}(\hat{f}) + \operatorname{supp}(\hat{g}) \subset [\mu+\lambda-\varepsilon, \infty).$$

The last set is disjoint from $\operatorname{supp}(\hat{h})$, and so $h*(f\cdot g_z) = 0$. Hence $U(h)\beta(f)(a)U(g) = 0$. □

One-parameter groups

2.4.3 Theorem (Borchers [16]) *Let $t \mapsto \alpha_t$ be a σ-weakly continuous one-parameter group of *-automorphisms of a W*-algebra $\mathscr{M}$ containing the identity operator in a Hilbert space $\mathscr{H}$. Then the following two conditions are equivalent:*

(1) *There is a strongly continuous one-parameter unitary group $t \mapsto U_t \in B(\mathscr{H})$ with non-negative spectrum* (namely, $U_t = \exp(itH)$, $H \geqslant 0$) *such that* $\alpha_t(a) = U_t a U_t^*$ $(a \in \mathscr{M},\ t \in \mathbb{R})$.

(2) *There is a strongly continuous one-parameter unitary group $t \mapsto V_t \in \mathscr{M}$ with non-negative spectrum such that* $\alpha_t(a) = V_t a V_t^*$ $(a \in \mathscr{M},\ t \in \mathbb{R})$.

Proof (2)⇒(1) is trivial. We shall show (1)⇒(2). Let $H = \int_{-\infty}^{\infty} \gamma\, dP(\gamma)$ be the spectral decomposition of H; then $P([t, \infty)) = 1$ when $t \leqslant 0$. For $t \in \mathbb{R}$, let e_1, e_2 be projections in $\mathscr{M}$ such that $e_1, e_2 \leqslant P([t, \infty))$ and let $e_1 \vee e_2$ be the supremum of e_1 and e_2 in $\mathscr{M}_p$, where $\mathscr{M}_p$ is the set of all projections in $\mathscr{M}$. Since $(1 - P([t, \infty)))(e_1\mathscr{H} + e_2\mathscr{H}) = 0$, $e_1 \vee e_2 \leqslant P([t, \infty))$. Therefore there is the largest projection P_t in $\mathscr{M}$ such that $P_t \leqslant P([t, \infty))$.

Since $U_s P_t U_s^* \leqslant P([t, \infty))$ and $U_s P_t U_s^* \in \mathscr{M}$, $U_s P_t U_s^* = P_t$ for $s \in \mathbb{R}$. Clearly $P_{t_1} \geqslant P_{t_2}$ if $t_1 \leqslant t_2$. For each $t \in \mathbb{R}$, define $q_t = \wedge_{s<t} P_s$ in $\mathscr{M}$. Then $\{q_t\}$ is a decreasing family of projections which is left continuous in t and $q_t \to 0$ strongly as $t \to \infty$. Moreover $q_t = 1$ for $t \leqslant 0$. Thus there is a unique projection-valued measure P_1 on $\mathbb{R}$ such that $P_1([t, \infty)) = q_t (t \in \mathbb{R})$.

Let $V_t = \int_{-\infty}^{\infty} \exp(it\gamma)\, dP_1(\gamma)$; then $V_t \in \mathscr{M}$ and $t \to V_t$ has non-negative spectrum. Since $U_t P_1(\gamma) U_t^* = P_1(\gamma)$ for $t \in \mathbb{R}$ and $\gamma \in \hat{\mathbb{R}}$, $P(\gamma)P_1(\mu) = P_1(\mu)P(\gamma)$ for $\gamma, \mu \in \hat{\mathbb{R}}$. Let $K = \int_0^{\infty} \gamma\, dP_1(\gamma)$; then $P_1([t, \infty)) \leqslant P([t, \infty))$ $(t \in \mathbb{R})$ and the commutativity of $\{P(\gamma)\}$ and $\{P_1(\mu)\}$ implies $\overline{H-K} \geqslant 0$, where $\overline{H-K}$ is the closure of $H-K$. Now let $W_t = U_t V_{-t}$; then $W_t = \exp it(\overline{H-K})$. Let $W_t = \int_{-\infty}^{\infty} \exp(i\mu t)\, dP_2(\mu)$ and $\gamma_t(a) = W_t a W_t^*$ $(a \in \mathscr{M}, t \in \mathbb{R})$. Let F_t be the largest projection in $\mathscr{M}$ such that $F_t \leqslant P_2([t, \infty))$; then $U_s F_t U_s^* \leqslant U_s P_2([t, \infty)) U_s^* = P_2([t, \infty))$ and so $U_s F_t U_s^* = F_t (s \in \mathbb{R})$. Similarly $V_s F_t V_s^* = F_t$. Since $tF_t \leqslant \overline{H-K}$, $H \geqslant K + tF_t$. Let $K + tF_t = \int_{-\infty}^{\infty} \gamma\, dP_3(\gamma)$ be the spectral decomposition of $K + tF_t$; then, if $t \geqslant 0$, $P_3([t, \infty)) \leqslant P([t, \infty))$ and $P_1([t, \infty)) \leqslant$

$P_3([t,\infty))$. By the maximality of $P_1([t,\infty))$, $P_1([t,\infty)) = P_3([t,\infty))$ $(t \geqslant 0)$; hence $K = K + tF_t$ $(t>0)$, and so $F_t = 0$ $(t>0)$. By Lemma 2.4.2, for $f \in L^1(\mathbb{R})$ with $\mathrm{supp}(\hat{f}) \subset [t,\infty)$ $(t>0)$, $\gamma(f)(a)\mathscr{H} = \gamma(f)(a)P_2([0,\infty))\mathscr{H} \subset P_2([t,\infty))\mathscr{H}$ for $a \in \mathscr{M}$. Let e be the orthogonal projection of $\mathscr{H}$ onto $[\gamma(f)(a)\mathscr{H}]$. Since $\gamma(f)(a) \in \mathscr{M}$, $e \in \mathscr{M}$ and so $e \leqslant P_2([t,\infty))$ $(t>0)$. Hence $e = 0$. Therefore $\gamma(f)(a) = 0$ for $a \in \mathscr{M}$ and $f \in L^1(R)$ with $\mathrm{supp}(\hat{f}) \subset [t,\infty)$ $(t>0)$.

$$\gamma(\bar{f})(a^*) = \int \overline{f(t)}\gamma_t(a^*)\,\mathrm{d}t = \left\{\int f(t)\gamma_t(a)\,\mathrm{d}t\right\}^* = 0.$$

Since $\mathrm{supp}(\hat{\bar{f}}) = -\,\mathrm{supp}\,\hat{f}$, $\gamma(g)(a) = 0$ for $a \in \mathscr{M}$ and $g \in L^1(R)$ with $\mathrm{supp}(\hat{g}) \subset (-\infty,0)\cup(0,\infty)$. For $g \in L^1(R)$ with $\int g(t)\,\mathrm{d}t = 0$, $\hat{g}(0) = 0$; any such g can be approximated by h with $\hat{h}$, zero in neighborhood of 0 and so $\gamma(g)(a) = \int g(t)\gamma_t(a)\,\mathrm{d}t = 0$, $\int_{\mathbb{R}} \langle \gamma_t(a), \xi\rangle g(t)\,\mathrm{d}t = 0$ $(\xi \in \mathscr{M}_*)$ and so $\langle \gamma_t(a), \xi \rangle =$ constant. Hence $\langle \gamma_t(a), \xi\rangle = \langle a, \xi\rangle$ $(a \in \mathscr{M}, \xi \in \mathscr{M}_*)$.

Therefore $\gamma_t(a) = a$ $(a \in \mathscr{M})$ and so $W_t \in \mathscr{M}'$. Finally we have $V_t a V_t^* = U_t a U_t^*$ $(a \in \mathscr{M})$. □

Actions by $\mathbb{R}^n$

Next we shall generalize Theorem 2.4.3 to higher dimensional cases. Let $\mathbb{R}^n$ be the n-dimensional euclidean group and let $t \mapsto u_t$ $(t \in \mathbb{R}^n)$ be a strongly continuous unitary representation of $\mathbb{R}^n$ into the group of unitary operators in a Hilbert space $\mathscr{H}$; then by Stone's theorem, there is a projection-valued measure P on the dual group $\hat{\mathbb{R}}^n (= \mathbb{R}^n)$ such that $u_t = \int_{\mathbb{R}^n} \exp(\mathrm{i}\langle t, x\rangle)\,\mathrm{d}P(x)$, where $\langle\ ,\ \rangle$ is the inner product of $\mathbb{R}^n$.

Now suppose that there exists a closed convex subset Λ in $\mathbb{R}^n$ such that no straight line is contained in Λ and $P(\Lambda) = 1_{\mathscr{H}}$, where $1_{\mathscr{H}}$ is the identity operator on $\mathscr{H}$. We shall call this condition the spectrum condition. Take an element $\omega \notin \Lambda$ and let $\Lambda_0 = \Lambda - \omega$. Since Λ_0 contains no straight line, there is a linearly independent family $\{f_1, f_2, \ldots, f_n\}$ of linear functionals such that $\Lambda_0 \subseteqq \{x \in \mathbb{R}^n \mid \text{all } f_j(x) \geqslant 0\}$. Let $f_j(x) = \langle t_j, x\rangle$ $(x \in \mathbb{R}^n)$; then

$$\begin{aligned} u_{\lambda t_j}\exp(-\mathrm{i}\langle \lambda t_j, \omega\rangle) &= \int_{\Lambda} \exp(\mathrm{i}\langle \lambda t_j, x - \omega\rangle)\,\mathrm{d}P(x) \\ &= \int_{\Lambda_0} \exp(\mathrm{i}\langle \lambda t_j, x\rangle)\mathrm{d}P(x+\omega) \qquad \text{for } \lambda \in \mathbb{R}. \end{aligned}$$

Since $\langle t_j, x\rangle \geqslant 0$ for $x \in \Lambda_0$, $u_{\lambda t_j}\exp(-\mathrm{i}\langle \lambda t_j, \omega\rangle) = \exp(\mathrm{i}\lambda H_j)$ for $\lambda \in \mathbb{R}$, where $H_j \geqslant 0$.

2.4.4 Proposition *Under the above spectrum condition, let $\mathscr{M}$ be a W^*-algebra containing the identity operator on a Hilbert space $\mathscr{H}$ such that $u_t \mathscr{M} u_t^* = \mathscr{M}$*

($t \in \mathbb{R}^n$); then there is a strongly continuous unitary representation $t \mapsto V_t (\in \mathcal{M})$ of $\mathbb{R}^n$ such that $V_t a V_t^* = u_t a u_t^*$ ($a \in \mathcal{M}, t \in \mathbb{R}^n$).

Proof By the above consideration, $u_{\lambda t_j} \exp(-i\langle \lambda t_j, \omega \rangle) = \exp(i\lambda H_j)$ with $H_j \geqslant 0$. Let $H_j = \int_{-\infty}^{\infty} \gamma \, dP(\gamma)$ be the spectral decomposition of H_j. Then by the consideration in the proof of Theorem 2.4.3, there is the largest projection $q_{\lambda,j}$ in $\mathcal{M}$ such that $q_{\lambda,j} \leqslant P_j([\lambda, \infty))$. Since $u_s q_{\lambda,j} u_s^* \leqslant P_j([\lambda, \infty))$ for $s \in \mathbb{R}^n$ and $u_s q_{\lambda,j} u_s^* \in \mathcal{M}$, $u_s q_{\lambda,j} u_s^* = q_{\lambda,j}$ for $s \in R^n$. By the considerations in the proof of Theorem 2.4.3 there is a projection-valued measure Q_j on $\mathbb{R}$ such that $Q_j([\lambda, \infty)) = \wedge_{\mu < \lambda} q_{\mu,j}$ in $\mathcal{M}$. Let $V_{\lambda t_j} = \int_{-\infty}^{\infty} \exp(i\lambda\gamma) \, dQ_j(\gamma)$; then

$$V_{\lambda t_j} a V_{\lambda t_j}^* = u_{\lambda t_j} a u_{\lambda t_j}^* \qquad (a \in \mathcal{M}).$$

Since

$$V_{\lambda t_k} q_{\mu,j} V_{\lambda t_k}^* = u_{\lambda t_k} q_{\mu,j} u_{\lambda, t_k}^* = q_{\mu,j} \qquad (\mu, \lambda \in \mathbb{R}),$$
$$V_{\lambda t_k} V_{\mu t_j} = V_{\mu t_j} V_{\lambda t_k} \qquad (\lambda, \mu \in \mathbb{R} \text{ and } j, k = 1, 2, \ldots, n).$$

Consider a strongly continuous unitary representation $\lambda_1 t_1 + \lambda_2 t_2 + \cdots + \lambda_n t_n \mapsto V_{\lambda_1 t_1} V_{\lambda_2 t_2} \cdots V_{\lambda_n t_n}$ of $\mathbb{R}^n$ into $\mathcal{M}$; then clearly

$$V_t a V_t^* = u_t a u_t^* \qquad (a \in \mathcal{M}) \text{ for } t \in \mathbb{R}^n.$$

Moreover, $V_t = \int_\Gamma \exp(i\langle t, x \rangle) \, dQ(x)$, where $\Gamma = \bigcap_{j=1}^n \{x \in \mathbb{R}^n \mid f_j(x) \geqslant 0\}$ and so $V_t \exp(i\langle t, \omega \rangle) = \int_\Gamma \exp(i\langle t, x + \omega \rangle) \, dQ(x) = \int_\Gamma \exp(i\langle t, x \rangle) \, dQ(x - \omega)$.

□

Now for $u_t = \int_{\mathbb{R}^n} \exp(i\langle t, x \rangle) \, dP(x)$, define $Sp(u)$ to be equal to the support of P which equals the complement of the largest open subset O such that $P(O) = 0$.

2.4.5 Theorem *Let $t \mapsto \alpha_t$ be a σ-weakly continuous representation of $\mathbb{R}^n$ by *-automorphisms on a factor $\mathcal{M}$ containing the identity operator in a Hilbert space $\mathcal{H}$. Then the following two conditions are equivalent:*

(1) *There is a strongly continuous unitary representation $t \mapsto u_t \in B(\mathcal{H})$ of $\mathbb{R}^n$ satisfying the spectrum condition '$Sp(u) \subset \Lambda$' such that $\alpha_t(a) = u_t a u_t^*$ ($a \in \mathcal{M}, t \in \mathbb{R}^n$).*
(2) *There is a strongly continuous unitary representation $t \mapsto V_t (\in \mathcal{M})$ of $\mathbb{R}^n$ such that $Sp(V) \subseteq Sp(u) \subset \Lambda$ and $\alpha_t(a) = V_t a V_t^*$ ($a \in \mathcal{M}, t \in \mathbb{R}^n$).*

To prove Theorem 2.4.5, we shall provide a lemma.

2.4.6 Lemma *Let $t \mapsto W_t$ be a strongly continuous unitary representation of $\mathbb{R}^n$ into the unitary group of $B(\mathcal{H})$; then $x_0 \in \mathbb{R}^n$ belongs to $Sp(w)$ if and only if there exists a sequence $\{\xi_n\}$ of elements in $\mathcal{H}$ such that $\|\xi_n\| = 1$ and $\|(W_t - \exp(i\langle t, x_0 \rangle)1)\xi_n\|^2 \to 0$ ($n \to \infty$) for each $t \in \mathbb{R}^n$.*

Proof Let $W_t = \int \exp(i\langle t, x\rangle)\, dP(x)$. Suppose that $x_0 \in Sp(W)$. Take a decreasing sequence $\{K_n\}$ of compact neighborhoods of x_0 such that $\bigcap_{n=1}^{\infty} K_n = \{x_0\}$. Let $P_n = \int_{K_n} dp(x)$; then $P_n \neq 0$. Take an element ξ_n in $P_n\mathscr{H}$ with $\|\xi_n\| = 1$; then

$$\|(W_t - \exp(i\langle t, x_0\rangle)1)\xi_n\|^2 = \int |\exp(i\langle t, x - x_0\rangle) - 1|^2\, d\|P(x)\xi_n\|^2$$

$$= \int_{K_n} |\exp(i\langle t, x - x_0\rangle) - 1|^2\, d\|P(x)\xi_n\|^2$$

$$\leqslant \max_{x \in K_n} |\exp(i\langle t, x - x_0\rangle) - 1|^2 \to 0 \qquad (n \to \infty).$$

Conversely suppose that $\|(W_t - \exp(i\langle t, x_0\rangle)1)\xi_n\| \to 0\ (n \to \infty)$ for each $t \in \mathbb{R}^n$ with $\|\xi_n\| = 1$. For $f \in L^1(\mathbb{R}^n)$,

$$\int \|(W_t - \exp(i\langle t, x_0\rangle)1)\xi_n\|^2 f(t)\, dt$$

$$= \int f(t)\, dt \int |\exp(i\langle t, x - x_0\rangle) - 1|^2\, d\|P(x)\xi_n\|^2$$

$$= \int d\|P(x)\xi_n\|^2 \int |\exp(i\langle t, x - x_0\rangle) - 1|^2 f(t)\, dt$$

$$= \int d\|P(x)\xi_n\|^2 \int \{2 - [\exp(i\langle t, x - x_0\rangle) + \exp(i\langle t, x_0 - x\rangle)]\} f(t)\, dt$$

$$= \int \{2\hat{f}(0) - (\hat{f}(x - x_0) + \hat{f}(x_0 - x))\}\, d\|P(x)\xi_n\|^2 \to 0 \qquad (n \to \infty).$$

$\{\hat{f} \mid f \in L^1(\mathbb{R}^n)\}$ is uniformly dense in $C_0(\mathbb{R}^n)$, so that $\int \{2g(0) - (g(x - x_0) + g(x_0 - x))\}\, d\|P(x)\xi_n\|^2 \to 0$ for each $g \in C_0(\mathbb{R}^n)$. Let K be a compact neighborhood of 0 with $K = -K$ and let $g \in C_0(\mathbb{R}^n)$ such that $g(0) = 1$, $g(K^c) = 0$. Then

$$\int_{(K + x_0)^c} \{2g(0) - (g(x - x_0) + g(x_0 - x))\}\, d\|P(x)\xi_n\|^2$$

$$= 2\|P((K + x_0)^c)\xi_n\|^2 \to 0 \qquad (n \to \infty).$$

Hence $\|P(K + x_0)\xi_n\|^2 \to 1$ and so $P(K + x_0) \neq 0$; since K is arbitrary, $x_0 \in Sp(W)$. □

Proof of Theorem 2.4.5 By Proposition 2.4.4, there is a strongly continuous unitary representation $t \mapsto V_t$ of $\mathbb{R}^n$ into $\mathscr{M}$ such that $u_t a u_t^* = V_t a V_t^*$ $(a \in \mathscr{M})$ and so $V_t^* u_t a u_t^* V_t = a$; hence $V_t^* u_t \in \mathscr{M}'$ and so $u_t = V_t V_t'$, where $t \mapsto V_t'$ is a strongly continuous unitary representation of $\mathbb{R}^n$ into $\mathscr{M}'$. Now we shall show that $Sp(u) \supset Sp(V) + Sp(V')$. Let $x_1 \in Sp(V)$, $x_2 \in Sp(V')$ and let K be any compact neighborhood of 0; then by Lemma 2.4.6, there exist two sequences $\{\xi_n\}$ and $\{\eta_n\}$ in $\mathscr{H}$ such that

$$\|[V_t - \exp(\mathrm{i}\langle t, x_1\rangle)1]\xi_n\| \to 0,$$
$$\|(V_t' - \exp(\mathrm{i}\langle t, x_2\rangle)1)\eta_n\| \to 0$$

and $P(x_1 + K_n)\xi_n = \xi_n$, $Q(x_2 + K_n)\eta_n = \eta_n$ with $\|\xi_n\| = \|\eta_n\| = 1$, where $V_t = \int \exp(\mathrm{i}\langle t, x\rangle)\,\mathrm{d}P(x)$ and $V_t' = \int \exp(\mathrm{i}\langle t, x\rangle)\,\mathrm{d}Q(x)$. Now suppose that $P(x_1 + K)Q(x_2 + K) = 0$, then $Z(P(x_1 + K_n))\cdot Z(Q(x_2 + K_n)) = 0$, where $Z(P(x_1 + K_n))$(resp. $Z(P(x_2 + K_n))$) is the central envelope of $P(x_1 + K_n)$ (resp. $Q(x_2 + K_n)$). Since $\mathscr{M}$ is a factor, $Z(P(x_1 + K_n)) = Z(P(x_2 + K_n)) = 1_{\mathscr{H}}$, a contradiction. Therefore we can take $\{\xi_n\}$ and $\{\eta_n\}$ as follows: $\xi_n = \eta_n$ and $\xi_n \in P(x_1 + K_n)Q(x_2 + K_n)\mathscr{H}$. Then

$$\begin{aligned}\|(u_t - \exp(\mathrm{i}\langle t, x_1 + x_2\rangle)1)\xi_n\| &= \|(V_t V_t' - \exp(\mathrm{i}\langle t, x_1 + x_2\rangle)1)\xi_n\| \\ &\leqslant \|V_t(V_t' - \exp(\mathrm{i}\langle t, x_2\rangle)1)\xi_n\| \\ &\quad + \|\exp(\mathrm{i}\langle t, x_2\rangle)(V_t - \exp(\mathrm{i}\langle t, x_1\rangle)1)\xi_n\| \to 0.\end{aligned}$$

Hence $x_1 + x_2 \in Sp(u)$ and so $Sp(V) + Sp(V') \subset Sp(u)$. Take $x_0 \in Sp(V')$ and let $W_t' = \exp(-\mathrm{i}\langle t, x_0\rangle)V_t'$ and $W_t = \exp(\mathrm{i}\langle t, x_0\rangle)V_t$; then $u_t = W_t W_t'$. By the above consideration, $Sp(u) \supseteq Sp(W) + Sp(W')$. Since $0 \in Sp(W')$, $Sp(W) \subset Sp(u)$. Moreover $W_t \in \mathscr{M}$ and $u_t a u_t^* = W_t a W_t^*$. □

Remark By more sophisticated discusions, Theorem 2.4.5 can be extended to a general W*-algebra $\mathscr{M}$. The theorem is due to H. Araki (unpublished).

2.4.7 Corollary (Araki [2]) *Let $t \mapsto \alpha_t$ be a σ-weakly continuous representation of $\mathbb{R}^n$ by *-automorphisms on a W*-algebra $\mathscr{M}$ containing the identity operator in a Hilbert space $\mathscr{H}$. Suppose that there exists a strongly continuous unitary representation $t \mapsto u_t \in B(\mathscr{H})$ satisfying the spectrum condition such that $u_t a u_t^* = \alpha_t(a)$ $(t \in \mathbb{R}^n, a \in \mathscr{M})$ and $\{u_t | t \in \mathbb{R}^n\}$ has one eigenvalue ξ_0 with ξ_0, a cyclic vector for $\mathscr{M}$. Then $u_t \in \mathscr{M}$ $(t \in \mathbb{R})$.*

Proof From the definition of V_t $(t \in \mathbb{R}^n)$ in the proof of Proposition 2.4.4, $V_t\xi_0 = \xi_0$ $(t \in \mathbb{R}^n)$. Define $V_t' = u_t V_t^*$, then $V_t' \in \mathscr{M}'$ and $V_t'\xi_0 = \xi_0$. Hence $aV_t'\xi_0 = V_t' a\xi_0 = a\xi_0$ for $a \in \mathscr{M}$. Since $[\mathscr{M}\xi_0] = \mathscr{H}$, $V_t' = 1_{\mathscr{H}}$ and so $u_t = V_t \in \mathscr{M}$ $(t \in \mathbb{R}^n)$. □

2.4.8 Corollary (Borchers [16]) *Let $t \mapsto \alpha_t$ be a σ-weakly continuous representation of $\mathbb{R}^4$ by *-automorphisms on a W*-algebra $\mathscr{M}$ containing the identity operator in a Hilbert space $\mathscr{H}$. Suppose that there exists a strongly continuous unitary representation $t \mapsto u_t (\in B(\mathscr{H}))$ of $\mathbb{R}^4$ such that $\alpha_t(a) = u_t a u_t^*$ $(a \in \mathscr{M}, t \in \mathbb{R}^4)$. Let $u_t = \int \exp(\mathrm{i}\langle t, x\rangle)\,\mathrm{d}P(x)$ be the Stone representation of $\{u_t | t \in \mathbb{R}^4\}$.*

Suppose that Sp(u) is contained in the forward light cone – i.e.

$$P(\{(x_0, x_1, x_2, x_3) | x_1{}^2 + x_2{}^2 + x_3{}^2 \leqslant c^2 x_0{}^2, x_0 \geqslant 0\}) = 1_{\mathscr{H}};$$

then there is a strongly continuous unitary representation $t \mapsto V_t (\in \mathscr{M})$ *of* $\mathbb{R}^4$ *such that* $u_t a u_t{}^* = V_t a V_t{}^*$ $(a \in \mathscr{M}, t \in \mathbb{R}^n)$ *– namely* $\{\alpha_t | t \in \mathbb{R}^n\}$ *is observable. Moreover Sp(V) is again contained in the forward light cone. If there also is a vacuum state – i.e. there is a vector* ξ_0 *with* $\|\xi_0\| = 1$ *such that* $u_t \xi_0 = \xi_0$ $(t \in \mathbb{R}^4)$ *and* $[\mathscr{M}\xi_0] = \mathscr{H}$; *then* $u_t \in \mathscr{M}$ $(t \in \mathbb{R}^4)$.

2.4.9 Proposition *Let G be a locally compact group and let* $g \mapsto \alpha_g$ $(g \in G)$ *be a σ-weakly continuous representation of G by *-automorphisms on a W*-algebra* $\mathscr{M}$ *containing the identity operator in a Hilbert space* $\mathscr{H}$. *Suppose that there exists a strongly continuous unitary representation* $g \mapsto u_g \in \mathscr{M}$ *such that* $\alpha_g(a) = u_g a u_g{}^*$ $(a \in \mathscr{M}, g \in G)$. Let $\mathscr{H}_0 = \{\xi \in \mathscr{H} | u_g \xi = \xi$ for all $g \in G\}$. If $\dim(\mathscr{H}_0) = 1$ and $[\mathscr{M}\mathscr{H}_0] = \mathscr{H}$; then $\mathscr{M} = B(\mathscr{H})$.

Proof Take $\xi_0 \in \mathscr{H}_0$ with $\|\xi_0\| = 1$. For $a' \in \mathscr{M}'$, $u_g a' \xi_0 = a' u_g \xi_0 = a' \xi_0$; hence $\mathscr{M}' \xi_0 \subset \mathscr{H}_0$. Since $\dim(\mathscr{H}_0) = 1$, $a' \xi_0 = \lambda \xi_0$ with $\lambda \in \mathbb{C}$; then $a a' \xi_0 = a' a \xi_0$ and so $a a' \xi_0 = a \lambda \xi_0 = \lambda a \xi_0$ for $a \in \mathscr{M}$. Hence $a' = \lambda 1_{\mathscr{H}}$ and so $\mathscr{M} = \mathscr{M}'' = B(\mathscr{H})$. □

2.4.10 Corollary (Araki [2]) *Let* $t \mapsto \alpha_t$ *be a σ-weakly continuous representation of* $\mathbb{R}^n$ *by *-automorphisms on a W*-algebra* $\mathscr{M}$ *containing the identity operator in a Hilbert space* $\mathscr{H}$. *Suppose that there exists a strongly continuous unitary representation* $t \mapsto u_t (\in B(\mathscr{H}))$ *of* $\mathbb{R}^n$ *satisfying the spectrum condition and* $\alpha_t(a) = u_t a u_t{}^*$ $(a \in \mathscr{M}, t \in \mathbb{R}^n)$. *Let* $\mathscr{H}_0 = \{\xi \in \mathscr{H} | u_t \xi = \xi$ *for all* $t \in \mathbb{R}^n\}$. *If* $\dim(\mathscr{H}_0) = 1$ *and* $[\mathscr{M}\mathscr{H}_0] = \mathscr{H}$ *(namely the vacuum state is unique), then* $\mathscr{M} = B(\mathscr{H})$.

General groups

2.4.11 Proposition *Let G be a group and let* $g \mapsto \alpha_g$ $(g \in G)$ *be a representation of G by *-automorphisms on a W*-algebra* $\mathscr{M}$ *containing the identity operator in a Hilbert space* $\mathscr{H}$. *Suppose that there is a unitary representation* $g \mapsto u_g \in B(\mathscr{H})$ *of G such that* $\alpha_g(a) = u_g a u_g{}^*$ $(a \in \mathscr{M}, g \in G)$. *Let* $\mathscr{H}_0 = \{\xi \in \mathscr{H} | u_g \xi = \xi$ *for all* $g \in G\}$ *and let P be the orthogonal projection of* $\mathscr{H}$ *onto* $\mathscr{H}_0$. *Suppose that* $[\mathscr{M}\mathscr{H}_0] = \mathscr{H}$ *and* $\{PaP | a \in \mathscr{M}\}$ *is a commutative family; then the W*-algebra* $\{\mathscr{M}, u_g | g \in G\}''$ *is of type I. Moreover if there is an element* $\xi_0 \in \mathscr{H}_0$ *such that* $[\mathscr{M}\xi_0] = \mathscr{H}$, *then* $\{\mathscr{M}, u_g | g \in G\}'$ *is commutative.*

Proof For $x \in \{\mathscr{M}, u_g | g \in G\}'$ and $\xi \in \mathscr{H}_0$, $u_g x \xi = x u_g \xi = x \xi$; hence

$$[\{\mathscr{M}, u_g | g \in G\}' \mathscr{H}_0] \subset \mathscr{H}_0,$$

and so $P\in\{\mathscr{M},u_g|g\in G\}''$. For $a\in\mathscr{M}$, $Pau_gP=PaP$ and $Pu_gaP=Pu_gau_g{}^*u_gP=P\alpha_g(a)u_gP=P\alpha_g(a)P$ and so $P\{\mathscr{M},u_g|g\in G\}''P=$ the σ-weak closure of $\{PaP|a\in\mathscr{M}\}$ in $\{\mathscr{M},u_g|g\in G\}''$, hence P is an abelian projection. $[\{\mathscr{M},u_g|g\in G\}''P\mathscr{H}]\supset[\mathscr{M}\mathscr{H}_0]=\mathscr{H}$; therefore the central support of P in $\{\mathscr{M},u_g|g\in G\}''$ is $1_{\mathscr{H}}$. Hence $\{\mathscr{M},u_g|g\in G\}''$ is of type I.

If there is an element ξ_0 in $\mathscr{H}_0$ such that $[\mathscr{M}\xi_0]=\mathscr{H}$, then $P\{\mathscr{M},u_g|g\in G\}'P=P[P\{\mathscr{M},u_g|g\in G\}P]'P=P\{P\mathscr{M}P\}'P=$ the σ-weak closure of $P\mathscr{M}P$, for $[P\mathscr{M}P\xi_0]=P\mathscr{H}$ and so the commutative W*-algebra $P\{P\mathscr{M}P\}''P$ has a cyclic vector ξ_0, so that $P\{P\mathscr{M}P\}''P$ is maximal. □

2.4.12 Proposition *Let G be a group and let $g\mapsto\alpha_g$ be a representation of G by *-automorphisms on a W*-algebra $\mathscr{M}$ containing the identity operator in a Hilbert space $\mathscr{H}$. Suppose that $[\mathscr{M}\mathscr{H}_0]=\mathscr{H}$ and there exists a $\sigma(\mathscr{M},\mathscr{M}_*)$-dense *-subalgebra $\mathscr{A}$ of $\mathscr{M}$ and a sequence $\{g_n\}$ in G such that $\|[\alpha_{g_n}(a),b]\|\to 0$ $(n\to\infty)$ for a, $b\in\mathscr{A}$; then $\{PxP|x\in\mathscr{M}\}$ is a commutative family.*

Proof For $\eta_1,\eta_2\in\mathscr{H}$, $P\eta_i$ $(i=1,2)$ is invariant under u_g $(g\in G)$. Consider the space $\mathscr{H}\oplus\mathscr{H}$, and let Γ be the closed convex hull of $\{(u_g\eta_1,u_g\eta_2)|g\in G\}$ in $\mathscr{H}\oplus\mathscr{H}$. Then Γ is invariant under $u_g\oplus u_g$ $(g\in G)$. Let (η',η'') be the unique element in Γ with $\|(\eta',\eta'')\|=\inf_{(\xi_1,\xi_2)\in\Gamma}\|(\xi_1,\xi_2)\|$. Then (η',η'') is invariant under $u_g\oplus u_g$ $(g\in G)$. Therefore for arbitrary $\varepsilon>0$, there exists $\sum_{i=1}^m\lambda_iu_{h_i}{}^{-1}$ with $\lambda_i\geqslant 0$, $\sum_{i=1}^m\lambda_i=1$, $h_i\in G$ $(i=1,2,\ldots,m)$ such that for $\eta\in\mathscr{H}_0$ and $a\in\mathscr{A}$, $\|(\sum_{i=1}^m\lambda_iu_{h_i}{}^{-1}-P)a^*\eta\|<\varepsilon$ and $\|(\sum_{i=1}^m\lambda_iu_{h_i}{}^{-1}-P)a\eta\|<\varepsilon$. Take an n such that for $b\in\mathscr{A}$, $|(\eta,[\alpha_{g_n}(\sum_{i=1}^m\lambda_i\alpha_{h_i}{}^{-1}(a)),b]\eta)|<\varepsilon$ $(n\geqslant n_0)$, where (,) is the inner product of $\mathscr{H}$. Then

$$\begin{aligned}
&\left|\left(\eta,\left[\alpha_{g_n}\left(\sum_{i=1}^m\lambda_i\alpha_{h_i}{}^{-1}(a)\right),b\right]\eta\right)\right|\\
&=\left|\left(\eta,\sum_{i=1}^m\lambda_iu_{g_n}u_{h_i}{}^{-1}au_{h_i}u_{g_n}{}^{-1}b\eta\right)-\left(\eta,\sum_{i=1}^m\lambda_ibu_{g_n}u_{h_i}{}^{-1}au_{h_i}u_{g_n}{}^{-1}\eta\right)\right|\\
&=\left|\left(\sum_{i=1}^m\lambda_iu_{h_i}{}^{-1}a^*\eta,u_{g_n}{}^{-1}b\eta\right)-\left(u_{g_n}{}^{-1}b^*\eta,\sum_{i=1}^m\lambda_iu_{h_i}{}^{-1}a\eta\right)\right|\\
&\geqslant|(Pa^*\eta,u_{g_n}{}^{-1}b\eta)-(u_{g_n}{}^{-1}b^*\eta,Pa\eta)|-\|b\eta\|\varepsilon-\|b^*\eta\|\varepsilon\\
&=|(Pa^*\eta,b\eta)-(b^*\eta,Pa\eta)|-\|b\eta\|\varepsilon-\|b^*\eta\|\varepsilon.
\end{aligned}$$

Hence $(\eta,bPa\eta)=(\eta,aPb\eta)$ $(a,b\in\mathscr{A})$, and so $\{PaP|a\in\mathscr{A}\}$ is a commutative family. Since $\mathscr{A}$ is $\sigma(\mathscr{M},\mathscr{M}_*)$-dense in $\mathscr{M}$, $\{PxP|x\in\mathscr{M}\}$ is a commutative family. □

Algebraic quantum field theory

The main application is to the theory of local algebras: We shall state the Haag–Kastler–Araki axioms for the C*-algebra A of quasi-local observables of a local quantum field theory. Let $\mathbb{R}^n$ ($n \geqslant 2$) be the n-dimensional euclidean space (n is the dimension of spacetime and c the speed of light).

(1) To every bounded open region $O \subset \mathbb{R}^n$, one assigns a C*-algebra $A(O)$ such that
  (i) $O_1 \subset O_2$ implies $1 \in A(O_1) \subset A(O_2)$ (isotopy);
  (ii) if the regions O_1 and O_2 are spacelike separated (namely for $p = (p_0, p_1, \ldots, p_{n-1}) \in O_1$, $q = (q_0, q_1, \ldots, q_{n-1}) \in O_2$, $c^2(p_0 - q_0)^2 - \sum_{j=1}^{n-1} (p_j - q_j)^2 < 0$, where c is a fixed positive number);

  then the elements of $A(O_1)$ commute with all elements of $A(O_2)$ (local commutativity). The algebra A of quasi-local observables is the C*-algebra generated by the union of $\{A(O)\}$.

(2) There exists a representation of the vector group $\mathbb{R}^n$ as *-automorphisms of A, $\alpha: \mathbb{R}^n \mapsto \mathrm{Aut}(A)$, and furthermore $\alpha_g A(O) = A(O + g)$ for every $g \in \mathbb{R}^n$.

(3) A representation π of A on a Hilbert space $\mathscr{H}$ is called a *-representation satisfying the spectrum condition if the following conditions hold:
  (i) there exists a strongly continuous unitary representation u of $\mathbb{R}^n$ in $\mathscr{H}$;
  (ii) the representation u_g implements the automorphisms α_g, that is $u_g \pi(a) u_g^{-1} = \pi(\alpha_g(a))$ ($a \in A, g \in \mathbb{R}^n$);
  (iii) the spectrum of the representation u is contained in the forward light cone – i.e. let $u_g = \int \exp(\mathrm{i}\langle g, x\rangle)\, \mathrm{d}P(x)$ be the Stone representation of u, where $\langle\ ,\ \rangle$ is the inner product of $\mathbb{R}^n$;

  then the support of the projection-valued measure is contained in the set $\{p \in R^n \mid c^2 {p_0}^2 - ({p_1}^2 + {p_2}^2 + \cdots + {p_{n-1}}^2) \geqslant 0 \text{ and } p_0 \geqslant 0\}$, where c is a fixed positive number. Such representations are said to be representations with positive energy.

(4) A state ϕ on A is called a vacuum state if it is invariant under α_g ($g \in \mathbb{R}^n$) – i.e. $\phi(\alpha_g(a)) = \phi(a)$ ($a \in A$) and the cyclic representation $\{\pi_\phi, \mathscr{H}_\phi\}$ (the GNS construction of A) constructed via ϕ is a representation satisfying the spectrum condition.

The W*-algebra $\pi_\phi(A)''$ is called the quasi-local W*-algebra of quantum field theory.

Then we have the following.

2.4.13 Corollary *In the quasi-local W*-algebra of quantum field theory, if one assumes the spectrum condition, local commutativity, and the existence of a vacuum state (not necessarily unique), then the algebra $\mathscr{M}$ is of type I with $\mathscr{M}'$ commutative.*

2.4.14 Notes and remarks In 1964, Araki [2] obtained Theorem 2.4.1. After this result, mathematical physicists conjectured that the observability of positive energy would be concluded without the assumption of 'a vacuum state'. This conjecture means mathematically the likelihood of the innerness of bounded derivations on a W*-algebra (i.e. everywhere-defined derivations (cf. Theorem 2.3.1)). Stimulated by this conjecture, mathematicians studied the innerness of derivations on a W*-algebra. In 1966, Kadison [90] proved that derivations on a W*-algebra are automatically spatial and by using Kadison's spatiality theorem, the author [162] proved the innerness of derivations on a W*-algebra (cf. Theorem 2.5.3). The result is now called the 'Derivation Theorem' and was used by Borchers in the original proof of Theorem 2.4.3. The proof of Theorem 2.4.3, presented here, is essentially part of Arveson's spectral theory on one-parameter automorphism groups [6], which gives an elegant proof of Derivation Theorem for W*-algebra Arveson's theory came much later than the original proof in 1966.

References [2], [6], [16], [56], [57], [71], [118], [158], [160].

2.5 Application to bounded derivations

In this section, we shall discuss the automatic innerness of derivations on W*-algebras and simple C*-algebras with an identity. There are few known examples of infinite-dimensional, non-commutative algebras except for operator algebras which have only inner derivations. The algebra $B(E)$ of all bounded linear operators on a Banach space E has only inner-bounded derivations. However, there are many examples of C*-algebras which have outer-bounded derivations. In C*-quantum physics, simple C*-algebras are often used. This is due to the philosophy that nature is unique. The good behavior of simple C*-algebras like the innerness of all derivations on the algebra suggests that C*-algebras are appropriate for quantum physics.

The derivation theorem

Let $\mathcal{M}$ be a W*-algebra and let A be a commutative W*-subalgebra of $\mathcal{M}$ containing the identity of $\mathcal{M}$.

2.5.1 Lemma *Let δ be a derivation on $\mathcal{M}$; then there is an element x_0 in $\mathcal{M}$ such that $\delta(a) = [x_0, a]$ for $a \in A$ and $\|x_0\| \leqslant \|\delta\|$.*

Proof Let A^u be the group of all unitary elements in A. For $u \in A^u$, put $Tu(x) = (ux + \delta(u))u^{-1}$ $(x \in \mathcal{M})$. Then if $u, v \in A^u$,

$$\begin{aligned} Tu\,Tv(x) &= \{u(vx+\delta(v))v^{-1}+\delta(u)\}u^{-1} \\ &= uvxv^{-1}u^{-1}+u\delta(v)v^{-1}u^{-1}+\delta(u)u^{-1} \\ &= \{uvx+\delta(uv)\}(uv)^{-1} = Tuv(x). \end{aligned}$$

Hence $Tu\,Tv = Tuv$. Let K be the σ-closed convex subset of $\mathscr{M}$ generated by $\{Tu(0)|u\in A^u\}$; then $Tu(K)\subset K$. Since $\|Tu(0)\| = \|\delta(u)u^{-1}\| \leqslant \|\delta\|$, $\sup_{x\in K}\|x\| \leqslant \|\delta\|$. Since $\{Tu|u\in A^u\}$ is commutative and Tu is σ-continuous, by the Markov–Kakutani fixed point theorem, there is an element x_0 such that $Tu(x_0)=x_0$ for $u\in A^u$, hence $(ux_0+\delta(u))u^{-1}=x_0$ and so $\delta(u)=[x_0,u]$ $(u\in A^u)$. Since any element of A is a finite linear combination of elements in A^u, $\delta(a)=[x_0,a]$ $(x\in A)$. Clearly $\|x_0\|\leqslant\|\delta\|$. □

2.5.2 Lemma *Let δ be a *-derivation on $\mathscr{M}$ and let ϕ be a normal state on $\mathscr{M}$; then there is a self-adjoint element h in $\mathscr{M}$ such that $\phi(\delta(x))=\phi(\delta_{ih}(x))$ for $x\in\mathscr{M}$, where $\delta_{ih}(x)=[ih,x]$ $(x\in\mathscr{M})$ and $\|h\|\leqslant\|\delta\|$.*

Proof In the proof of Lemma 2.5.1, if δ is a *-derivation, then $(\delta(u)u^{-1})^* = u\delta(u^*)=\{\delta(uu^*)-\delta(u)u^*\} = -\delta(u)u^{-1}$. Hence we can choose a self-adjoint element k in $\mathscr{M}$ such that $[ik,a]=\delta(a)$ $(a\in A)$, and $\|k\|\leqslant\|\delta\|$.

Now suppose that there exists a normal state ϕ on $\mathscr{M}$ such that $\delta^*(\phi)\notin \{\delta_{ih}{}^*(\phi)|h^*=h,\ \|h\|\leqslant\|\delta\|,\ h\in\mathscr{M}\}=\mathscr{L}$, where $(\delta^*\phi)(x)=\phi(\delta(x))$ $(x\in\mathscr{M})$. Since the mapping $h\to\delta_{ih}{}^*\phi$ of $\mathscr{M}^s$ with $\sigma(\mathscr{M},\mathscr{M}_*)$ onto $\mathscr{M}_*{}^s$ with $\sigma(\mathscr{M}_*,\mathscr{M})$ is continuous, $\mathscr{L}$ is $\sigma(\mathscr{M}_*,\mathscr{M})$-compact in $\mathscr{M}_*{}^s$. Therefore by the bipolar theorem there is a self-adjoint element a in $\mathscr{M}$ such that $\sup_{\delta_{ih}{}^*(\phi)\in\mathscr{L}}|\delta_{ih}{}^*(\phi)(a)| < |\delta^*(\phi)(a)|$. Hence $\{\delta_{ih}(a)|h^*=h,\ \|h\|\leqslant\|\delta\|,\ h\in\mathscr{M}\}\not\ni\delta(a)$. This contradicts Lemma 2.5.1. □

2.5.3 Theorem *Let $\mathscr{M}$ be a W^*-algebra and let δ be a derivation on $\mathscr{M}$; then there is an element a in $\mathscr{M}$ such that $\delta(x)=[a,x]$ $(x\in\mathscr{M})$ and $\|a\|\leqslant\|\delta\|$.*

Proof Let $\delta^*(x)=\delta(x^*)^*$ $(x\in\mathscr{M})$; then δ^* is also a derivation on $\mathscr{M}$ and $\delta=(\delta+\delta^*)/2+\mathrm{i}(\mathrm{i}\delta^*-\mathrm{i}\delta)/2$ with two *-derivations on the right.

Assume that δ is a *-derivation. By Lemma 2.5.2, for $\phi\in\mathscr{M}_*{}^+$ there is a self-adjoint element h in $\mathscr{M}$ such that $\phi(\delta(x))=\phi(\delta_{ih}(x))$ $(x\in\mathscr{M})$ and $\|h\|\leqslant\|\delta\|$. Let $\delta_1=\delta-\delta_{ih}$; then δ_1 is a *-derivation. Let $\alpha_t=\exp(t\delta_1)$; then $t\mapsto\alpha_t$ is a uniformly continuous one-parameter group of *-automorphism on $\mathscr{M}$ and $\phi(\alpha_t(x))=\phi(x)$ $(x\in\mathscr{M},\ t\in\mathbb{R})$ – i.e. ϕ is an invariant normal state on $\mathscr{M}$ under α. Let $\{\pi_\phi, U_\phi, \mathscr{H}_\phi\}$ be the covariant representation of the dynamical system $\{\mathscr{M},\alpha\}$ constructed via ϕ (cf. page 49); then

$$\pi_\phi(\alpha_t(x)) = U_\phi(t)\pi_\phi(x)U_\phi(t)^{-1} \qquad (x\in\mathscr{M}).$$

Let $U_\phi(t)=\exp(itH_\phi)$ be the Stone representation; then H_ϕ is a *bounded* self-adjoint operator on $\mathscr{H}_\phi$. In fact,

$$\begin{aligned}\|\pi_\phi(\delta_1(a))1_\phi\|^2 &= \|H_\phi\pi_\phi(a)1_\phi\|^2 = (H_\phi\pi_\phi(a)1_\phi, H_\phi\pi_\phi(a)1_\phi)\\ &= (\pi_\phi(a)1_\phi, H_\phi{}^2\pi_\phi(a)1_\phi) \leqslant \|\pi_\phi(a)1_\phi\|\,\|H_\phi{}^2\pi_\phi(a)1_\phi\|\\ &= \phi(a^*a)^{1/2}(H_\phi{}^2\pi_\phi(a)1_\phi, H_\phi{}^2\pi_\phi(a)1_\phi)^{1/2}\\ &\leqslant \phi(a^*a)^{1/2+1/4}(H_\phi{}^4\pi_\phi(a)1_\phi, H_\phi{}^4\pi_\phi(a)1_\phi)^{1/4}\\ &\leqslant \cdots \leqslant \phi(a^*a)^{2^{-1}+4^{-1}\cdots+(2^n)^{-1}}(H_\phi{}^{2^n}\pi_\phi(a)1_\phi, H_\phi{}^{2^n}\pi_\phi(a)1_\phi)^{(2^n)^{-1}}\\ &= \phi(a^*a)^{1-(2^n)^{-1}}\phi(\delta_1{}^{2^n}(a)^*\delta_1{}^{2^n}(a))^{(2^n)^{-1}}\\ &\leqslant \phi(a^*a)^{1-(2^n)^{-1}}(\|\delta_1{}^{2^n}\|^2\|a\|^2)^{(2^n)^{-1}}\end{aligned}$$

Hence $\|H_\phi a_\phi\| \leqslant \|a_\phi\|\,\|\delta_1\|$ and so $\|H_\phi\| \leqslant \|\delta_1\|$.

Let

$$w_\phi(t) = \exp it(\|H_\phi\|1 + H_\phi);$$

then

$$w_\phi(t)\pi_\phi(x)w_\phi(t)^{-1} = U_\phi(t)\pi_\phi(x)U_\phi(t)^{-1} \qquad (x\in\mathscr{M}).$$

Hence by Theorem 2.4.3 there is a positive element k in $\mathscr{M}$ such that

$$U_\phi(t)\pi_\phi(x)U_\phi(t)^{-1} = \pi_\phi\exp(itk)\pi_\phi(x)\pi_\phi\exp(-itk).$$

Let $\mathscr{I}$ be the kernel of π_ϕ; then there is a central projection z in $\mathscr{M}$ such that $\mathscr{M}(1-z)=\mathscr{I}$. Since $\delta_1(1-z)=0$,

$$\alpha_t(xz) = \exp(itkz)xz\exp(-itkz) \qquad (x\in\mathscr{M})$$

and so

$$\delta_1(xz) = \delta_1(x)z = [ik, xz] \qquad (x\in\mathscr{M}).$$

Therefore δ is inner on $\mathscr{M}z$. From this we can conclude that δ is inner on $\mathscr{M}z$ without the assumption of *-ness. Namely there is an element b in $\mathscr{M}$ such that $\delta(xz)=[bz, xz]$ $(x\in\mathscr{M})$. Therefore

$$\delta(uz)u^{-1}z = [bz, uz]u^{-1}z = bz - uzbzu^{-1}z = bz - ubu^{-1}z \qquad (u\in\mathscr{M}^u)$$

and

$$\|bz - ubu^{-1}z\| \leqslant \|\delta\|.$$

Let $C(b)$ be the uniformly closed convex span of $\{ubu^{-1}|u\in\mathscr{M}^u\}$; then $C(b)\cap Z \neq \phi$, where Z is the center of $\mathscr{M}$. Hence $\delta(xz)=[bz-lz, xz]$ $(x\in\mathscr{M})$, where $l\in C(b)\cap Z$ and so there is an element az in $\mathscr{M}$ such that $\|az\|\leqslant\|\delta\|$ and $\delta(xz)=[az, xz]$ $(x\in\mathscr{M})$. Let $\{z_\alpha\}$ be a maximal family of mutually orthogonal central projections in $\mathscr{M}$ such that $\delta(xz_\alpha)=[a_\alpha z_\alpha, xz_\alpha]$ $(x\in\mathscr{M})$ with $\|a_\alpha z_\alpha\|\leqslant\|\delta\|$ and let $a=\sum_\alpha a_\alpha z_\alpha$; then $a\in\mathscr{M}$ and $\delta(x)=[a,x]$ $(x\in\mathscr{M})$ with $\|a\|\leqslant\|\delta\|$. □

2.5.4 Corollary *Let A be a C^*-algebra on a Hilbert space $\mathscr{H}$ and let δ be a* *derivation on A; then there is an element a in the weak closure of A in* $B(\mathscr{H})$ *with* $\|a\| \leqslant \|\delta\|$ *such that* $\delta(x) = [a, x]$ $(a \in A)$.

Proof Let A^{**} be the second dual of A; then A^{**} is a W*-algebra. Let δ^{**} be the second dual of δ; then δ^{**} is a derivation on A^{**}; hence there is an element b in A^{**} such that $\delta^{**}(x) = [b, x]$ $(x \in A^{**})$.

The weak closure of A on $\mathscr{H}$ is *-isomorphic to $A^{**}/\mathscr{I}$, where $\mathscr{I}$ is a σ-closed ideal of A^{**}. Let $\mathscr{I} = A^{**}z$, where z is a central projection of A^{**}; then $\delta^{**}(z) = 0$ and so $\delta^{**}(xz) = [b, xz] = [bz, xz]$ $(x \in A^{**})$. □

For *-derivations, we have more precise expressions.

2.5.5 Proposition *Let A be a C*-algebra on a Hilbert space* $\mathscr{H}$ *and let* δ *be a *-derivation on A; then there is a positive element h in the weak closure* $\bar{A}$ *of A in* $B(\mathscr{H})$ *with* $\|h\| = \|\delta\|$ *such that* $\delta(a) = \mathrm{i}[h, a]$ $(a \in A)$ *and moreover if* $\delta(a) = \mathrm{i}[k, a]$ $(a \in A)$ *for a positive element k in* $\bar{A}$, *then* $h \leqslant k$.

Proof Without loss of generality, one may assume that A has an identity. Let $\mathscr{F} = \{k \in \bar{A} \mid k \geqslant 0 \text{ and } \delta(a) = \mathrm{i}[k, a]\ (a \in A)\}$. We shall introduce a partial ordering $\lesssim$ in $\mathscr{F}$ as follows: $k_1 \lesssim k_2$ if and only if $k_1 z \geqslant k_2 z$ for every $z \in Z_p$, where Z_p is the set of all central projections of $\bar{A}$. Suppose that $\mathscr{F}_1 = \{k_\alpha\}$ is a linearly ordered subset of $\mathscr{F}$ and let k_0 be an accumulation point of $\mathscr{F}_1$ in the weak operator topology on $\bar{A}$; then clearly $k_0 z \leqslant k_\alpha z$ for all $k_\alpha \in \mathscr{F}_1$ and $z \in Z_p$.

Therefore by Zorn's lemma, $\mathscr{F}$ has a maximal element h. We shall show that $hz \leqslant kz$ for all $z \in Z_p$ and $k \in \mathscr{F}$. In fact, suppose that $hz_0 \nleqslant k_0 z_0$ for some $z_0 \in Z_p$ and $k_0 \in \mathscr{F}$. Let $(h - k_0)z_0 = a_1 - a_2$ be the orthogonal decomposition: then $a_1 > 0$, $a_2 \geqslant 0$ and $a_1 - a_2 \in Z$, where Z is the center of $\bar{A}$. Let $h_0 z_0 = hz_0 - a_1$ and $h_0 = h_0 z_0 + h(1 - z_0)$; then

$$h_0 z = h_0 z_0 z + h(1 - z_0)z \leqslant hz_0 z + h(1 - z_0)z = hz \qquad \text{for } z \in Z_p$$

and

$$h_0 z_0 = h_0 z_0 - a_1 z_0 = hz_0 - a_1 < hz_0,$$

a contradiction. By a similar method, one can easily show that hz is not invertible in $\bar{A}z$ for all $z \in Z_p$.

Now let A^{**} be the second dual of A; then it is a W*-algebra and A is a C*-subalgebra of A^{**}, when it is canonically embedded into A^{**}. Let $\{z_\beta \mid \beta \in \Pi\}$ be the set of all minimal central projections in A^{**} such that $A^{**}z_\beta$ is a type I factor and let $z_0 = \bigvee_{\beta \in \Pi} z_\beta$. Then A is *-isomorphic to Az_0. By the previous discussions, there is a positive element l in $A^{**}z_0$ such that $\delta(a)z_0 = \mathrm{i}[l, a]$ $(a \in A)$, and for any positive element k in $A^{**}z_0$ with

$\delta(a)z_0 = \mathrm{i}[k, a]$ $(a \in A)$, $lz_\beta \leqslant kz_\beta$ for $\beta \in \Pi$. Let B be a C^*-subalgebra of $A^{**}z_0$ generated by Az_0 and l; then we shall show that $\| l \| \leqslant \| \delta \|$. In fact, if $\| l \| > \| \delta \|$, there is a z_β such that $\| lz_\beta \| > \| \delta \|$.

$$\delta(a)z_\beta = \mathrm{i}[lz_\beta, a] \qquad (a \in A)$$

and so

$$\| laz_\beta - alz_\beta \| \leqslant \| \delta \| \, \| a \|.$$

Since $A^{**}z_\beta$ is a type-I factor, there is a pure state ϕ on A such that the mapping $az_\beta \mapsto \pi_\phi(a)$ $(a \in A)$ is uniquely extended to a $*$-isomorphism ρ of $A^{**}z_\beta$ onto $B(\mathscr{H}_\phi)$. Therefore

$$\| \rho(l)\pi_\phi(a) - \pi_\phi(a)\rho(l) \| \leqslant \| \delta \| \, \| a \|.$$

Since l is not invertible, there is a sequence $\{\xi_n\}$ in $\mathscr{H}_\phi$ such that $\| \xi_n \| = 1$ and $\| \rho(l)\xi_n \| \to 0$. Since $\{\pi_\phi, \mathscr{H}_\phi\}$ is irreducible, for an arbitrary η with $\|\eta\| = 1$, there is an element a_n in A such that $\| a_n \| \leqslant 1$ and $\pi_\phi(a_n)\xi_n = \eta$. Therefore

$$\| \rho(l)\pi_\phi(a_n)\xi_n - \pi_\phi(a_n)\rho(l)\xi_n \| = \| \rho(l)\eta - \pi_\phi(a_n)\rho(l)\xi_n \| \leqslant \| \delta \|.$$

Hence

$$\| \rho(l)\eta \| \leqslant \| \delta \| + \| \pi(a_n)\rho(l)\xi_n \| \leqslant \| \delta \| + \| \rho(l)\xi_n \|.$$

This implies that $\| \rho(l) \| = \| l \| \leqslant \| \delta \|$, a contradiction.

Now we shall identify A with the subalgebra Az_0 of B. Let ψ be a state on B and consider the $*$-representation $\{\pi_\psi, \mathscr{H}_\psi\}$ of B. Then there is a positive element m in $\pi_\psi(A)''$ such that $\pi_\psi(\delta(a)) = \mathrm{i}[m, \pi_\psi(a)]$ $(a \in A)$, and for any positive k in $\pi_\psi(A)''$ with $\pi_\psi(\delta(a)) = \mathrm{i}[k, \pi_\psi(a)]$ $(a \in A)$, $mz \leqslant kz$ for all central projections z of $\pi_\psi(A)''$.

$$\pi_\psi(\delta(a)) = \pi_\psi([\mathrm{i}l, a]) = \mathrm{i}[\pi_\psi(l), \pi_\psi(a)] \qquad (a \in A);$$

hence $m - \pi_\psi(l) \in \pi_\psi(A)'$. By the proof of 2.4.3(2), one can construct a positive element K in $\pi_\psi(A)''$ such that $\pi_\psi(l) \geqslant K$ and $\pi_\psi(\delta(a)) = \mathrm{i}[K, \pi_\psi(a)]$ $(a \in A)$. Then $m - K \in \pi_\psi(A)'' \cap \pi_\psi(A)'$. By the minimality of m, $mz \leqslant Kz$ for all central projections z of $\pi_\psi(A)''$. Therefore $m \leqslant K \leqslant \pi_\psi(l)$.

Since $\| \pi_\psi(l) \| \leqslant \| \delta \|$, $\|m\| \leqslant \| \delta \|$. Now let ϕ be a state on A; then take a state extension $\tilde{\phi}$ of ϕ to B and follow the above procedure and then reduce the obtained m to the central envelope $z(p)$ of p in $\pi_{\tilde{\phi}}(A)''$, where p is the orthogonal projection of $\mathscr{H}_{\tilde{\phi}}$ onto $[\pi_{\tilde{\phi}}(A)1_{\tilde{\phi}}]$; then one can conclude the existence of h in the proposition. It is a routine matter to extend this statement to an arbitrary $*$-representation of A.

Therefore we have proved that there is a positive element h in $\bar{A}$ such that $\delta(a) = \mathrm{i}[h, a]$ $(a \in A)$, and for any positive element k in $\bar{A}$ with $\delta(a) = \mathrm{i}[k, a]$ $(a \in A)$, $h \leqslant k$, and moreover $\| h \| \leqslant \| \delta \|$. Finally we shall show that $\| h \| = \| \delta \|$.

Suppose that $\|h\| < \|\delta\|$ and let $l = h - (\|h\|/2)1$; then $\|l\| = (\|h\|/2)$ and $\delta(a) = \mathrm{i}[l, a]$. Hence $\|\delta\| \leqslant 2\|l\| \leqslant \|h\| < \|\delta\|$, a contradiction. □

Let A be a C*-algebra, A^{**} the second dual of A; then A^{**} is a W*-algebra and A is a C*-subalgebra of A^{**}, when it is canonically embedded into A^{**}. Let $M(A) = \{x \in A^{**} \mid Ax, xA \subset A\}$; then $M(A)$ is a C*-subalgebra of A^{**} containing the identity, and A is a closed two-sided ideal of $M(A)$. $M(A)$ is called the multiplier algebra of A. If A has an identity, then $M(A) = A$. Let δ be a derivation on A; then there is an element d in A^{**} such that $\delta(a) = [d, a]$ $(a \in A)$. For $x \in M(A)$ and $a \in A$,

$$[d, x]a = (dx - xd)a = dxa - xda = dxa - xad + xad - xda = \delta(xa) - x\delta(a) \in A$$

and

$$a[d, x] = adx - axd = adx - dax + dax - axd = -\delta(a)x + \delta(ax) \in A.$$

Therefore the derivation δ on A can be extended to a derivation $\tilde{\delta}$ on $M(A)$ with $\tilde{\delta}(x) = [d, x]$ $(x \in M(A))$.

Since A is $\sigma(A^{**}, A^{*})$-dense and any derivation on $M(A)$ is $\sigma(A^{**}, A^{*})$-continuous, any extension on $M(A)$ is unique. Also, any derivation on $M(A)$ is an extension of its restriction to A. Hence there is a one-to-one correspondence between derivations on A and derivations on $M(A)$.

Now let h be a positive element in A^{**} such that $\|h\| = \|\delta\|$, $\delta(a) = \mathrm{i}[h, a]$ $(a \in A)$ and moreover for any positive element k in A^{**} with $\delta(a) = \mathrm{i}[k, a]$ $(a \in A)$, $h \leqslant k$. Let B be a C*-subalgebra of A^{**} generated by A and h. Suppose that $A \subsetneqq B$ and let S be the set of all self-adjoint linear functionals on B such that $f(A) = 0$, and $\|f\| \leqslant 1$. S is a $\sigma(B^{*}, B)$-compact convex set. Let g be an extreme point in S and let $g = g_1 - g_2$ be the orthogonal decomposition of g with $g_1, g_2 \geqslant 0$ and $\|g\| = \|g_1\| + \|g_2\|$. Put $\xi = g_1 + g_2$ and let $\{\pi_\xi, \mathscr{H}_\xi\}$ be the GNS representation of B constructed via ξ.

2.5.6 Lemma *Let $\overline{\pi_\xi(A)}$ be the weak closure of $\pi_\xi(A)$ on $\mathscr{H}_\xi$. If $\overline{\pi_\xi(A)} \neq 0$, then it is a factor containing $1_{\mathscr{H}_\xi}$.*

Proof Let $\xi(y) = (\pi_\xi(y)1_\xi, 1_\xi)$ $(y \in B)$. Then there are two elements η_1, η_2 in $\mathscr{H}_\xi$ with $g_i(y) = (\pi_\xi(y)\eta_i, \eta_i)$ $(i = 1, 2)$ for $y \in B$. Since $g_1 = g_2$ on A, a mapping $\pi_\xi(a)\eta_1 \mapsto \pi_\xi(a)\eta_2 (a \in A)$ defines a partial isometry u' on $\mathscr{H}_\xi$ with $u' \in \pi_\xi(A)'$. Suppose that $\overline{\pi_\xi(A)}$ contains a non-trivial central projection z. Set $l_i(y) = (\pi_\xi(y)z\eta_i, \eta_i)$ and

$$k_i(y) = (\pi_\xi(y)(1_{\mathscr{H}_\xi} - z)\eta_i, \eta_i) \ (i = 1, 2) \qquad \text{for } y \in B.$$

Then

$$l_2(x) = (\pi_\xi(x)z\eta_2, \eta_2) = (\pi_\xi(x)zu'\eta_1, u'\eta_1) = (\pi_\xi(x)z\eta_1, \eta_1) \qquad (x \in A).$$

Analogously $k_2 = k_1$ on A. On the other hand, there is a positive element b in $\overline{\pi_\xi(A)}$ with $\pi_\xi(\delta(a)) = [ib, \pi_\xi(a)]$ $(a \in A)$, and so

$$[ib, \pi_\xi(a)] = \pi_\xi([ih, a]) = [\pi_\xi(ih), \pi_\xi(a)] \qquad (a \in A).$$

Hence $b - \pi_\xi(h) \in \pi_\xi(A)'$. This implies that z commutes with $\pi_\xi(h)$ and so z belongs to the center of $\overline{\pi_\xi(B)}$. Hence $l_i, k_i \geqslant 0$ $(i = 1, 2)$ and

$$\begin{aligned} 1 = \|g\| &= \|l_1 - l_2 + k_1 - k_2\| \leqslant \|l_1 - l_2\| + \|k_1 - k_2\| \\ &\leqslant \|l_1\| + \|l_2\| + \|k_1\| + \|k_2\| = \|\eta_1\|^2 + \|\eta_2\|^2 \\ &= \|g_1\| + \|g_2\| = 1. \end{aligned}$$

Since z is a non-trivial central projection in $\overline{\pi_\xi(B)}$ and $[\pi_\xi(B)1_\xi] = \mathscr{H}_\xi$, we have $0 < \|l_1 - l_2\| < 1$ and so $0 < \|k_1 - k_2\| < 1$. Therefore

$$g = \|l_1 - l_2\| \frac{l_1 - l_2}{\|l_1 - l_2\|} + \|k_1 - k_2\| \frac{k_1 - k_2}{\|k_1 - k_2\|}$$

and so by the extremality of g

$$\frac{l_1 - l_2}{\|l_1 - l_2\|} = \frac{k_1 - k_2}{\|k_1 - k_2\|} = g_1 - g_2.$$

But $s(l_1 + l_2)s(k_1 + k_2) = 0$, a contradiction. □

Now we are ready to prove the following theorem.

2.5.7 Theorem *Let A be a simple C^*-algebra and let $M(A)$ be the multiplier algebra of A; then for each $*$-derivation δ on A, there is a positive element d in $M(A)$ such that $\delta(a) = [d, a]$ $(a \in A)$, $\|d\| = \|\delta\|$. In particular, any derivation on A is realized by an element b in the multiplier algebra $M(A)$ such that $\delta(a) = [b, a]$ $(a \in A)$.*

Proof First, assume that δ is a $*$-derivation on A, and let $\mathscr{I}$ be the closed two-sided ideal of B generated by A. If $A \subsetneqq \mathscr{I}$, then there is an extreme point g in S such that $g(\mathscr{I}) \neq 0$. If $\pi_\xi(A) = 0$, then $\pi_\xi(\mathscr{I}) = 0$ and so $g(\mathscr{I}) = 0$. Therefore $\pi_\xi(A) \neq 0$ and so $\pi_\xi(A)$ is faithful, for A is simple. By 2.5.6, $\overline{\pi_\xi(A)}$ is a factor. Take a positive element k in $\overline{\pi_\xi(A)}$ such that

$$\pi_\xi(\delta(a)) = [ik, \pi_\xi(a)] \qquad (a \in A), \ \|\delta\| = \|k\| \text{ and } k \leqslant \pi_\xi(h);$$

then

$$[\pi_\xi(ih), \pi_\xi(a)] = [ik, \pi_\xi(a)] \ (a \in A)$$

and so $\pi_\xi(h) - k \in \{\overline{\pi_\xi(A)}\}'$. Since $\overline{\pi_\xi(A)}$ is a factor,

$$\|\pi_\xi(h)\| = \|k + (\pi_\xi(h) - k)\| = \|k\| + \|\pi_\xi(h) - k\| = \|\delta\| + \|\pi_\xi(h) - k\|.$$

Hence

$$\|\delta\| = \|h\| \geqslant \|\pi_\xi(h)\| = \|\delta\| + \|\pi_\xi(h) - k\|$$

and so $\pi_\xi(h) - k = 0$. This contradicts $\overline{\pi_\xi(A)} \subsetneqq \overline{\pi_\xi(B)}$. Hence $\mathscr{I} = A$. In particular, hA, $Ah \subset A$ and so h is an element of $M(A)$. The rest is clear.

□

2.5.8 Corollary *Any derivation δ on a simple C*-algebra with identity is inner. Furthermore, for any *-derivation δ on a simple C*-algebra A with identity, one can choose a positive element h in A such that*

$$\delta(a) = \mathrm{i}[h, a] (a \in A) \text{ and } \|\delta\| = \|h\|.$$

In general, a C*-algebra may have bounded derivations which cannot be realized by elements in its multiplier algebra. In fact, one can prove the following theorem.

2.5.9 Theorem ([1], [61]) *Let A be a separable C*-algebra. Then A has only bounded derivations which are realized by elements in its multiplier algebra $M(A)$, if and only if it is a C*-direct sum of simple C*-algebras and a C*-algebra with a continuous trace.*

2.5.10 Notes and remarks The proof of Lemma 2.5.1 is due to Johnson and Ringrose [212]. The original development of fundamental results on derivations is as follows. In 1966, the innerness of derivations on a W^*-algebra was obtained (cf. [90] [162]); by using it, Borchers obtained Theorem 2.4.3 ([16]) in 1966. Here we present a proof of the innerness in the opposite direction. This proof is somewhat new. It would be interesting to find a proof of the innerness directly from Araki's theorem (2.4.1).

The innerness of derivations on a simple C*-algebra was proved by the author [163], [164]. The norm of a good 'h' with $\delta(x) = \mathrm{i}[h, x]$ $(x \in \mathscr{M})$ was found by Kadison, Lance and Ringrose [92].

References [90], [92], [93], [96], [162], [163], [164].

2.6 Uniformly continuous dynamical systems

In §2.4, we showed that if a W*-dynamical system $\{\mathscr{M}, \mathbb{R}^n, \alpha\}$ satisfies the spectrum condition, then it is observable. In particular, if a W*-dynamical system $\{\mathcal{M}, \mathbb{R}, \alpha\}$ has a corresponding hamiltonian which is lower bounded, then it is observable. In §2.5, we have shown that a bounded *-derivation on a W*-algebra always has a corresponding bounded hamiltonian, so that it is observable (more strongly, inner).

In this section, we shall extend these results to uniformly continuous W*-dynamical systems $\{\mathcal{M}, G, \alpha\}$, where G is a real Lie group. We shall also formulate the lower boundedness conditions for W*-dynamical systems $\{\mathcal{M}, G, \alpha\}$ with a Lie group G and raise a problem which seems interesting.

Let $\mathcal{H}$ be a Hilbert space. A bounded operator a on $\mathcal{H}$ is said to be skew-self-adjoint if $a^* = -a$. Let L be the set of all skew self-adjoint elements in $B(\mathcal{H})$; then L is a real Lie algebra with product $[a,b] = ab - ba$ $(a, b \in L)$.

2.6.1 Lemma *Let $\mathfrak{g}$ be a finite-dimensional real Lie subalgebra of L; then $\mathfrak{g}$ is reductive.*

Proof By 2.2.9, $[[x, y], y] = 0$ implies $[x, y] = 0$ for $x, y \in \mathfrak{g}$. Let $\mathfrak{r}$ be the radical of $\mathfrak{g}$. Then we shall show that $\mathfrak{r}$ is commutative. Let $\mathscr{D}^0\mathfrak{r} = \mathfrak{r}$, $\mathscr{D}^1\mathfrak{r} = [\mathfrak{r},\mathfrak{r}]$, $\mathscr{D}^2\mathfrak{r} = \mathscr{D}(\mathscr{D}\mathfrak{r}), \ldots, \mathscr{D}^n\mathfrak{r} = \mathscr{D}(\mathscr{D}^{n-1}\mathfrak{r})$ and let k be the greatest number such that $\mathscr{D}^{k-1}\mathfrak{r} \neq \{0\}$ and $\mathscr{D}^k\mathfrak{r} = \{0\}$. If $k - 2 \geqslant 0$, then $[\mathscr{D}^{k-2}\mathfrak{r}, \mathscr{D}^{k-1}\mathfrak{r}] = 0$, for $\mathscr{D}^{k-1}\mathfrak{r}$ is a commutative ideal of $\mathscr{D}^{k-2}\mathfrak{r}$; hence $[\mathscr{D}^{k-2}\mathfrak{r}, [\mathscr{D}^{k-2}\mathfrak{r}, \mathscr{D}^{k-2}\mathfrak{r}]] = 0$ and so by 2.2.9, $[\mathscr{D}^{k-2}\mathfrak{r}, \mathscr{D}^{k-2}\mathfrak{r}] = \mathscr{D}^{k-1}\mathfrak{r} = 0$, a contradiction. Therefore $k = 1$ and so $\mathfrak{r}$ is commutative. This implies that the radical of $\mathfrak{g}$ is the center of $\mathfrak{g}$. □

2.6.2 Proposition *Let $\mathfrak{g}$ be a finite-dimensional real Lie subalgebra of L; then $\mathfrak{g}$ is a direct sum of a compact semi-simple Lie algebra and a commutative Lie algebra (called a compact Lie algebra).*

Proof By 2.6.1, $\mathfrak{g}$ is the direct sum of a semi-simple Lie algebra and a commutative Lie algebra. Therefore it is sufficient to assume that $\mathfrak{g}$ is semi-simple. Let $\tilde{\mathfrak{g}}$ be the complex linear subspace of $B(\mathcal{H})$ spanned by $\mathfrak{g}$. For $x \in \mathfrak{g}$, let $(ad(x) - \lambda 1)y = 0$ for some non-zero $y \in \tilde{\mathfrak{g}}$ and some number λ; then

$$\begin{aligned}\{(ad(x) - \lambda 1)y\}^* &= \{[x, y] - \lambda y\}^* = [y^*, x^*] - \bar{\lambda} y^* \\ &= -[y^*, x] - \bar{\lambda} y^* = [x, y^*] - \bar{\lambda} y^* \\ &= (ad(x) - \bar{\lambda} 1)y^* = 0.\end{aligned}$$

Therefore,

$$ad(x)y^*y = (ad(x)y^*)y + y^*(ad(x)y) = \bar{\lambda} y^*y + \lambda y^*y = (\bar{\lambda} + \lambda)y^*y.$$

Hence,

$$[y^*y, [y^*y, x]] = -[y^*y, (\bar{\lambda} + \lambda)y^*y] = 0,$$

and so by 2.2.9,

$$[y^*y, x] = -(\bar{\lambda} + \lambda)y^*y = 0.$$

This implies $\bar{\lambda} + \lambda = 0$ – i.e. λ is purely imaginary. Hence $\Phi(x^2) < 0$ for $x \in \mathscr{g}$, where Φ is the Killing form of $\mathscr{g}$. □

2.6.3 Corollary (Kadison and Singer [94]) *Let G be a connected finite-dimensional real Lie group and suppose that G has a faithful uniformly continuous unitary representation into the unitary group of $B(\mathscr{H})$; then G is a direct product of a compact group and a vector group.*

Proof Let $\mathscr{g}$ be the real Lie algebra of G; then there is a faithful representation du of $\mathscr{g}$ into L. By 2.6.2, $\mathscr{g}$ is compact. □

Now let $\mathscr{A}$ be a C*-algebra on a Hilbert space $\mathscr{H}$ and let $\mathscr{M}$ be the weak closure of $\mathscr{A}$ in $B(\mathscr{H})$.

2.6.4 Proposition *Let M be the real Lie algebra of all skew-self-adjoint elements in $\mathscr{M}$, and let N be the Lie subalgebra of elements m in M such that $[m, \mathscr{A}] \subset \mathscr{A}$, and let D be the set of all bounded *-derivations on $\mathscr{A}$. For each $m \in N$, let $f(m)$ be a bounded *-derivation on $\mathscr{A}$ such that $f(m)(x) = [m, x]$ $(x \in \mathscr{M})$ (by 2.5.3, $f(N) = D$). Let $\mathscr{g}$ be a finite-dimensional real Lie algebra and let λ be a homomorphism of $\mathscr{g}$ into D; then there is a homomorphism μ of $\mathscr{g}$ into N such that $\lambda = f \circ \mu$.*

Proof Let $\mathscr{h}$ be a real linear subspace of N such that $f|\mathscr{h}$ is a one-to-one linear mapping of $\mathscr{h}$ onto $\lambda(\mathscr{g})$. Let $(h_1, h_2, \ldots, h_n)$ be a base of $\mathscr{h}$ and let Z be the center of $\mathscr{M}$. Then for i, j, there is an element h in $\mathscr{h}$ such that $f([h_i, h_j]) = f(h)$ and so $[h_i, h_j] \in \mathscr{h} + z_{ij}$, where $z_{ij} \in Z$. Let $\mathscr{h}'$ be the linear subspace of Z generated by $\{z_{ij} | i, j = 1, 2, \ldots, n\}$; then

$$[\mathscr{h} + \mathscr{h}', \mathscr{h} + \mathscr{h}'] = [\mathscr{h}, \mathscr{h}] \subset \mathscr{h} + \mathscr{h}'.$$

Therefore $\mathscr{h} + \mathscr{h}'$ is a finite-dimensional real Lie algebra. By Proposition 2.6.2, it is easily seen $\mathscr{h} + \mathscr{h}'$ is compact. Since $\mathscr{h}'$ is a central ideal of $\mathscr{h} + \mathscr{h}'$, there is an ideal $\mathscr{h}_1$ in $\mathscr{h} + \mathscr{h}'$ such that $\mathscr{h} + \mathscr{h}' = \mathscr{h}_1 + \mathscr{h}'$ and $\mathscr{h}_1 \cap \mathscr{h}' = \{0\}$. Then $f|\mathscr{h}_1$ is an isomorphism γ of $\mathscr{h}_1$ onto $\lambda(\mathscr{g})$. Put $\mu = \gamma^{-1} \circ \lambda$; then $\lambda = f \circ \mu$. □

2.6.5 Theorem (Dixmier [53]) *Let G be a simply connected finite-dimensional real Lie group and let $g \mapsto \alpha_g$ be a norm-continuous representation of G by *-automorphisms on $\mathscr{A}$. Then there is a uniformly continuous unitary representation $g \mapsto u_g$ of G into the unitary group of $\mathscr{M}$ such that $\alpha_g(a) = u_g a u_g^*$*

$(a \in \mathscr{A}, g \in G)$. *Moreover if* $\mathscr{A}$ *is a simple* C*-*algebra with an identity, then one can take* u_g $(g \in G)$ *in* $\mathscr{A}$.

Proof Let $\mathscr{g}$ be the Lie algebra of G. Then, α defines a homomorphism $d\alpha$ of $\mathscr{g}$ into D. By 2.6.4, there is a homomorphism μ of $\mathscr{g}$ into N. Since G is simply connected, there is a unitary representation $g \mapsto u_g$ $(\in \mathscr{M})$ of G such that $\mu = du$. Since $\mu(\mathscr{g})$ is a set of bounded operators in $\mathscr{M}$, $g \mapsto u_g$ is uniformly continuous.

Moreover

$$\alpha_{\exp tl}(a) = \exp t[\mu(l), a] = \exp t\mu(l) a \exp t\mu(l)^{-1} \qquad (l \in \mathscr{g}, a \in \mathscr{A}) \qquad \square$$

Let $\mathscr{M}$ be a W*-algebra on a Hilbert space $\mathscr{H}$ containing the identity operator $1_{\mathscr{H}}$, and let G be a locally compact group. Let $\{\mathscr{M}, G, \alpha\}$ be a W*-dynamical system. In general, we cannot expect there to be a strongly continuous unitary representation $t \mapsto u_t$ of G in the Hilbert space $\mathscr{H}$ such that $\alpha_t(a) = u_t a u_t^*$ $(a \in \mathscr{M}, t \in G)$. However there are many cases in which we can be assured of the existence of such a representation.

2.6.6 Proposition *Suppose that there is a cyclic vector* ξ_0 *in* $\mathscr{H}$ *such that* $(\alpha_t(a)\xi_0, \xi_0) = (a\xi_0, \xi_0)$ $(a \in \mathscr{M})$; *then there exists a strongly continuous unitary representation* $t \mapsto u_t$ *of* G *in* $\mathscr{H}$ *such that* $u_t a u_t^* = \alpha_t(a)$ $(a \in \mathscr{M}, t \in G)$.

Proof Put $u_t a\xi_0 = \alpha_t(a)\xi_0$ $(a \in \mathscr{M}, t \in G)$; then

$$\begin{aligned} \|u_t a\xi_0\|^2 = \|\alpha_t(a)\xi_0\|^2 &= (\alpha_t(a)\xi_0, \alpha_t(a)\xi_0) \\ &= (\alpha_t(a^*)\alpha_t(a)\xi_0, \xi_0) = (\alpha_t(a^*a)\xi_0, \xi_0) \\ &= (a^*a\xi_0, \xi_0) = \|a\xi_0\|^2. \end{aligned}$$

Hence u_t is isometric on $\mathscr{M}\xi_0$. Since $[\mathscr{M}\xi_0] = \mathscr{H}$, u_t can be uniquely extended to a unitary operator (denoted again by u_t)

$$u_{t_1t_2} a\xi_0 = \alpha_{t_1t_2}(a)\xi_0 = \alpha_{t_1}(\alpha_{t_2}(a))\xi_0 = u_{t_1}u_{t_2}a\xi_0;$$

hence $u_{t_1t_2} = u_{t_1}u_{t_2}$ $(t_1, t_2 \in G)$. Suppose that $t \to e$, where e is the unit element of G; then $\alpha_t(a)\xi_0 \to a\xi_0$ (weakly) and so $u_t \to 1_{\mathscr{H}}$ (weakly). Since u_t and $1_{\mathscr{H}}$ are unitary, $u_t \to 1_{\mathscr{H}}$ (strongly). Moreover

$$u_t a u_t^* b\xi_0 = u_t a u_t^{-1} b\xi_0 = u_t a \alpha_t^{-1}(b)\xi_0 = \alpha_t(a) b\xi_0;$$

hence

$$u_t a u_t^* = \alpha_t(a) \qquad (a \in \mathscr{M}, t \in G) \qquad \square$$

2.6.7 Theorem (Araki [4]). *Suppose that there exists a cyclic and separating vector* ξ_0 *in* $\mathscr{H}$ – i.e. $[\mathscr{M}\xi_0] = [\mathscr{M}'\xi_0] = \mathscr{H}$; *then there exists a strongly* *continuous unitary representation* $t \mapsto u_t$ *of* G *in* $\mathscr{H}$ *such that* $u_t a u_t^* = \alpha_t(a)$ $(a \in \mathscr{M}, t \in G)$.

Next assume that α_t is inner for each $t \in G$ – i.e. for each $t \in G$, there is a unitary operator v_t in $\mathscr{M}$ such that $\alpha_t(a) = v_t a v_t^*$ $(a \in \mathscr{M})$. In general, one cannot deduce from this assumption that there exists a strongly continuous unitary representation $t \mapsto u_t$ of G in $\mathscr{H}$ such that $u_t \in \mathscr{M}$ and $\alpha_t(a) = u_t a u_t^*$ $(a \in \mathscr{M}, t \in G)$.

If $G = \mathbb{R}^1$ and $\mathscr{H}$ is separable, then there exists a strongly continuous unitary representation $t \mapsto u_t$ of $\mathbb{R}^1$ in $\mathscr{H}$ such that $u_t \in \mathscr{M}$ and $\alpha_t(a) = u_t a u_t^*$ $(a \in \mathscr{M}, t \in \mathbb{R}^1)$ ([95], [129]). On the other hand, if $G = \mathbb{R}^2$, it is not true. We shall show this in the following.

A negative result

Let $\mathscr{H} = L^2(\mathbb{R}^1)$ be the Hilbert space of all square Lebesgue integrable functions on $\mathbb{R}^1$. Let $U(\lambda)$ $(\lambda \in \mathbb{R}^1)$ be the strongly continuous one-parameter group of unitary operators in $L^2(\mathbb{R}^1)$ such that $\{U(\lambda)f\}(x) = f(x+\lambda)$ $(f \in L^2(\mathbb{R}^1)$ and $x \in \mathbb{R}^1)$, and let $V(\mu)$ $(\mu \in \mathbb{R}^1)$ be the strongly continuous one-parameter group of unitary operators in $L^2(\mathbb{R}^1)$ such that $\{V(\mu)f\}(x) = \exp(i\mu x)f(x)$ $(f \in L^2(\mathbb{R}^1)$ and $x \in \mathbb{R}^1)$.

Then,

$$\{U(\lambda)V(\mu)f\}(x) = \exp(\mathrm{i}\mu(x+\lambda))f(x+\lambda) \text{ and } \{V(\mu)U(\lambda)f\}(x) = \exp(\mathrm{i}\mu x)f(x+\lambda).$$

Hence

$$U(\lambda)V(\mu) = \exp(\mathrm{i}\mu\lambda)V(\mu)U(\lambda).$$

Now let $\mathscr{M} = B(L^2(\mathbb{R}^1))$ and put

$$\alpha(\lambda, \mu)(a) = U(\lambda)V(\mu)aV(\mu)^*U(\lambda)^* \qquad (a \in B(L^2(\mathbb{R}^1));$$

then

$$\begin{aligned} \alpha(\lambda_1+\lambda_2, \mu_1+\mu_2)(a) &= U(\lambda_1)U(\lambda_2)V(\mu_1)V(\mu_2)aV(\mu_2)^*V(\mu_1)^*. \\ U(\lambda_2)^*U(\lambda_1)^* \\ &= U(\lambda_1)V(\mu_1)U(\lambda_2)\exp(\mathrm{i}\mu_1\lambda_2)V(\mu_2)aV(\mu_2)^*\exp(-\mathrm{i}\mu_1\lambda_2)U(\lambda_2)^*V(\mu_1)^*U(\lambda_1)^* \\ &= \alpha(\lambda_1, \mu_1)\alpha(\lambda_2, \mu_2)(a) \qquad (a \in B(L^2(\mathbb{R}^1)) \text{ and } (\lambda_1, \mu_1), (\lambda_2, \mu_2) \in \mathbb{R}^2). \end{aligned}$$

Therefore $(\lambda, \mu) \mapsto \alpha(\lambda, \mu)$ is a σ-weakly continuous representation of $\mathbb{R}^2$ by *-automorphisms on $B(L^2(\mathbb{R}^1))$.

We now show that the automorphism group α is not unitaily implemented.

Proof (*indirect*) Now suppose that there is a strongly continuous representation $(\lambda, \mu) \mapsto W(\lambda, \mu)$ of $\mathbb{R}^2$ in $L^2(\mathbb{R}^1)$ such that

$$\alpha(\lambda, \mu)(a) = W(\lambda, \mu) a W(\lambda, \mu)^* \qquad (a \in B(L^2(\mathbb{R}^1)));$$

then

$$W(\lambda, \mu) = x(\lambda, \mu) U(\lambda) V(\mu), \qquad \text{where } |x(\lambda, \mu)| = 1.$$

$$W(\lambda, 0) = x(\lambda, 0) U(\lambda) = \exp(\mathrm{i} a\lambda) U(\lambda) \qquad \text{for some real number } a.$$

$$W(0, \mu) = x(0, \mu) V(\mu) = \exp(\mathrm{i} b\mu) V(\mu) \qquad \text{for some real number } b.$$

Since

$$\begin{aligned} W(\lambda, \mu) &= W(\lambda, 0) W(0, \mu) = \exp(\mathrm{i} a\lambda) U(\lambda) \exp(\mathrm{i} b\mu) V(\mu) \\ &= \exp(\mathrm{i}(a\lambda + b\mu)) U(\lambda) V(\mu), \qquad x(\lambda, \mu) = \exp(\mathrm{i}(a\lambda + b\mu)). \end{aligned}$$

We now compute the product $W(\lambda, \mu_1) W(\lambda, \mu_2)$ in two ways:

$$\begin{aligned} W(\lambda_1, \mu_1) W(\lambda_2, \mu_2) &= \exp(\mathrm{i}(a\lambda_1 + b\mu_1)) V(\lambda_1) V(\mu_1) \exp(\mathrm{i}(a\lambda_2 + b\mu_2)) U(\lambda_2) V(\mu_2) \\ &= \exp(\mathrm{i}(a\lambda_1 + a\lambda_2 + b\mu_1 + b\mu_2)) U(\lambda_1) V(\mu_1) U(\lambda_2) V(\mu_2). \end{aligned}$$

On the other hand,

$$\begin{aligned} &W(\lambda_1, \mu_1) W(\lambda_2, \mu_2) \\ &\quad = W(\lambda_1 + \lambda_2, \mu_1 + \mu_2) = \exp(\mathrm{i}\{a(\lambda_1 + \lambda_2) + b(\mu_1 + \mu_2)\}) U(\lambda_1 + \lambda_2) V(\mu_1 + \mu_2) \\ &\quad = \exp(\mathrm{i}\{a(\lambda_1 + \lambda_2) + b(\mu_1 + \mu_2)\}) U(\lambda_1) U(\lambda_2) V(\mu_1) V(\mu_2) \\ &\quad = \exp(\mathrm{i}\{a(\lambda_1 + \lambda_2) + b(\mu_1 + \mu_2)\}) U(\lambda_1) V(\mu_1) U(\lambda_2) \exp(\mathrm{i}\mu_1\lambda_2) V(\mu_2) \\ &\quad = \exp(\mathrm{i}\{a(\lambda_1 + \lambda_2) + b(\mu_1 + \mu_2)\}) \exp(\mathrm{i}\mu_1\lambda_2) U(\lambda_1) V(\mu_1) U(\lambda_2) V(\mu_2). \end{aligned}$$

Hence $\exp(\mathrm{i}\mu_1\lambda_2) = 1$ for $\mu_1, \lambda_2 \in \mathbb{R}^1$, a contradiction. Hence α is not implemented.

Lie groups

Now we shall discuss the problem of extending the spectrum conditions in §2.4 to Lie groups.

Let G be a simply connected finite-dimensional real Lie group and let $\mathfrak{g}$ be the Lie algebra of G. Let $t \mapsto u_t$ be a strongly continuous unitary representation of G in a Hilbert space $\mathscr{H}$. Let V be the linear subspace of $\mathscr{H}$ consisting of all infinitely differentiable vectors in $\mathscr{H}$ with respect to u_G; then V is a dense subspace of $\mathscr{H}$ and there is a homomorphism $\mathrm{d}u$ of $\mathfrak{g}$ into essentially skew-self-adjoint operators defined on V – i.e. the closure $\overline{\mathrm{d}u(x)}$ of $\mathrm{d}u(x)$ in $\mathscr{H}$ is skew-self-adjoint for each $x \in \mathfrak{g}$.

Now suppose that there exists a base $(x_1, x_2, \ldots, x_n)$ of $\mathfrak{g}$ such that $-\mathrm{i}\,\overline{\mathrm{d}u(x_j)}$ $(j = 1, 2, \ldots, n)$ is lower bounded – i.e. there is a real number c_j such that $-\mathrm{i}\,\overline{\mathrm{d}u(x_j)} \geqslant c_j 1_{\mathscr{H}}$ $(j = 1, 2, \ldots, n)$. We shall call this the lower boundedness condition. Let $\mathscr{M}$ be a W*-algebra containing the identity operator $1_{\mathscr{H}}$ in $\mathscr{H}$ such that $u_t\mathscr{M}u_t^* = \mathscr{M}(t\in G)$. Put $\alpha_t(a) = u_t a u_t^*$ $(a\in\mathscr{M}, t\in G)$; then $t\mapsto\alpha_t$ is a σ-weakly continuous representation of G by *-automorphisms on $\mathscr{M}$. Consider a one-parameter group of *-automorphisms as follows.

$$\lambda\mapsto\exp(\mathrm{i}\lambda(-\mathrm{i}\,\overline{du(x_j)} - c_j 1_{\mathscr{H}}))a\exp(-\mathrm{i}\lambda(-\mathrm{i}\,\overline{du(x_j)} - c_j 1_{\mathscr{H}}))$$
$$= \exp(\lambda\,\overline{du(x_j)})a\exp(-\lambda\,\overline{du(x_j)}) = \alpha_{\exp(\lambda x_j)}(a) \qquad (a\in\mathscr{M}, \lambda\in\mathbb{R}^1).$$

Then by 2.4.3 there is a positive operator k_j in $\mathscr{H}$ such that $k\eta\mathscr{M}$ and $\alpha_t(a) = \exp(\mathrm{i}\lambda k_j)a\exp(-\mathrm{i}\lambda k_j)$ $(a\in\mathscr{M}, \lambda\in\mathbb{R}^1)$.

A solution to the following open question would be interesting.

2.6.8 Problem *Let $\mathscr{M}$ be a W*-algebra containing the identity operator $1_{\mathscr{H}}$ in a Hilbert space $\mathscr{H}$ and let G be a simply connected finite-dimensional real Lie group. Let $t\mapsto\alpha_t$ be a σ-weakly continuous representation of G by *-automorphisms on $\mathscr{M}$. Suppose that there exists a strongly continuous unitary representation $t\mapsto u_t$ of G in $\mathscr{H}$ such that $\alpha_t(a) = u_t a u_t^*$ $(a\in\mathscr{M}, t\in G)$ and the representation u satisfies the lower boundedness condition. Then can we conclude that there exists a strongly continuous unitary representation $t\mapsto v_t$ of G in $\mathscr{H}$ such that $\alpha_t(a) = v_t a v_t^*$ $(a\in\mathscr{M}, t\in G)$ and $v_t\in\mathscr{M}$ $(t\in G)$?*

In general, let G be a simply connected finite-dimensional real Lie group and let $t\mapsto u_t$ be a strongly continuous unitary representation of G in a Hilbert space $\mathscr{H}$. If u is uniformly continuous, then it is lower bounded. Moreover let $N = \{t\in G \mid \alpha_t = 1\}$; then by 2.6.2 G/N is a direct sum of a vector group and a compact group. It would be interesting to inquire what happens to the group G under the assumption of the existence of a lower-bounded unitary representation.

2.6.9 Problem *Let $\mathscr{H}$* be a Hilbert space and let V be a dense linear subspace of $\mathscr{H}$. Let $\mathfrak{g}$ be a finite-dimensional real Lie algebra of essentially skew-self-adjoint operators on V – i.e. for $x\in\mathfrak{g}$, the closure $\bar{x}$ in $\mathscr{H}$ is skew-self-adjoint. Suppose that $\mathfrak{g}$ is lower bounded – i.e. there is a base $(x_1, x_2, \ldots, x_n)$ in $\mathfrak{g}$ such that $\{-\mathrm{i}\bar{x}_j \mid j = 1, 2, \ldots, n\}$ are lower-bounded self-adjoint operators in $\mathscr{H}$. Then, determine the Lie algebra $\mathfrak{g}$.

One may formulate a more general spectrum condition problem as follows. Let G be a simply connected finite-dimensional real Lie group and let $\mathfrak{g}$ be its Lie algebra. Then $\mathfrak{g}$ is a finite-dimensional real linear space. Let Λ be a closed convex subset of $\mathfrak{g}^*$ $(=\mathfrak{g})$ satisfying the spectrum condition – i.e. Λ contains no straight line.

Take an inner product $\langle\ ,\ \rangle$ on $\mathfrak{g}$. By the bipolar theorem on convex subsets, we can write $\Lambda = \cap_{\alpha\in\Pi}\{x\in\mathfrak{g} \mid \langle x_\alpha, x\rangle \geqslant c_\alpha\}$ where c_α is a real number for each α. Since Λ contains no straight line, the family $\{x_\alpha | a \in \Pi\}$ contains a linearly independent family $\{x_{\alpha_1}, x_{\alpha_2}, \ldots, x_{\alpha_n}\}$, where $n = \dim(\mathcal{g})$. Let $t \mapsto u_t$ be a strongly continuous unitary representation of G in a Hilbert space $\mathscr{H}$, and let $u_{\exp(\lambda x_\alpha)} = \exp(i\lambda H_\alpha)(\lambda \in R^1)$, where H_α is a self-adjoint operator in $\mathscr{H}$. If $H_\alpha \geqslant c_\alpha 1_{\mathscr{H}}$ for each $\alpha \in \Pi$, then we denote it by $\widetilde{Sp}(u) \subset \Lambda$.

Let $\mathscr{M}$ be a W*-algebra containing the identity operator $1_{\mathscr{H}}$ in $\mathscr{H}$ such that $u_t \mathscr{M} u_t^* = \mathscr{M}$ $(t \in G)$. Let $\alpha_t(a) = u_t a u_t^*$ $(a \in \mathscr{M}, t \in G)$. Suppose that $\widetilde{Sp}(u) \subset \Lambda$; then can we conclude that there exists a strongly continuous unitary representation $t \mapsto v_t$ of G in $\mathscr{H}$ such that $v_t \in \mathscr{M}$, $\alpha_t(a) = v_t a v_t^*$ $(a \in \mathscr{M}, t \in G)$ and $\widetilde{Sp}(v) \subset \Lambda$?

2.6.10 Notes and remarks The original proof of the Kadison–Singer theorem (2.6.3) is different from that presented here. In this section, we have used Putnam's theorem (2.2.9) repeatedly. It would be interesting to extend it to lower-bounded self-adjoint operators which are closely related to the lower boundedness condition.

References [4], [53], [94], [129].

2.7 C*-dynamical systems and ground states

In previous sections, we have discussed W*-dynamical systems as a generalization of quantum field-theoretic models. However, since a W*-dynamical system is in a fixed state, it is not suitable for treating phase transitions in quantum statistical mechanics and quantum field theory. It is also not suitable for treating equilibrium states at different temperatures in quantum statistical mechanics. For these purposes, we have to use C*-dynamical systems. In this section, we shall discuss an algebraic condition which enables a C*-dynamical system to have a state (vacuum state, ground state) which satisfies the spectrum condition.

2.7.1 From a W*-dynamical system to a C*-dynamical system Let $\{\mathscr{M}, \mathbb{R}^n, \alpha\}$ be a W^*-dynamical system on a Hilbert space $\mathscr{H}$ and suppose that there exists a strongly continuous unitary representation $t \mapsto u_t$ of $\mathbb{R}^n$ on $\mathscr{H}$ such that $\alpha_t(a) = u_t a u_t^*$ $(a \in \mathscr{M}, t \in \mathbb{R}^n)$. Suppose that u satisfies the spectrum condition – i.e. there exists a closed convex subset C of $\mathbb{R}^n$ satisfying the spectrum condition such that C contains no straight line and $Sp(u) \subset C$. Moreover we assume that there exists an invariant cyclic vector ξ_0 with $\|\xi_0\| = 1$ in $\mathscr{H}$ – i.e. $u_t \xi_0 = \xi_0$ $(t \in \mathbb{R}^n)$ and $[\mathscr{M}\xi_0] = \mathscr{H}$. Then $0 \in Sp(u)$ by 2.4.7, $u_t \in \mathscr{M}$ $(t \in \mathbb{R}^n)$.

Let $u_t = \int_{\mathbb{R}^n} \exp(i\langle t, x\rangle)\, dp(x)$ be the Stone representation and let

$$S_m = \{(t_1, t_2, \ldots, t_n) \in \mathbb{R}^n | t_1^2 + t_2^2 + \cdots + t_n^2 \leqslant m\}.$$

Put $E_m = \int_{S_m} \mathrm{d}p(x)$, and let

$$u_{m,t} = u_t E_m + (1_{\mathscr{H}} - E_m), \qquad \alpha_{m,t}(a) = u_{m,t} a u_{m,t}{}^* \qquad (a \in \mathscr{M});$$

then $t \mapsto \alpha_{m,t}$ $(t \in \mathbb{R}^n)$ is a norm-continuous representation of $\mathbb{R}^n$ by *-automorphisms on $\mathscr{M}$, and the dynamical system $\{\mathscr{M}, \mathbb{R}^n, \alpha_m\}$ satisfies the spectrum condition and has an invariant cyclic ξ_0. Let $\mathscr{A}$ be the C*-subalgebra of $\mathscr{M}$ generated by $1_{\mathscr{H}}$ and $\{E_m \mathscr{M} E_m | m = 1, 2, \ldots, \}$. Then $\{\mathscr{A}, \mathbb{R}^n, \hat{\alpha}\}$ is a C*-dynamical system, where $\hat{\alpha}$ is the restriction of α to $\mathscr{A}$, and $\mathscr{A}$ is σ-weakly dense in $\mathscr{M}$. Moreover, $\| \hat{\alpha}_{m,t}(x) - \hat{\alpha}_t(x) \| \to 0$ $(m \to \infty)$ for each $x \in \mathscr{A}$, where $\hat{\alpha}_{m,t}$ is the restriction of $\alpha_{m,t}$ to $\mathscr{A}$. Let $\phi_0(x) = (x\xi_0, \xi_0)$ $(x \in \mathscr{A})$; then ϕ_0 is invariant under $\hat{\alpha}$ – i.e. $\phi_0(\hat{\alpha}_t(x)) = \phi_0(x)$ $(x \in \mathscr{A}, t \in \mathbb{R}^n)$ and also $\hat{\alpha}_t(x) = u_t x u_t{}^*$ $(x \in \mathscr{A}, t \in \mathbb{R}^n)$.

Now we shall consider a more general converse problem as follows. Let $\{\mathscr{A}, \mathbb{R}^n, \alpha\}$ be a C*-dynamical system, and let C be a closed convex subset of $\mathbb{R}^n$ satisfying the spectrum condition with an additional condition $0 \in C$. Under what conditions can we conclude that $\{\mathscr{A}, \mathbb{R}^n, \alpha\}$ has an invariant state ϕ_0 satisfying the given spectrum condition?

Let ϕ be an invariant state on $\mathscr{A}$ for $\{\mathscr{A}, \mathbb{R}^n, \alpha\}$ and let $\{\pi_\phi, \mathscr{H}_\phi\}$ be the GNS representation of $\mathscr{A}$ constructed via ϕ. Let $u_\phi(t) a_\phi = (\alpha_t(a))\phi$ for $t \in \mathbb{R}^n$ and $a \in \mathscr{A}$; then $\|u_\phi(t) a_\phi\|^2 = \phi(\alpha_t(a^*)\alpha_t(a)) = \phi(\alpha_t(a^* a)) = \phi(a^* a) = \|a_\phi\|^2$. Therefore $u_\phi(t)$ can be uniquely extended to a unitary operator (denoted again by $u_\phi(t)$) on $\mathscr{H}_\phi$. Then $t \mapsto u_\phi(t)$ is a strongly continuous unitary representation of $\mathbb{R}^n$ and moreover $u_\phi(t) 1_\phi = 1_\phi$ $(t \in \mathbb{R}^n)$ and

$$\begin{aligned} \pi_\phi(\alpha_t(a)) b_\phi &= (\alpha_t(a) b)\phi = (\alpha_t(a \alpha_{-t}(b)))_\phi \\ &= u_\phi(t) \pi_\phi(a) u_\phi(-t) b_\phi \qquad (a, b \in \mathscr{A} \text{ and } t \in \mathbb{R}^n). \end{aligned}$$

Hence

$$\pi_\phi(\alpha_t(a)) = u_\phi(t) \pi_\phi(a) u_\phi(t)^*.$$

Such a representation of $\{\mathscr{A}, \mathbb{R}^n, \alpha\}$ is said to be a covariant representation of $\{\mathscr{A}, \mathbb{R}^n, \alpha\}$, and is denoted by $\{\pi_\phi, u_\phi, \mathscr{H}_\phi\}$.

Therefore the problem is equivalent to the following problem: under what conditions, can we find an invariant state ϕ for $\{\mathscr{A}, \mathbb{R}^n, \alpha\}$ such that the corresponding covariant representation $\{\pi_\phi, u_\phi, \mathscr{H}_\phi\}$ satisfies $Sp(u_\phi) \subset C$?

Let $\mathscr{F}_C$ be the set of all Lebesgue integrable functions f on $\mathbb{R}^n$ such that the Fourier transform $\hat{f}$ has a compact support $s(\hat{f})$ with $s(\hat{f}) \subset \mathbb{R}^n \backslash C$, where $\hat{f}(x) = \int_{\mathbb{R}^n} \exp(\mathrm{i}\langle t, x\rangle) f(t)\,\mathrm{d}t$. For $f \in \mathscr{F}_C$, $a \in \mathscr{A}$, let $T_f(a) = \int_{\mathbb{R}^n} \alpha_t(a) f(t)\,\mathrm{d}t$, and let $\mathscr{L}_C$ be the least closed left ideal of $\mathscr{A}$ containing $\{T_f(a) | f \in \mathscr{F}_C, a \in \mathscr{A}\}$.

2.7.2 Theorem (Doplicher [55]) *A C*-dynamical system $\{\mathscr{A}, \mathbb{R}^n, \alpha\}$ has an invariant state ϕ which satisfies the spectrum condition $Sp(u_\phi) \subset C$ if and only if $\mathscr{L}_C \subsetneqq \mathscr{A}$.*

Proof Suppose that there exists a state ϕ satisfying the conditions. Then for $f \in \mathscr{F}_C$ and $a \in \mathscr{A}$,

$$\begin{aligned}\pi_\phi(T_f(a))1_\phi &= \int_{\mathbb{R}^n} f(t)(t)\,\mathrm{d}t \int_{\mathbb{R}^n} \exp(\mathrm{i}\langle t, x\rangle)\,\mathrm{d}P(x)(\pi_\phi(a)1_\phi) \\ &= \int_{\mathbb{R}^n} \mathrm{d}P(x)(\pi_\phi(a)1_\phi) \int_{\mathbb{R}^n} f(t)\exp(\mathrm{i}\langle t, x\rangle)\,\mathrm{d}t \\ &= \int_{\mathbb{R}^n} \hat{f}(x)\,\mathrm{d}P(x)(\pi_\phi(a)1_\phi) \\ &= \int_C \hat{f}(x)\,\mathrm{d}P(x)\pi_\phi(a)1_\phi = 0,\end{aligned}$$

where

$$u_\phi(t) = \int_{\mathbb{R}^n} \exp(\mathrm{i}\langle t, x\rangle)\,\mathrm{d}P(x).$$

Hence $T_f(a) \in \mathscr{L}$, where $\mathscr{L} = \{x \in \mathscr{A} \mid \phi(x^*x) = 0\}$, and so $\mathscr{L}_C \subset \mathscr{L}$.

Conversely, suppose that $\mathscr{L}_C \subsetneqq \mathscr{A}$; then there is a state ψ on $\mathscr{A}$ such that $\psi(\mathscr{L}_C) = 0$. Let μ be a finitely additive invariant probability measure on $\mathbb{R}^n$ (by the amenability of $\mathbb{R}^n$, such a measure does exist), and let $\phi(a) = \int_{\mathbb{R}^n} \psi(\alpha_t(a))\,\mathrm{d}\mu(t)$ for $a \in \mathscr{A}$. (This is well-defined, because $t \mapsto \psi(\alpha_t(a))$ is bounded continuous.) Then ϕ is an invariant state on $\mathscr{A}$. Moreover $\alpha_t(T_f(a)) = T_g(a)$, where $g(s) = f(t+s)$ $(s \in \mathbb{R}^n)$, so that $\hat{g}(x) = \exp(\mathrm{i}\langle t, x\rangle)\hat{f}(x)$ and so $s(\hat{g}) \subset \mathbb{R}^n \backslash C$. Therefore $\alpha_t(\mathscr{L}_C) \subset \mathscr{L}_C$ for $t \in \mathbb{R}^n$; hence $\phi(\mathscr{L}_C) = 0$. Since

$$\begin{aligned}\pi_\phi(T_f(a))1_\phi &= \int_{\mathbb{R}^n} u_\phi(t)\pi_\phi(a)f(t)1_\phi \\ &= \int \hat{f}(x)\,\mathrm{d}P(x)\pi_\phi(a)1_\phi = 0\end{aligned}$$

for $f \in \mathscr{F}_C$ and $[\pi_\phi(\mathscr{A})1_\phi] = \mathscr{H}_\phi$, $\int_{\mathbb{R}^n} \hat{f}(x)\,\mathrm{d}P(x) = 0$; hence

$$\int_{\mathbb{R}^n \backslash C} \mathrm{d}P(x) = 0$$

i.e. $Sp(u_\phi) \subset C$. □

2.7.3 Definition *Let $\{\mathscr{A}, G, \alpha\}$ be a C*-dynamical system. $\{\mathscr{A}, G, \alpha\}$ is said to be approximately uniform if there exists a sequence of uniformly continuous (namely, norm-continuous) C*-dynamical systems $\{\mathscr{A}, G, \alpha_m\}$ such that $\|\alpha_{m,t}(a) - \alpha_t(a)\| \to 0$ $(m \to \infty)$ for $a \in \mathscr{A}$ and $t \in G$.*

2.7.4 Proposition *Suppose that $\{\mathscr{A}, \mathbb{R}^n, \alpha\}$ is approximately uniform and*

$\{\mathscr{A}, \mathbb{R}^n, \alpha_m\}$ has an invariant state ϕ_m on $\mathscr{A}$ such that $Sp(u_{\phi_m}) \subset C$ $(m=1,2,\ldots,)$; then $\{\mathscr{A}, \mathbb{R}^n, \alpha\}$ has an invariant state ϕ on $\mathscr{A}$ such that $Sp(u_\phi) \subset C$.

Proof Let ϕ be an accumulation point of $\{\phi_m\}$ in the state space of $\mathscr{A}$. For $f \in \mathscr{F}_C$, $a, b \in \mathscr{A}$,

$$\begin{aligned}|\phi(aT_f(b)) - \phi_m(aT_{m,f}(b))| &\leqslant |\phi(aT_f(b)) - \phi_m(aT_f(b))| \\ &\quad + |\phi_m(aT_f(b)) - \phi_m(aT_{m,f}(a))| \\ &\leqslant |\phi(aT_f(b)) - \phi_m(aT_f(b))| \\ &\quad + \|a\| \, \|T_f(b) - T_{m,f}(b)\|.\end{aligned}$$

On the other hand,

$$\begin{aligned}\|T_f(b) - T_{m,f}(b)\| &= \left\| \int \alpha_t(b) f(t)\,\mathrm{d}t - \int \alpha_{m,t}(b) f(t)\,\mathrm{d}t \right\| \\ &\leqslant \int \|\alpha_t(b) - \alpha_{m,t}(b)\| \, |f(t)|\,\mathrm{d}t \to 0 \qquad (m \to \infty).\end{aligned}$$

Hence $\phi(aT_f(b)) = 0$ and so $\phi(\mathscr{L}_c) = 0$. Moreover $\phi_m(\alpha_{m,t}(a)) = \phi_m(a)$ $(a \in \mathscr{A}, t \in \mathbb{R}^n)$, so that $\phi(\alpha_t(a)) = \phi(a)$. □

2.7.5 Proposition *Suppose that $\{\mathscr{A}, \mathbb{R}^1, \alpha\}$ is approximately uniform; then $\{\mathscr{A}, \mathbb{R}^1, \alpha\}$ has an invariant state ϕ such that $Sp(u_\phi) \subset [0, \infty)$.*

Proof Let $\{\mathscr{A}, \mathbb{R}^1, \alpha_m\}$ be a sequence of uniformly continuous C*-dynamical systems such that $\|\alpha_{m,t}(a) - \alpha_t(a)\| \to 0$ $(m \to \infty)$ for $t \in \mathbb{R}^1$ and $a \in \mathscr{A}$. Let us represent $\mathscr{A}$ as a C*-algebra containing the identity $1_{\mathscr{H}}$ in a Hilbert space $\mathscr{H}$. Since $\mathrm{d}\alpha_m$ is a bounded *-derivation on $\mathscr{A}$, by 2.5.5, there is a positive element h_m in $\mathscr{A}''$ such that $\mathrm{d}\alpha_m(a) = \mathrm{i}[h_m, a]$ $(a \in \mathscr{A})$ and $\|h_m\| = \|d\alpha_m\|$. By the minimality of h_m, h_m is not invertible. Let $\mathscr{B}$ be the C*-subalgebra of $B(\mathscr{H})$ generated by $\mathscr{A}$ and h_m, and let A be a C*-subalgebra of $\mathscr{B}$ generated by $1_{\mathscr{H}}$ and h_m. Then $A = C(K)$, where $C(K)$ is the C*-algebra of all continuous functions on a compact space K. Since h_m is not invertible, there is a point p_0 in K such that $h_m(p_0) = 0$. A functional f: $a \mapsto a(p_0)$ $(a \in A)$ is a state on A. Let ψ_m be a state on $\mathscr{B}$ such that $\psi_m | A = f$. Then,

$$\begin{aligned}|\psi_m(\mathrm{i}[h_m, a])| &\leqslant |\psi_m(h_m a)| + |\psi_m(a h_m)| \\ &\leqslant \psi_m(h_m)^{1/2} \psi_m(a^* h_m a)^{1/2} + \psi_m(a h_m a^*)^{1/2} \psi(h_m)^{1/2} \\ &= 0 \qquad \text{for } a \in \mathscr{B}.\end{aligned}$$

Hence

$$\psi_m(\exp(t\delta_{\mathrm{i}h_m})(a)) = \psi_m(a) \qquad \text{for } a \in \mathscr{B} \text{ and } t \in \mathbb{R}^1,$$

where $\delta_{\mathrm{i}h_m}(a) = \mathrm{i}[h_m, a]$ $(a \in \mathscr{B})$. Let $\beta_t = \exp(t\delta_{\mathrm{i}h_m})$; then ψ_m is an invariant state on $\mathscr{B}$ for $\{\mathscr{B}, \mathbb{R}^1, \beta\}$. Consider the covariant representation $\{\pi_{\psi_m}, u_{\psi_m}, \mathscr{H}_{\psi_m}\}$ of $\{\mathscr{B}, \mathbb{R}^1, \beta\}$; then

$$\begin{aligned}\pi_{\psi_m}(\exp t\delta_{ihm}(a)) &= \pi_{\psi_m}(\exp(ith_m)a\exp(ith_m)^*)\\ &= \pi_{\psi m}(\exp(ith_m))\pi_{\psi_m}(a)\pi_{\psi_m}(\exp(ith_m)^*)\\ &= u_{\psi_m}(t)\pi_{\psi_m}(a)u_{\psi_m}(t)^* \qquad (a\in\mathscr{B}).\end{aligned}$$

Hence

$$u_{\psi_m}(t)\pi_{\psi_m}(\exp(-ith_m))\in\pi_{\psi}(\mathscr{B})'.$$

On the other hand,

$$u_{\psi_m}(t)1_{\psi_m} = \pi_{\psi_m}(\exp(ith_m))1_{\psi_m} = 1_{\psi_m} \qquad (t\in\mathbb{R}^1).$$

Hence

$$\pi_{\psi_m}(b)u_{\psi_m}(t)\pi_{\psi_m}(\exp(-ith_m))1_{\psi_m} = u_{\psi_m}(t)\pi_{\psi_m}(\exp(-ith_m))\pi_{\psi_m}(b)1_{\psi_m} \qquad (b\in\mathscr{B})$$

and

$$\pi_{\psi_m}(b)u_{\psi_m}(t)\pi_{\psi_m}\exp(-ith_m)1_{\psi_m} = \pi_{\psi_m}(b)1_{\psi_m} \qquad (b\in\mathscr{B});$$

hence

$$u_{\psi_m}(t)\pi_{\psi_m}(\exp(-ith_m)) = 1_{\mathscr{H}_{\psi_m}}$$

and so

$$u_{\psi_m}(t) = \pi_{\psi_m}(\exp(ith_m)) = \exp it\pi_{\psi_m}(h_m).$$

Therefore $Sp(u_{\psi_m})\subset[0,\infty)$. Let $K = [\pi_{\psi_m}(\mathscr{A})1_{\psi_m}]$; then

$$u_{\psi_m}(t)\pi_{\psi_m}(\mathscr{A})1_{\psi_m} = u_{\psi_m}(x)\pi_{\psi_m}(\mathscr{A})u_{\psi_m}(-t)1_{\psi_m}\subset\pi_{\psi_m}(\mathscr{A})1_{\psi_m}$$

and so K is an invariant closed subspace of $\mathscr{H}$ under u_{ψ_m}. Let $\varphi_m = \psi_m|\mathscr{A}$; then ϕ_m is an invariant state on $\mathscr{A}$ such that $Sp(u_{\phi_m}) = Sp(u_{\psi_m}|K)\subset[0,\infty)$. Hence by 2.7.4, there exists an invariant state ϕ on $\mathscr{A}$ under α such that $Sp(u_\phi)\subset[0,\infty)$, □

Let G be a simply connected finite-dimensional real Lie group and let a C*-dynamical system $\{\mathscr{A}, G, \alpha\}$ be an approximately uniform C*-dynamical system; then there exists a sequence of uniformly continuous C*-dynamical systems $\{\mathscr{A}, G, \alpha_m\}$ such that $\|\alpha_{m,t}(a)-\alpha_t(a)\|\to 0$ for $a\in\mathscr{A}$. Let $N_m = \{t\in G|\alpha_{m,t}(a) = a \text{ for all } a\in\mathscr{A}\}$; then N_m is a closed normal subgroup of G. By 2.6.3 and 2.6.5, G/N_m is a direct product of a vector group and a compact group, and so it is an amenable group; hence it has a left invariant finitely additive probability measure ν_m. Let ψ_m be a state on $\mathscr{A}$ and put

$$\phi_m(a) = \int\psi_m(\alpha_{m,t}(a))\,\mathrm{d}\nu_m(t) \qquad \text{for } a\in\mathscr{A};$$

then $\phi_m(\alpha_{m,t}(a)) = \phi_m(a)$; hence ϕ_m is an invariant state on $\mathscr{A}$ under α_m. Let $\{\pi_{\phi_m}, u_{\phi_m}, \mathscr{H}_{\phi_m}\}$ be the covariant representation of $\{\mathscr{A}, G, \alpha_m\}$ constructed via ϕ_m. Let $\mathscr{g}$ be the Lie algebra of G and let Λ be a closed convex subset of $\mathscr{g}$ which contains no straight line.

2.7.6 Theorem *Suppose that* $\widetilde{Sp}(u_{\phi_m}) \subset \Lambda$ *for all m; then there is an invariant state* ϕ *on* $\mathscr{A}$ *under* α *such that* $\widetilde{Sp}(u_\phi) \subset \Lambda$, *where* $\{\pi_\phi, u_\phi, \mathscr{H}_\phi\}$ *is the covariant representation of* $\{\mathscr{A}, G, \alpha\}$ *constructed via* ϕ.

Proof Let $\Lambda = \bigcap_{\beta\in\pi}\{x\in\mathscr{g} \mid \langle x_\beta, x\rangle \geqslant C_\beta\}$; then $H_{m,\beta} \geqslant C_\beta 1_{\mathscr{H}\phi_m}$, where $U_{\phi_m, \exp\lambda x_\beta} = \exp(i\lambda H_{m,\beta})(\lambda\in\mathbb{R}^1)$.

Let ϕ be an accumulation point of $\{\phi_n\}$ in the state space of $\mathscr{A}$; then one can easily see that ϕ is an invariant state on $\mathscr{A}$ under α. Let $\{\pi_\phi, U_\phi, \mathscr{H}_\phi\}$ be the covariant representation of $\{\mathscr{A}, G, \alpha\}$ constructed via ϕ, and let

$$U_{\phi, \exp(\lambda x_\beta)} = \exp(i\lambda H_\beta)(\lambda\in\mathbb{R}^1);\text{ then for } a\in\mathscr{D}\,(\mathrm{d}\alpha),$$

$$(H_\beta a_\phi, a_\phi) = -i\frac{\mathrm{d}}{\mathrm{d}\lambda}(\exp(i\lambda H_\beta)a(\exp - i\lambda H_\beta)1_\phi, a_\phi)|_{\lambda=0}$$

$$= -i\phi(a^*\,\mathrm{d}\alpha(x_\beta)a).$$

Similarly

$$(H_{m,\beta} b\phi_m, b\phi_m) = -i\phi_m(b^*\,\mathrm{d}\alpha_m(x_\beta)b) \qquad \text{for } b\in\mathscr{D}(\mathrm{d}\alpha_m).$$

Let $C_c^\infty(G)$ be the space of all infinitely differentiable functions on G with compact support. For $f\in C_c^\infty(G)$, and $a\in\mathscr{A}$, let $T_{f,m}(a) = \int_G \alpha_{m,t}(a) f(t)\,\mathrm{d}v(t)$ and $T_f(a) = \int_G \alpha_t(a) f(t)\,\mathrm{d}v(t)$, where v is a left invariant Haar measure on G. Then

$$\mathrm{d}\alpha_m(x_\beta)T_{f,m}(a) = \frac{\mathrm{d}}{\mathrm{d}\lambda}\alpha_{m,\exp(\lambda x_\beta)}(T_{f,m}(a))|_{\lambda=0}$$

$$= \frac{\mathrm{d}}{\mathrm{d}\lambda}\int_G \alpha_{m,\exp(\lambda x_\beta)t}(a) f(t)\,\mathrm{d}v(t)|_{\lambda=0}$$

$$= \frac{\mathrm{d}}{\mathrm{d}\lambda}\int_G \alpha_{m,t}(a) f((\exp\lambda x\beta)^{-1}t)\,\mathrm{d}v(t)|_{\lambda=0}$$

$$= -\int_G \alpha_{m,t}(a)\frac{\mathrm{d}}{\mathrm{d}\lambda} f((\exp\lambda x_\beta)^{-1}t)|_{\lambda=0}\,\mathrm{d}v(t).$$

Similarly,

$$\frac{\mathrm{d}}{\mathrm{d}\lambda}\alpha_{\exp(\lambda X_\beta)}(T_f(a))|_{\lambda=0} = -\int_G \alpha_t(a)\frac{\mathrm{d}}{\mathrm{d}\lambda} f((\exp\lambda x_\beta)^{-1}t)|_{\lambda=0}\,\mathrm{d}v(t).$$

Since $\| \alpha_{m,t}(a) - \alpha_t(a) \| \to 0$ $(m \to \infty)$ and $\| \alpha_{m,t}(a) - \alpha_t(a) \| \leqslant 2$,

$$\| \mathrm{d}\alpha_m(x_\beta) T_{f,m}(a) - \mathrm{d}\alpha(x_\beta) T_f(a) \| \to 0 \qquad (m \to \infty).$$

Without loss of generality, we may assume that $\phi_m \to \phi$ in the state space of $\mathscr{A}$. Therefore,

$$\begin{aligned} &|\phi_m(T_{f,m}(a)^* \mathrm{d}\alpha_m(x_\beta)(T_{f,m}(a))) - \phi(T_f(a)^* \mathrm{d}\alpha(x_\beta)(T_f(a)))| \\ &\quad \leqslant |\phi_m(T_{f,m}(a)^* \mathrm{d}\alpha_m(x_\beta)(T_{f,m}(a))) - \phi_m(T_f(a)^* \mathrm{d}\alpha(x_\beta)(T_f(a)))| \\ &\qquad + |\phi_m(T_f(a)^* \mathrm{d}\alpha(x_\beta)(T_f(a))) - \phi(T_f(a)^* \mathrm{d}\alpha(x_\beta)(T_f(a)))| \\ &\quad \leqslant \| T_{f,m}(a)^* - T_f(a)^* \| \, \| \mathrm{d}\alpha_m(x_\beta)(T_{f,m}(a)) \| \\ &\qquad + \| T_f(a)^* \| \, \| \mathrm{d}\alpha_m(x_\beta)(T_{f,m}(a)) - \mathrm{d}\alpha(x_\beta)(T_f(a)) \| \\ &\qquad + |(\phi_m - \phi)(T_f(a)^* \mathrm{d}\alpha(x_\beta)(T_f(a)))| \to 0 \qquad (m \to \infty) \end{aligned}$$

Hence

$$(H_\beta T_f(a)1_\phi, T_f(a)1_\phi) \geqslant C_\beta (T_f(a)1_\phi, T_f(a)1_\phi),$$

Suppose that $a \in \mathscr{D}(\mathrm{d}\alpha)$ and $\{f_n\}$ $(\subset C_c^\infty(G))$ is an approximate identity of the group algebra $L^1(G, \gamma)$ with support in small neighbourhoods and integral one; then $T_{f_n}(a) \to a$ and $\mathrm{d}\alpha(x_\beta) T_{f_n}(a) = T_{f_n}(\mathrm{d}\alpha(x_\beta)(a)) \to \mathrm{d}\alpha(x_\beta)(a)$ and so

$$(H_\beta a 1_\phi, a 1_\phi) \geqslant C_\beta (a 1_\phi, a 1_\phi) \qquad \text{for } a \in \mathscr{D}(\mathrm{d}\alpha).$$

Since $H_\beta = \overline{H_\beta | \mathscr{D}(d\alpha) 1_\phi}$, $H_\beta \geqslant C_\beta 1_{\mathscr{H}_\beta}$. □

2.7.7 Notes and remarks Proposition 2.7.5 is due to Powers and Sakai [145]. In later chapters, the reader will find that all C*-dynamical systems $\{\mathscr{A}, \mathbb{R}, \alpha\}$ appearing in quantum lattice systems are approximately uniform, so that they always have ground states.

References [55], [145].

3

Unbounded derivations

Introduction

In Chapter 2, we mainly discussed bounded derivations on C*-algebras and special unbounded derivations which arise from a fixed physical state (and consequently associated with a W*-dynamical system). However, in quantum physics, we have to study various physical states like ground states, vacuum states and equilibrium states on a C*-algebra of quasi-local observables. In addition a partial differentiation on a mainfold is not bounded. Therefore, unbounded derivations in C*-algebras are perhaps much more important than those previously discussed.

In the previous chapter, we have seen that an everywhere-defined derivation on a C*-algebra is automatically bounded; for such derivations, the purely algebraic property implies continuity. On the other hand, for a densely defined derivation, the purely algebraic property (Leibnitz's rule) does not imply closability. Therefore it is quite difficult to study densely defined derivations from a purely algebraic point of view. Fortunately all densely defined derivations which arise from analysis, geometry and quantum physics turn out to be closable. They are also self-adjoint with respect to the *-operation. In this chapter, we shall study closable *-derivations in C*-algebras. The study of general unbounded derivations in C*-algebras can be divided into three steps: (i) closability, (2) the domain of a closed *-derivation and (3) generators.

Closable *-derivations are also of special interest, because they include partial differentiations in manifolds. In this chapter, we shall discuss these topics.

3.1 Definition of derivations

Let A be a C*-algebra. A derivation δ in a C*-algebra A is a linear mapping in A satisfying the following conditions:

(1) the domain $\mathscr{D}(\delta)$ of δ is a dense subalgebra of A;
(2) $\delta(ab) = \delta(a)b + a\delta(b)\,(a, b\in\mathscr{D}(\delta))$.

δ is said to be a *-derivation if it also satisfies:

(3) $a\in\mathscr{D}(\delta)$ implies $a^*\in\mathscr{D}(\delta)$ and $\delta(a^*) = \delta(a)^*$.

δ is said to be *closed* if $x_n\in\mathscr{D}(\delta), x_n \to x$ and $\delta(x_n)\to y$ implies $x\in\mathscr{D}(\delta)$ and $\delta(x) = y$. δ is said to be *closable* if $x_n\in\mathscr{D}(\delta), x_n\to 0$ and $\delta(x_n)\to y$ implies $y = 0$. A closable derivation δ can be extended to a closed derivation $\bar{\delta}$, the least closed derivation with $\bar{\delta}\supset\delta$, called the *closure* of δ.

Examples

(1) Inner derivations $\delta_a(x) = [a, x](x\in A)$. Inner derivations are bounded.
(2) Let $A = C([0,1])$, $\mathscr{D}(\delta) = C^{(1)}([0,1])$ = the *-algebra of all complex-valued, continuously differentiable functions on the unit interval $[0,1]$. Define $\delta(f) = (\mathrm{d}/\mathrm{d}t)f(t)(f\in C^{(1)}([0,1]))$; then δ is a closed unbounded *-derivation in $C([0,1])$.
(3) Let $A = C_0(\mathbb{R})$ be the C*-algebra of all complex-valued continuous functions on the real line $\mathbb{R}$ vanishing at infinity, $C_0{}^{(1)}(\mathbb{R}) = \mathscr{D}(\delta)$. Define $\delta(f)(t) = (\mathrm{d}/\mathrm{d}t)f(t)(f\in C_0{}^{(1)}(\mathbb{R}))$; then δ is a closed unbounded *-derivation in $C_0(\mathbb{R})$.

(4) Ising models Let $\mathbb{Z}$ be the group of all integers and let $\mathbb{Z}^n = \underbrace{\mathbb{Z}\times\mathbb{Z}\times\cdots\times\mathbb{Z}}$. Let B be the matrix algebra of all 2×2 matrices over the complex field, and let $B_p = B\ (p\in\mathbb{Z}^n)$.

Let $A = \bigotimes_{p\in\mathbb{Z}^n} B_p$ be the infinite C^*-tensor product of $\{B_p | p\in\mathbb{Z}^n\}$. Consider the spin matrix $\sigma = \begin{pmatrix} 1 & 0 \\ 0 & -1\end{pmatrix}$ in B and let σ_p be the corresponding matrix in B_p to σ. Define the total potential H as follows: $H = \sum_{p,q\in\mathbb{Z}^n} - J(|p-q|)\sigma_p\sigma_q$, where $J(1) = 1$ and $J(n) = 0$, otherwise, and $|p-q| = \sum_{i=1}^n |p_i - q_i|$ with $p = (p_1, p_2, \ldots, p_n)$ and $q = (q_1, q_2, \ldots, q_n)$. By using H, we shall define a *-derivation $\delta_{\mathrm{i}H}$ in A as follows: $\delta_{iH}(a) = \mathrm{i}[H, a](a\in A_0)$ where $A_0 = \bigcup_{m=1}^\infty \bigotimes_{p\in\Lambda_m} B_p \subset A$ with $\Lambda_m = \{x\in\mathbb{Z}^n | |x|\leqslant m\}$. ($\bigotimes_{p\in\Lambda_m} B_p$ is embbed canonically into A by identifying it with $\bigotimes_{p\in\Lambda_m} B_p \otimes \bigotimes_{p\in\mathbb{Z}^n\setminus\Lambda_m} 1_p$, where 1_p is the identity of B_p).

Then one can easily see that for

$$a\in\bigotimes_{p\in\Lambda_m} B_p,\ \delta_{iH}(a) = \mathrm{i}\left[\sum_{p,q\in\Lambda_{m+1}} - J(|p-q|)\sigma_p\sigma_q, a\right];$$

hence δ_{iH} is well-defined. Then δ is a closable unbounded *-derivation in A.

(5) Heisenberg models Consider the three Pauli spin matrices in B:

$$\sigma_1 = \begin{pmatrix} 0 & 1 \\ 1 & 0 \end{pmatrix}, \qquad \sigma_2 = \begin{pmatrix} 0 & -i \\ i & 0 \end{pmatrix}, \qquad \sigma_3 = \begin{pmatrix} 1 & 0 \\ 0 & -1 \end{pmatrix}.$$

Let $\sigma_j(p)$ be the matrix corresponding to σ_j in $B_p(j = 1, 2, 3)$. Define the total potential H as follows: $H = \sum_{p,q \in \mathbb{Z}^n} - J(|p - q|) \sum_{j=1}^{3} \sigma_i(p)\sigma_i(q)$. By using H, we shall define a *-derivation δ_{iH} in A as follows: $\delta_{iH}(a) = i[H, a](a \in A_0)$. Then δ is a closable unbounded *-derivation in A.

(6) Quasi-free derivations in the canonical anti-commutation relation algebra Let $\mathscr{H}$ be a Hilbert space. For $f, g \in \mathscr{H}$ define $\langle f, g \rangle = \overline{(f, g)}$ and denote a new Hilbert space having the scalar product $\langle \ , \ \rangle$ by $\bar{\mathscr{H}}$. Let $f \to a(f)$ be the identical mapping of $\mathscr{H}$ onto $\bar{\mathscr{H}}$; then it is complex-conjugate linear. Let $f \to a^*(f)$ be the identical mapping of $\mathscr{H}$ onto $\mathscr{H}$; then $f \to a^*(f)$ is linear. Now let $\mathscr{F}$ be a free algebra with its identity over the complex field generated by a linear space $\mathscr{H} \oplus \bar{\mathscr{H}}$. Let $\mathscr{I}$ be the two-sided ideal of $\mathscr{F}$ generated by $\{a(f)a(g) + a(g)a(f), a^*(f)a^*(g) + a^*(g)a^*(f),\ a(f)a^*(g) + a^*(g) a(f) - \overline{(f, g)}1 \mid f, g \in \mathscr{H}\}$; then the quotient algebra $\mathscr{F}/\mathscr{I} = \mathscr{A}_0(\mathscr{H})$ satisfies the following relations:

$$\{a(f), a(g)\} = a(f)a(g) + a(g)a(f) = 0, \{a^*(f), a^*(g)\} = 0$$

and

$$\{a(f), a^*(g)\} = \overline{(f, g)}1 = \langle f, g \rangle 1,$$

where $a(\cdot)$ and $a^*(\cdot)$ in $\mathscr{F}/\mathscr{I}$ are denoted by the canonical images of $a(\cdot)$ and $a^*(\cdot)$ in $\mathscr{F}$. Define $a(f)^* = a^*(f)(f \in \mathscr{H})$ and $\{a^*(f)\}^* = a(f)(f \in \mathscr{H})$; then the *-operation can be uniquely extended to a *-operation in $\mathscr{F}/\mathscr{I} = \mathscr{A}_0(\mathscr{H})$ and $\mathscr{A}_0(\mathscr{H})$ becomes a *-algebra. Let $\mathscr{H}$ be a finite-dimensional Hilbert space and let $f_1, f_2, \ldots, f_n$ be a base of $\mathscr{H}$. Since $a(f)^2 = 0, a^*(f)^2 = 0$ and $a(f)a(g)^* = \langle f, g \rangle 1 - a(g)^*a(f)$,

$$\{a(f_{i_1})a(f_{i_2}) \cdots a(f_{i_m})a(f_{j_i})^* a(f_{j_2})^* \cdots a(f_{jl})^*, 1 \mid i_1 < i_2 < \cdots < i_m;$$
$$j_1 < j_2 \cdots < j_l; 1 \leqslant i_m \leqslant n; 1 \leqslant j_l \leqslant n\}$$

is a base of $\mathscr{A}_0(\mathscr{H})$; therefore the dimension of

$$\mathscr{A}_0(\mathscr{H}) = \sum_{m=0}^{2n} {}_{2n}C_m = (1 + 1)^{2n} = 2^n \times 2^n = 2^{\dim(\mathscr{H})} \times 2^{\dim(\mathscr{H})}.$$

$$a(f)^*a(f) + a(f)a(f)^* = \|f\|^2 1 \qquad (f \in \mathscr{H}).$$

If $\|f\| = 1$, then $a(f)^*a(f) + a(f)a(f)^* = 1$. Since

$$a(f)^2 = 0, \qquad (a(f)^*a(f))^2 = a(f)^*a(f)(1 - a(f)a(f)^*) = a(f)^*a(f).$$

Hence $a(f)^*a(f)$ is an idempotent. Similarly $(a(f)a(f)^*)^2 = a(f)a(f)^*$.

Let $\mathscr{H}_0$ be the one-dimensional subspace of $\mathscr{H}$ generated by f; then $\mathscr{A}_0(\mathscr{H}_0)$ is isomorphic to the full matrix algebra of 2×2. Moreover if we represent $a(f) = \begin{pmatrix} 0 & 1 \\ 0 & 0 \end{pmatrix}$ and $a(f)^* = \begin{pmatrix} 0 & 0 \\ 1 & 0 \end{pmatrix}$, then one can easily see that $\mathscr{A}_0(\mathscr{H}_0)$ is *-isomorphic to the full matrix algebra of 2×2 over the complex field. It is also not so difficult to see that $\mathscr{A}_0(\mathscr{H})$ is *-isomorphic to the full matrix algebra of $2^n \times 2^n$ over the complex field. If $\mathscr{H}$ is infinite-dimensional, then arbitrary element ψ in $\mathscr{A}_0(\mathscr{H})$ can be expressed as follows:

$$\psi = \sum a(g_1)a(g_2)\cdots a(g_m)a(g_{m+1})^* \cdots a(g_{n+m})^* \in \mathscr{A}_0(K),$$

where K is a finite-dimensional subspace of $\mathscr{H}$ depending on ψ and so $\|\psi\|$ is defined by the norm of ψ in $\mathscr{A}_0(K)$. Since the C*-norm of a C*-algebra is unique, $\|\psi\|$ is uniquely defined. Therefore $\mathscr{A}_0(\mathscr{H})$ is a normed *-algebra with $\|x^*x\| = \|x\|^2$ $(x \in \mathscr{A}_0(\mathscr{H}))$. Let $\mathscr{A}(\mathscr{H})$ be the completion of the normed *-algebra $\mathscr{A}_0(\mathscr{H})$; then $\mathscr{A}(\mathscr{H})$ is a C*-algebra. $\mathscr{A}(\mathscr{H})$ is called the canonical anti-commutation relation algebra. It is clear that for $f \in \mathscr{H}$, $\|a(f)\| = \|a(f)^*\| = \|f\|$.

Now we shall define quasi-free derivations in $\mathscr{A}(\mathscr{H})$. Let H be a symmetric operator in $\mathscr{H}$ and define $\delta_{iH}(a(f)) = a(\mathrm{i}Hf)$; then δ_{iH} can be uniquely extended to a *-derivation with $\mathscr{D}(\delta_{iH}) = \mathscr{A}_0(\mathscr{D}(H))$ in $\mathscr{A}(\mathscr{H})$, where $\mathscr{D}(H)$ is the definition domain of H. δ_{iH} is a closable *-derivation (cf. 3.2.15).

Therefore its closure (again denoted by δ_{iH}) is a closed *-derivation. This closed *-derivation is called a quasi-free derivation in the canonical anti-commutation relation algebra $\mathscr{A}(\mathscr{H})$.

References [142][159].

3.2 Closability of derivations

Let A be a C*-algebra, δ a *-derivation in A. In the previous chapter, we have learned that if $\mathscr{D}(\delta) = A$, then δ is bounded. But if $\mathscr{D}(\delta) \subsetneq A$, then δ is not necessarily closable – in fact, we shall prove the following proposition.

3.2.1 Proposition *Let K be a totally disconnected compact space and let δ be a closed *-derivation in $C(K)$; then $\delta = 0$.*

Proof Let $a(=a^*) \in \mathscr{D}(\delta)$ and let $p(x)$ be a polynomial of one variable x. Define $p(a)(t) = p(a(t))(t \in K)$; then it is easily seen that $p(a) \in \mathscr{D}(\delta)$ and $\delta(p(a)) = p'(a)\delta(a)$, where p' is the derivative of p. For $f \in C^{(1)}([-\|a\|, \|a\|])$ take a sequence $\{p_n\}$ of polynomials on $[-\|a\|, \|a\|]$ such that $\|p_n - f\| \to 0$

and $\|p_n' - f'\| \to 0$; then $\|p_n(a) - f(a)\| \to 0$ and $\|\delta(p_n(a)) - f'(a)\delta(a)\| \to 0$. By the closedness of δ, $f(a) \in \mathcal{D}(\delta)$ and $\delta(f(a)) = f'(a)\delta(a)$.

Let G be a proper, open and closed, subset of K; then there is an element $e \in C(K)$ such that $e(t) = 1$ $(t \in G)$ and $e(t) = 0$ $(t \in K \backslash G)$. Since $\mathcal{D}(\delta)$ is a dense *-subalgebra of $C(K)$, there is an element $h \in \mathcal{D}(\delta)$ such that $h^* = h$, $\|e - h\| < \frac{1}{3}$; hence $h(t) \in [-\frac{1}{3}, \frac{1}{3}] \cup [\frac{2}{3}, \frac{4}{3}]$. Take an $f \in C^{(1)}(\mathbb{R})$ such that $f(x) = 0$ $(x \in [-\frac{1}{3}, \frac{1}{3}])$ and $f(x) = 1$ $(x \in [\frac{2}{3}, \frac{4}{3}])$; then $f(h) = e$ and so $f(h) = e \in \mathcal{D}(\delta)$. $\delta(e) = \delta(ee) = \delta(e)e + e\delta(e) = 2\delta(e)e$. Hence $\delta(e) = 0$. From this one can easily conclude that $\delta \equiv 0$. □

3.2.2 Example ([39]) Let K be the Cantor set of $[0, 1]$; then K is a compact subset. By Tietze's theorem, $C(K) = \{f|K \mid f \in C([0, 1])\}$, where $f|K$ is the restriction of f to K. For $f \in C^{(1)}([0, 1])$, define $\delta(f|K) = f'|K$; then δ is a *-derivation in $C(K)$. In fact, if $f|K = g|K$, then $f'|K = g'|K$, because K is a perfect set.

Clearly $\delta \not\equiv 0$ and so it is not closable.

In the proof of Proposition 3.2.1, we have shown the following.

3.2.3 Proposition *Let δ be a closed derivation in a commutative C*-algebra $C_0(\Omega)$ with a locally compact space Ω. Then, for $a\ (= a^*) \in \mathcal{D}(\delta)$ and $f \in C^{(1)}$ $([-\|a\|, \|a\|])$, $f(a) \in \mathcal{D}(\delta)$ and $\delta(f(a)) = f'(a)\delta(a)$.*

A solution to the following problem would be interesting.

3.2.4 Problem *Let K be a compact space and suppose that C(K) has a non-zero closed *-derivation. Then can we conclude that K has a one-dimensional differential structure? In this problem, we do not require that a conjectured differential structure should be directly related to a given derivation. In fact, even in C([0, 1]), there is a closed *-derivation δ which is not directly related to the differential structure* d/dt (*cf.* [112]).

In general, the assumption that K has a one-dimensional differential structure does not ensure that K has a non-zero closed *-derivation. In fact, one can easily extend Proposition 3.2.1 as follows: Suppose that a compact space K contains a dense open totally disconnected subset Ω; then any closed *-derivation is identically zero.

Now let $\Omega = \{(x, n^{-1}) | x = m/n^{-1} (m = 0, \pm 1, \pm 2, \ldots, \pm n);\ n = 1, 2, 3, \ldots\}$ in $\mathbb{R}^2$ and $K =$ the closure of Ω in $\mathbb{R}^2$; then Ω is a dense open totally disconnected subset of K and K has a one-dimensional differential structure on $[-1, 1] \times 0$.

3.2.5 Remark Let $C(K)$ be a commutative W*-algebra; then K is extremally disconnected. Hence there is no non-zero closed *-derivation in $C(K)$. Therefore to study unbounded derivations in a W*-algebra, one has to replace the norm topology by the σ-weak topology.

Let δ be a closable *-derivation in a C*-algebra A with an identity. If $\mathscr{D}(\delta)$ does not contain the identity, then define $\delta(1)=0$; δ can be uniquely extended to a closable *-derivation $\tilde{\delta}$ in A and $1 \in \mathscr{D}(\tilde{\delta})$.

In fact, suppose that $x_n + \lambda_n 1 \to 0$ and $\delta(x_n) \to y$.

Let $\bar{\delta}$ be the closure of δ. If $|\lambda_n| \to \infty$, then $x_n/\lambda_n \to -1$ and $\delta(x_n/\lambda_n) = \delta(x_n)/\lambda_n \to 0$; hence $-1 \in \mathscr{D}(\bar{\delta})$ and $\bar{\delta}(-1)=0$ and so $\tilde{\delta} \subset \bar{\delta}$. If $\lambda_n \to \lambda_0$ with $\lambda_0 \neq 0$, then

$$\frac{x_n}{\lambda_n} \to -1 \qquad \text{and} \qquad \delta\left(\frac{x_n}{\lambda_n}\right) = \frac{\delta(x_n)}{\lambda_n} \to \frac{y}{\lambda_0};$$

hence $-1 \in \mathscr{D}(\bar{\delta})$ and $\bar{\delta}(-1) = y/\lambda_0$. Clearly $\bar{\delta}(-1)=0$ and so $\tilde{\delta} \subset \bar{\delta}$.

If $\lambda_n \to 0$, then $x_n \to 0$ and so $y=0$; hence $\tilde{\delta}$ is closable. From the above considerations, also if δ is a closable *-derivation in a C*-algebra without identity, then consider the C*-algebra A_1, adjoining an identity to A and define $\delta(1)=0$; then δ can be uniquely extended to a closable *-derivation in A_1. Therefore for simplicity, in most cases we shall assume that A has an identity and $\delta(1)=0$.

3.2.6 Definition *Let δ be a *-derivation in a C*-algebra A. A self-adjoint element x in $\mathscr{D}(\delta)$ is said to be well-behaved with respect to δ if there is a state ϕ_x on A such that $|\phi_x(x)| = \|x\|$ and $\phi_x(\delta(x))=0$.*

3.2.7 Definition *A *-derivation δ in A is said to be well behaved if every self-adjoint element in $\mathscr{D}(\delta)$ is well-behaved with respect to δ.* (If δ has this property it is also called conservative. Both $\pm\delta$ are dissipative then.)

3.2.8 Definition *A *-derivation δ in A is said to be quasi-well-behaved if the self-adjoint portion in $\mathscr{D}(\delta)$ contains a dense open subset consisting of well-behaved elements.*

3.2.9 Theorem *If a *-derivation δ in A is quasi-well-behaved then δ is closable and its closure $\bar{\delta}$ is again quasi-well-behaved.*

Proof Assume that δ is not closable. Consider the graph $G = \{(x, \delta(x)) | x \in A\}$ in the C*-algebra $A \oplus A$, and let $\bar{G}$ be the norm closure of G in $A \oplus A$; then there is an element $a(\neq 0)$ in A such that $(0,a) \in \bar{G}$. If $a_n \to 0$ with $a_n \in \mathscr{D}(\delta)$ and $\delta(a_n) \to a$, then for $x, y \in \mathscr{D}(\delta)$, $xa_ny \to 0$ and $\delta(xa_ny) = \delta(x)a_ny + x\delta(a_n)y + xa_n\delta(y) \to xay$; hence $\bar{G}$ contains a set $\{(0,b) | b \in \mathscr{I}\}$, where $\mathscr{I}$ is a non-zero closed ideal of A. Hence there is a positive element y with $\|y\|=1$ such that $x_n(=x_n^*)\in\mathscr{D}(\delta)$, $x_n\to 0$ and $\delta(x_n)\to y$. Take $u\in W(\delta)^0$ ($W(\delta)^0$ is the interior of the set $W(\delta)$ of all well-behaved elements) such that $\|u-y\|<1/10$. For real λ, since $W(\delta)^0$ is open, there is an $n(\lambda)$ such that $u+\lambda x_n (n\geqslant n(\lambda))$ are well behaved. Hence $\phi_{u+\lambda x_n}(\delta(u+\lambda x_n))=0$ $(n\geqslant n(\lambda))$. Let ϕ_λ be an accumulation point $\{\phi_{u+\lambda x_n}\}$ in the state space of A; then

$$|\phi_\lambda(u)-\phi_{u+\lambda x_n}(u+\lambda x_n)|\leqslant|\phi_\lambda(u)-\phi_{u+\lambda x_n}(u)|+|\phi_{u+\lambda x_n}(u-(u+\lambda x_n))|$$
$$\leqslant|\phi_\lambda(u)-\phi_{u+\lambda x_n}(u)|+\|\lambda x_n\|.$$

Hence $|\phi_\lambda(u)|=\|u\|$ and analogously $\phi_\lambda(\delta(u)+\lambda y)=0$.

$\phi_\lambda(y)=\phi_\lambda(u)+\phi_\lambda(y-u)$; hence

$$|\phi_\lambda(y)|\geqslant\|u\|-\|y-u\|>\tfrac{9}{10}-\tfrac{1}{10}=\tfrac{8}{10}.$$

Therefore

$$\phi_\lambda(\delta(u)+\lambda y)=\phi_\lambda(\delta(u))+\lambda\phi_\lambda(y)\leqslant\|\delta(u)\|+\lambda\phi_\lambda(y)$$

and $\phi_\lambda(y)\geqslant 0$. Let $\lambda=-\frac{10}{8}\|\delta(u)\|$; then

$$\phi_\lambda(\delta(u)+\lambda y)\leqslant\|\delta(u)\|+\lambda\phi_\lambda(y)<\|\delta(u)\|-\|\delta(u)\|=0,$$

a contradiction. Hence δ is closable. For $y\in W(\delta)^0$, there is a positive number r_y such that

$$S_{r_y}(y)=\{x\mid\|x-y\|<r_y, x\in\mathscr{D}(\delta)\}\subset W(\delta)^0.$$

Let

$$v_{r_y}(y)=\{z\in\mathscr{D}(\bar{\delta})\mid\|z-y\|<r_y\}.$$

Then for $z\in v_{r_y}(y)$, there is a sequence (x_n) in $\mathscr{D}(\delta)$ such that $x_n\to z$ and $\delta(x_n)\to\bar{\delta}(z)$. Take n_0 such that $\|x_n-y\|<r_y$ $(n\geqslant n_0)$; then $x_n\in S_{r_y}(y)$. Hence from the above considerations, $z\in W(\bar{\delta})$ and so $W(\bar{\delta})^0$ is dense in the self-adjoint portion of $\mathscr{D}(\bar{\delta})$. □

3.2.10 Corollary *If δ is well-behaved, then its closure $\bar{\delta}$ is again well-behaved.*

3.2.11 Examples

(1) Let $A=C([0,1])$ and $\delta=\mathrm{d}/\mathrm{d}t$ with $\mathscr{D}(\delta)=C^{(1)}([0,1])$; then δ is quasi-well-behaved, but not well-behaved. $W(\delta)^0$ contains the set of all continuously differentiable functions f containing their maximum values inside the open interval $(0,1)$ with $\|f\|>\max(f(0),f(1))$.

(2) Let $A=C(T)$ (T is a one-dimensional torus group), and $\delta=\mathrm{d}/\mathrm{d}t$ with $\mathscr{D}(\delta)=C^{(1)}(T)$; then δ is well-behaved.

(3) Let $A=C_0(\mathbb{R})$ and $\delta=\mathrm{d}/\mathrm{d}t$ with $\mathscr{D}(\delta)=C_0^{(1)}(\mathbb{R})$; then δ is well-behaved.

3.2.12 Proposition *Let δ be a *-derivation in a C*-algebra A with an identity* *and suppose that* $1 \in \mathscr{D}(\delta)$. *Then* δ *is well-behaved if and only if for* $x(>0) \in \mathscr{D}(\delta)$ *there is a state* ϕ_x *on* A *such that* $\phi_x(x) = \|x\|$ *and* $\phi_x(\delta(x)) = 0$.

Proof Let y be a self-adjoint element of $\mathscr{D}(\delta)$ and let $y = y^+ - y^-$ be the orthogonal decomposition of y.

$$\phi_{\|y\|1 \pm y}(\|y\|1 \pm y) = \|\,\|y\|1 \pm y\| \qquad \text{and} \qquad \phi_{\|y\|1 \pm y}(\delta(y)) = 0.$$

If $\|y\| = \|y^+\|$, then

$$\phi_{\|y\|1+y}(\|y\|1 + y) = 2\|y\|;$$

hence

$$\phi_{\|y\|1+y}(y) = \|y\|.$$

If $\|y\| = \|y^-\|$, then

$$\phi_{\|y\|1-y}(\|y\|1 - y) = 2\|y^-\|;$$

hence

$$\phi_{\|y\|1-y}(y) = -\|y^-\| = -\|y\|. \qquad \square$$

3.2.13 Definition *A *-derivation* δ *in* A *is said to be approximately inner if there is a sequence* (h_n) *of self-adjoint elements in* A *such that* $\delta(a) = \lim_{n\to\infty} \mathrm{i}[h_n, a]$ $(a \in \mathscr{D}(\delta))$.

3.2.14 Proposition *If* δ *is approximately inner, then* δ *is well-behaved.*

Proof For $a(>0) \in \mathscr{D}(\delta)$, take a state ϕ_a such that $\phi(a) = \|a\|$. Then

$$\begin{aligned}\phi(\delta(a)) &= \lim_n \phi(\mathrm{i}h_n a - a\mathrm{i}h_n)\\ &= -\lim_n \phi(\mathrm{i}h_n(\|a\|1 - a) - (\|a\|1 - a)\mathrm{i}h_n).\end{aligned}$$

$$\begin{aligned}|\phi(h_n(\|a\|1 - a))| &= |\phi(h_n(\|a\|1 - a)^{1/2}(\|a\|1 - a))^{1/2})|\\ &\leqslant \phi(h_n(\|a\|1 - a)h_n)^{1/2}\phi(\|a\|1 - a)^{1/2} = 0.\end{aligned}$$

Analogously $|\phi((\|a\|1 - a)h_n)| = 0$; hence $\phi(\delta(a)) = 0$. $\square$

3.2.15 Examples Clearly, Ising and Heisenberg models are approximately inner. In addition, quasi-free derivations in the canonical anti-commutation relation algebra have approximately inner *-derivations as cores (i.e. they are the closures of approximately inner *-derivations. In fact, let H be a self-adjoint operator in a Hilbert space $\mathscr{H}$. By Weyl's theorem, one can express $H = H_1 + C$, where H_1 is a diagonalizable self-adjoint operator and C is a compact self-adjoint operator. Let $H_1 = \sum_{i=1}^{\infty} \lambda_i E_i$, where $\lambda_i \in R, \dim(E_i) = 1$ and $E_i E_j = 0$ $(i \neq j)$. Let $P_n = \sum_{i=1}^{n} E_i$; then $\mathscr{A}(P_n(\mathscr{H}))$ is a finite-dimensional full matrix algebra. Let $\{e_{ij} | ij = 1, 2, \ldots, 2^n\}$ be a matrix unit of $\mathscr{A}(P_n(\mathscr{H}))$. Set $ih_n = \sum_{i=1}^{2^n} \delta_{iH_1}(e_{i1})e_{1i}$; then we can easily see that

$$\delta_{iH1}(a) = [ih_n, a] \qquad (a \in \mathscr{A}(P_n(\mathscr{H})).$$

Since

$$\left(\sum_{i=1}^{2^n} \delta_{iH}(e_{i1})e_{1i} \right)^* = \sum_{i=1}^{2^n} e_{1i}{}^* \delta(e_{i1}{}^*) = \sum_{i=1}^{2^n} e_{i1} \delta(e_{1i})$$

and $e_{i1} e_{1i} = e_{ii}$,

$$0 = \delta\left(\sum_{i=1}^{2^n} e_{i1} e_{1i} \right) = \sum_{i=1}^{2^n} \delta(e_{i1}) e_{1i} + \sum_{i=1}^{2^n} e_{i1} \delta(e_{1i})$$

hence

$$\sum_{i=1}^{2^n} \delta(e_{i1}) e_{1i} = - \sum_{i=1}^{2^n} e_{i1} \delta(e_{1i}).$$

Therefore h_n is self-adjoint. $\| CP_n - C \| \to 0$, for C is compact. Let $\tilde{H}_1$ (resp. $\tilde{C}$) be the restriction of H_1 (resp. C) to $\bigcup_{n=1}^{\infty} P_n \mathscr{H}$; then $\bar{\tilde{H}}_1 = H_1$ and $\bar{\tilde{C}} = C$. Hence $\overline{\tilde{H}_1 + \tilde{C}} = H_1 + C = H$.

Let $\check{\delta}_{iH}$ be the restriction of δ_{iH} to $\bigcup_{n=1}^{\infty} \mathscr{A}(P_n(\mathscr{H}))$; then clearly $\bar{\check{\delta}}_{iH} \supset \delta_{iH}$. If S is symmetric in $\mathscr{H}$, then use the polar decomposition $S = V|S|$; then one can easily see that δ_{iS} has an approximately inner *-derivation as a core.

3.2.16 Proposition ([7]) *Let δ be a *-derivation in a C^*-algebra A. If $a \in W(\delta)^0$, then for any state ϕ on A with $|\phi(a)| = \|a\|, \phi(\delta(a)) = 0$.*

Proof For b $(= b^*) \in \mathscr{D}(\delta)$, there is a sequence (λ_n) of real numbers such that $\lambda_n \to 0, |\phi_{a + \lambda_n b}(a + \lambda_n b)| = \|a + \lambda_n b\|$ and $\phi_{a+\lambda_n b}(\delta(a) + \lambda_n \delta(b)) = 0$.

$$\begin{aligned} \|a\| + |\lambda_n| \, \|b - \delta(a)\| + \lambda_n{}^2 \|\delta(b)\| &\geq \|a + \lambda_n b - \lambda_n \delta(a + \lambda_n b)\| \\ &\geq |\phi_{a+\lambda_n b}(a + \lambda_n b - \lambda_n \delta(a + \lambda_n b))| \\ &= \|a + \lambda_n b\| \geq \|a + \lambda_n \delta(a)\| - |\lambda_n| \, \|\delta(a) - b\|. \end{aligned}$$

On the other hand,

$$|\phi(a + \lambda_n \delta(a))| \leq \|a + \lambda_n \delta(a)\|.$$

Hence

$$\begin{aligned} |\phi(a + \lambda_n \delta(a))| &= \|a\| \pm \lambda_n \phi(\delta(a)) \quad \text{(for large } n\text{)} \\ &\leq \|a\| + 2|\lambda_n| \, \|b - \delta(a)\| + \lambda_n{}^2 \|\delta(b)\| \end{aligned}$$

and so

$$\pm \lambda_n \phi(\delta(a)) \leq 2|\lambda_n| \, \|b - \delta(a)\| + \lambda_n{}^2 \|\delta(b)\|.$$

Taking λ_n such that $\pm\lambda_n = |\lambda_n|$

$$\phi(\delta(a)) \leqslant 2\|b - \delta(a)\| + |\lambda_n|\,\|\delta(b)\|$$

assuming that $\phi(\delta(a)) \neq 0$. We may assume that $\phi(\delta(a)) > 0$; otherwise consider $-a$. Take $b(=b^*)\in\mathscr{D}(\delta)$ such that $\|b - \delta(a)\| < (\phi(\delta(a)))/3$ and let $|\lambda_n| < \phi(\delta(a))/3\|\delta(b)\|$; then

$$0 < \phi(\delta(a)) < \frac{2\phi(\delta(a))}{3} + \frac{\phi(\delta(a))}{3},$$

a contradiction. □

3.2.17 Proposition *Let δ be a *-derivation in A. An element $x(>0)\in\mathscr{D}(\delta)^s$ is well-behaved with respect to δ if and only if $\|(1+\lambda\delta)(x)\| \geqslant \|x\|$ for all real λ.*

Proof Suppose that x is well-behaved; then

$$\|(1+\lambda\delta)(x)\| \geqslant |\phi_x((1+\lambda\delta)(x))| = |\phi(x)| = \|x\|.$$

Conversely if $\|(1+\lambda\delta)(x)\| \geqslant \|x\|$ for all real λ, then

$$\|(1+(\lambda+\mathrm{i}\mu)\delta)(x)\| = \|(1+\lambda\delta)(x) + \mathrm{i}\mu\delta(x)\| \geqslant \|(1+\lambda\delta)(x)\| \geqslant \|x\|$$

for $\lambda, \mu\in\mathbb{R}$. Hence there is a linear functional f on the linear subspace of A spanned by x and $\delta(x)$ such that $\|f\| = 1$, $f(\delta(x)) = 0$ and $f(x) = \|x\|$. By the Hahn–Banach theorem, one can extend f to a bounded linear functional $\tilde{f}$ on A such that $\|\tilde{f}\| = \|f\|$.

Since $x > 0$, $\tilde{f}$ is a state. □

3.2.18 Proposition *If a *-derivation δ in A with identity is well-behaved, then there is a state ϕ on A such that $\phi(\delta(x)) = 0$ for $x\in\mathscr{D}(\delta)$.*

Proof Suppose that there is an element $x(=x^*)$ in $\mathscr{D}(\delta)$ such that $\|1 - \delta(x)\| < 1$; then $|\phi_x(1) - \phi_x(\delta(x))| < 1$ and so $\phi_x(1) < 1$, a contradiction. Therefore $\|1 - \delta(a)\| \geqslant 1$ for all $a\in\mathscr{D}(\delta)$; hence there is a bounded linear functional f on A such that $\|f\| = 1$, $f(1) = 1$ and $f(\delta(a)) = 0$ for all $a\in\mathscr{D}(\delta)$.

□

3.2.19 Proposition *Let δ be a *-derivation in A and suppose that δ is well-behaved; then $\|(1+\lambda\delta)(a)\| \geqslant \|a\|$ for $a\in\mathscr{D}(\delta)$ and real λ.*

Proof Take a state ϕ_{a^*a}; then

$$\begin{aligned}\phi_{a^*a}((1+\lambda\delta)(a^*)(1+\lambda\delta)(a)) &= \phi_{a^*a}(a^*a + \lambda\delta(a^*)a + \lambda a^*\delta(a) + \lambda^2\delta(a^*)\delta(a)) \\ &\geqslant \phi_{a^*a}((a^*a) + \lambda\phi_{a^*a}(\delta(a^*a))) = \phi_{a^*a}(a^*a) = \|a\|^2.\end{aligned}$$

□

3.2.20 Proposition *Let δ be a *-derivation in A with identity and suppose that there is a sequence (δ_n) of well-behaved *-derivations in A such that $\mathscr{D}(\delta_n) \supset \mathscr{D}(\delta)$ $(n = 1, 2, \ldots)$ and $\delta(a) = \lim_{n \to \infty} \delta_n(a)$ for $a \in \mathscr{D}(\delta)$; then δ is again well-behaved.*

Proof For $a(\geqslant 0) \in \mathscr{D}(\delta)$, $\|(1 + \lambda\delta_n)(a)\| \geqslant \|a\|$ for real λ and so $\|(1 + \lambda\delta)(a)\| \geqslant \|a\|$; hence a is well-behaved with respect to δ (3.2.17). Since A has an identity, this implies δ is well-behaved (3.2.12). □

3.2.21 Definition *A *-derivation δ in A is said to be approximately bounded if there is a sequence (δ_n) of bounded *-derivations in A such that $\lim_n \delta_n(a) = \delta(a)$ for each $a \in \mathscr{D}(\delta)$.*

3.2.22 Corollary *Let δ be an approximately bounded *-derivation in A; then δ is well-behaved.*

Proof If A has no identity, then adjoin a unit element and define $\delta_n(1) = 0$ and $\delta(1) = 0$; then we may assume that A has an identity. Let $\alpha_{n,t}(a) = (\exp t\delta_n)(a)$ $(a \in A)$; then $t \to \alpha_{n,t}$ is a uniformly continuous one-parameter group of *-auto-morphisms on A.

Let ϕ be a state on A such that $|\phi(a)| = \|a\|$. Then $|\phi(\alpha_{n,t}(a))| \leqslant \|\alpha_{n,t}(a)\| = \|a\|$; hence $(\mathrm{d}/\mathrm{d}t)\phi(\alpha_{n,t}(a)) = \phi(\delta_n(a)) = 0$ $(a \in \mathscr{D}(\delta)^s)$, and so δ_n is well-behaved. Hence by 3.2.20, δ is well-behaved. □

3.2.23 Notes and remarks Proposition 3.2.1 is due to the author [166]. The well-behavedness of a *-derivation δ is equivalent to the dissipativity of $\pm\delta$. Kishimoto [103] first used the dissipativity in the theory of *-derivations. Example 3.2.2 is due to Bratteli and Robinson [39]. The notion of approximate innerness is important for quantum lattice systems and Fermi statistics as shown in 3.2.15. Theorem 3.2.9 is due to the author [170] and Batty [7].

References [7], [103], [170].

3.3 The domain of closed *-derivations

The domain problem, to be described later, of closed *-derivations in C*-algebras (functional calculus) plays a central role in the theory. In this section, we shall show that if $a(= a^*)$ belongs to the domain of a closed *-derivation δ and if $f \in C^{(2)}$ $([-\|a^-\|, \|a^+\|])$, then $f(a)$ belongs to the domain of δ (cf. Theorem 3.3.7). However one cannot replace $C^{(2)}$ by $C^{(1)}$ in a non-commutative C*-algebra, though it is possible for a commutative

C*-algebra. We shall propose some related problems to this interesting situation.

Let δ be a closed *-derivation in a C*-algebra A. Without loss of generality, we may assume that A has an identity and $\mathscr{D}(\delta)$ contains it. Define

$$\|a\|_\delta = \left\| \begin{pmatrix} a & \delta(a) \\ 0 & a \end{pmatrix} \right\| \qquad (a \in \mathscr{D}(\delta)),$$

where $\begin{pmatrix} x & y \\ z & w \end{pmatrix}$ $(x, y, z, w \in A)$ is the matrix of 2×2 over A and $\left\| \begin{pmatrix} x & y \\ z & w \end{pmatrix} \right\|$ is the C*-norm of $\begin{pmatrix} x & y \\ z & w \end{pmatrix}$. Then we can easily see that a mapping $a \to \begin{pmatrix} a & \delta(a) \\ 0 & a \end{pmatrix}$ is an isomorphism of A into $A \otimes B_2$, where B_2 is the full matrix algebra of 2×2. Consequently, $\mathscr{D}(\delta)$ becomes a normed *-algebra under $\|\cdot\|_\delta$, and the closedness of δ implies that $\mathscr{D}(\delta)$ is a Banach *-algebra under the norm $\|\cdot\|_\delta$ with $\|a^*\|_\delta = \|a\|_\delta$. $(a \in \mathscr{D}(\delta))$.

Let $a \in \mathscr{D}(\delta)$; then one may consider three spectra of a. $Sp(a)$ = the spectrum of a in the C*-algebra A; $Sp(a)_{\mathscr{D}(\delta)}$ = the spectrum of a in the Banach *-algebra $\mathscr{D}(\delta)$; $Sp(a)_{A \otimes B_2}$ = the spectrum of $\begin{pmatrix} a & \delta(a) \\ 0 & a \end{pmatrix}$ in the C*-algebra $A \otimes B_2$. Clearly $Sp(a)_A \subset Sp(a)_{\mathscr{D}(\delta)}$ and $Sp(a)_{A \otimes B_2} \subset Sp(a)_{\mathscr{D}(\delta)}$.

3.3.1 Proposition $Sp(a)_{A \otimes B_2} = Sp(a)_A \qquad (a \in \mathscr{D}(\delta))$.

Proof Let $\lambda \notin Sp(a)_A$; then there is a $(\lambda 1 - a)^{-1}$ in A. Let

$$X = \begin{pmatrix} (\lambda 1 - a)^{-1} & (\lambda 1 - a)^{-1}\delta(a)(\lambda 1 - a)^{-1} \\ 0 & (\lambda 1 - a)^{-1} \end{pmatrix};$$

then one can easily see that

$$X \cdot \begin{pmatrix} (\lambda 1 - a) & -\delta(a) \\ 0 & (\lambda 1 - a) \end{pmatrix} = \begin{pmatrix} (\lambda 1 - a) & -\delta(a) \\ 0 & (\lambda 1 - a) \end{pmatrix} \cdot X = \begin{pmatrix} 1 & 0 \\ 0 & 1 \end{pmatrix}.$$

Hence $x = \begin{pmatrix} \lambda 1 - a & -\delta(a) \\ 0 & \lambda 1 - a \end{pmatrix}^{-1}$ in $A \otimes B_2$, and so $\lambda \notin Sp(a)_{A \otimes B_2}$. Conversely let $\lambda \notin Sp(a)_{A \otimes B_2}$; then there is an element $Y = \begin{pmatrix} x & y \\ z & w \end{pmatrix}$ in $A \otimes B_2$ such that

$$Y \cdot \begin{pmatrix} \lambda 1 - a & -\delta(a) \\ 0 & \lambda 1 - a \end{pmatrix} = \begin{pmatrix} \lambda 1 - a & -\delta(a) \\ 0 & \lambda 1 - a \end{pmatrix} \cdot Y = \begin{pmatrix} 1 & 0 \\ 0 & 1 \end{pmatrix}$$

From this, $x(\lambda 1 - a) = (\lambda 1 - a)w = 1$; hence $(\lambda 1 - a)$ is invertible and $(\lambda 1 - a)^{-1} = x = w$. Hence $\lambda \notin Sp(a)_A$. □

3.3.2 Corollary *The boundary of* $Sp(a)_A \supseteq$ *the boundary of* $Sp(a)_{\mathscr{D}(\delta)}$. *In particular* $Sp(a)_A = Sp(a)_{\mathscr{D}(\delta)}$ *if* $a^* = a$.

Proof Since $\mathscr{D}(\delta)$ is a closed subalgebra of $A \otimes B_2$, by the general theory of Banach algebras [151], the boundary of $Sp(a)_{\mathscr{D}(\delta)} \subset$ the boundary of $Sp(a)_{A \otimes B_2}$. □

Henceforth, we shall denote $Sp(a)_A = Sp(a)_{\mathscr{D}(\delta)}$ by $Sp(a)$ for $a(=a^*) \in \mathscr{D}(\delta)$.

3.3.3 Problem *Is it true that* $Sp(a)_A = Sp(a)_{\mathscr{D}(\delta)}$ *for every* $a \in \mathscr{D}(\delta)$?

Let $a \in \mathscr{D}(\delta)$ and let f be a complex analytic function on a neighborhood of $Sp(a)_{\mathscr{D}(\delta)}$; then by the general theory of Banach algebras, $f(a) \in \mathscr{D}(\delta)$. Moreover let $\{f_n\}$ be a sequence of analytic functions on a fixed neighborhood U of $Sp(a)_{\mathscr{D}(\delta)}$ and suppose that $\{f_n\}$ converges uniformly to an f_0 on every compact subset of U; then f_0 is analytic on U and $\| f_n(a) - f_0(a) \|_\delta \to 0$.

3.3.4 Proposition *Let* δ *be a closed derivation in* A *and suppose that* A *has an identity* 1. *Then* $\mathscr{D}(\delta)$ *contains* 1 *automatically.*

Proof If $1 \notin \mathscr{D}(\delta)$, then define $\tilde{\delta}(1) = 0$ and extend δ to a derivation $\tilde{\delta}$. First of all, we shall show that $\tilde{\delta}$ is closed. In fact, let $x_n = y_n + \lambda_n 1 \to 0$ ($y_n \in \mathscr{D}(\delta)$) and $\tilde{\delta}(y_n) = \delta(y_n) \to z$. If $|\lambda_{n_j}| \to \infty$, then

$$\frac{x_{n_j}}{\lambda_{n_j}} = \frac{y_{n_j}}{\lambda_{n_j}} + 1 \to 0;$$

hence $y_{n_j}/\lambda_{n_j} \to -1$ and $\delta(y_{n_j}/\lambda_{n_j}) \to 0$. Therefore $-1 \in \mathscr{D}(\delta)$ and so $1 \in \mathscr{D}(\delta)$, a contradiction. Next suppose that $\lambda_{n_j} \to \lambda (\neq 0)$; then

$$\frac{x_{n_j}}{\lambda_{n_j}} = \frac{y_{n_j}}{\lambda_{n_j}} + 1 \to 0$$

and so $-1 \in \mathscr{D}(\delta)$, a contradiction. Hence $\lambda_n \to 0$, and so $y_n \to 0$ and $z = 0$; hence $\tilde{\delta}$ is closed. Take $a \in \mathscr{D}(\delta)$ such that $\| a - 1 \| < \frac{1}{2}$; then $Sp(a)_A \subset$ the disk with radius $\frac{1}{2}$ at the center 1. Take an open set U such that $0 \notin \bar{U}$ and $Sp(a)_{\mathscr{D}(\delta)} \subset U$. (This is possible, because the boundary of $Sp(a)_{\mathscr{D}(\delta)} \subset$ the boundary of $Sp(a)_A$.) Consider an analytic function f on $0 \cup U$ such that $f(0) = 0$, $f(\lambda) = 1$ on U; then by Runge's approximation theorem [77], there is a sequence of polynomials such that $p_n(0) = 0$ and $p_n(\lambda) \to f(\lambda)$ uniformly on every compact subset of U; hence $\| f(a) - p_n(a) \|_\delta = \| 1 - p_n(a) \|_{\tilde{\delta}} \to 0$. Since $p_n(0) = 0$, $p_n(a) \in \mathscr{D}(\delta)$; hence $1 \in \mathscr{D}(\delta)$. □

Let δ be a closed derivation in a C*-algebra A. For $a \in \mathscr{D}(\delta)$, let

$U(z)=\exp(za)(z\in\mathbb{C})$. Then $U(z)\in\mathscr{D}(\delta)$.

$$\frac{\mathrm{d}U(z)}{\mathrm{d}z}=\lim_{h\to 0}\frac{U(z+h)-U(z)}{h}=\lim_{h\to 0}\frac{U(z)\{U(h)-1\}}{h}=a\exp(za).$$

This limit is uniformly convergent on every compact subset of the complex plane; hence.

$$\frac{\mathrm{d}\delta(U(z))}{\mathrm{d}z}=\delta\left(\frac{\mathrm{d}U(z)}{\mathrm{d}z}\right)=\delta(aU(z))=\delta(a)U(z)+a\delta(U(z)).$$

Therefore for any arbitrary positive number ε, there exists a positive number α such that $|\delta(U(z))-\delta(U(0))/z-\delta(a)|<\varepsilon$ for $|z|<\alpha$. For $b\in\mathscr{D}(\delta), \delta(b^n)=\sum_{k=0}^{n-1}b^k\delta(b)b^{n-k-1}$. Hence

$$\delta(U(z))=\delta\left(U\left(\frac{z}{n}\right)^n\right)=\sum_{k=0}^{n-1}U\left(\frac{z}{n}\right)^k\delta\left(U\left(\frac{z}{n}\right)\right)U\left(\frac{z}{n}\right)^{n-k-1}$$
$$=\sum_{k=0}^{n-1}U\left(\frac{kz}{n}\right)\delta\left(U\left(\frac{z}{n}\right)\right)U\left(\left(\frac{n-k-1}{n}\right)z\right).$$

Let $|z/n|<\alpha$; then $|\delta(U(z/n))-(z/n)\delta(a)|<\varepsilon|z|/n$ and so

$$\left\|\sum_{k=0}^{n-1}U\left(\frac{z}{n}\right)^k\delta\left(U\left(\frac{z}{n}\right)\right)U\left(\frac{z}{n}\right)^{n-k-1}-\left(\sum_{k=0}^{n-1}U\left(\frac{z}{n}\right)^k\delta(a)U\left(\frac{z}{n}\right)^{n-k-1}\right)\frac{z}{n}\right\|$$
$$\leqslant\frac{|z|\varepsilon}{n}\sum_{k=0}^{n-1}\left\|U\left(\frac{z}{n}\right)^k\right\|\left\|U\left(\frac{z}{n}\right)^{n-k-1}\right\|$$
$$=\frac{|z|\varepsilon}{n}\sum_{k=0}^{n-1}\left\|U\left(\frac{kz}{n}\right)\right\|\left\|U\left(\frac{n-k-1}{n}z\right)\right\|$$
$$\leqslant\frac{\varepsilon|z|}{n}n\cdot\max_{|w|\leqslant|z|}\|U(w)\|^2=\varepsilon|z|\max_{|w|\leqslant|z|}\|U(w)\|^2.$$

Hence

$$\left\|\delta(U(z))-\sum_{k=0}^{n-1}U\left(\frac{z}{n}\right)^k\delta(a)U\left(\frac{z}{n}\right)^{n-k-1}\frac{z}{n}\right\|\leqslant\varepsilon|z|\max_{|w|\leqslant|z|}\|U(w)\|^2.$$

On the other hand,

$$\sum_{k=0}^{n-1}U\left(\frac{z}{n}\right)^k\delta(a)U\left(\frac{z}{n}\right)^{n-k-1}\frac{z}{n}=z\sum_{k=0}^{n-1}U\left(\frac{kz}{n}\right)\delta(a)U\left(\frac{n-k-1}{n}z\right)\frac{1}{n}$$
$$\to z\int_0^1 U(sz)\delta(a)U((1-s)z)\mathrm{d}s$$

in the uniform norm of A, as $n \to \infty$. Hence we have the following equality.

$$\delta(U(z)) = z \int_0^1 U(sz)\delta(a)U((1-s)z)\mathrm{d}s.$$

3.3.5 Proposition *Let δ be a closed derivation in a C^*-algebra A. For a $(=a^*)\in\mathscr{D}(\delta)$, $\|\delta\exp(ita)\| \leqslant |t|\,\|\delta(a)\|\,(t\in\mathbb{R})$.*

Proof

$$\|\delta\exp(ita)\| = |t| \left\| \int_0^1 \exp(itsa)\delta(a)\exp\mathrm{i}(1-s)ta\,\mathrm{d}s \right\|$$

$$\leqslant |t|\,\|\delta(a)\| \int_0^1 \mathrm{d}s = |t|\,\|\delta(a)\|. \qquad \square$$

3.3.6 Proposition *For a $(=a^*)\in\mathscr{D}(\delta)$, let f be a continuous function $[-\|a^-\|, \|a^+\|]$ such that $\int_{-\infty}^{\infty}|x\hat{f}(x)|\,\mathrm{d}x < +\infty$, where $\hat{f}$ is the Fourier transform of f on $(-\infty,\infty)$ (let $f(x)=0$ outside $[-\|a^-\|, \|a^+\|]$) and let $f(a)(\lambda) = f(a(\lambda))$ for $\lambda\in S_p(a)$; then $f(a)\in\mathscr{D}(\delta)$ and $\|\delta(f(a))\| \leqslant (1/(2\pi)^{1/2}\int_{-\infty}^{\infty}|x\hat{f}(x)|\,\mathrm{d}x)$ $\|\delta(a)\|$.*

Proof Since $\hat{f}$ is continuous on $(-\infty,\infty)$ vanishing at infinity, $\int_{-\infty}^{\infty}|x\hat{f}(x)|\,\mathrm{d}x < +\infty$ implies $\int_{-\infty}^{\infty}|\hat{f}(x)|\,\mathrm{d}x < +\infty$. Hence $\hat{\hat{f}}(y) = 1/(2\pi)^{1/2}\int_{-\infty}^{\infty} f(x)\exp(\mathrm{i}yx)\,\mathrm{d}x$ exists. Since $\hat{\hat{f}}$ is continuous on $\mathbb{R}$ and f is continuous on $[-\|a^-\|, \|a^+\|]$ and $\hat{\hat{f}} = f$ a.e., $\hat{\hat{f}} = f$. Therefore,

$$f(a)(\lambda) = f(a(\lambda)) = \frac{1}{(2\pi)^{1/2}} \int_{-\infty}^{\infty} \hat{f}(x)\exp(\mathrm{i}a(\lambda)x)\,\mathrm{d}x \qquad \text{for } \lambda\in Sp(a).$$

Since

$$\int_{-\infty}^{\infty} |\hat{f}(x)\exp(\mathrm{i}xa)|\,\mathrm{d}x = \int_{-\infty}^{\infty} |\hat{f}(x)|\,\mathrm{d}x < +\infty,$$

$\int_{-\infty}^{\infty}\hat{f}(x)\exp(\mathrm{i}xa)\,\mathrm{d}x$ is absolutely convergent, so that $\int_{-\infty}^{\infty}\hat{f}(x)\exp(\mathrm{i}xa)\,\mathrm{d}x\in A$. Since

$$\int_{-\infty}^{\infty} \|\hat{f}(x)\delta\exp(\mathrm{i}xa)\|\,\mathrm{d}x \leqslant \int_{-\infty}^{\infty} |\hat{f}(x)|\,|x|\,\|\delta(a)\|\,\mathrm{d}x < +\infty,$$

$$\delta\left(\int_{-\infty}^{\infty} \hat{f}(x)\exp(\mathrm{i}xa)\,\mathrm{d}x\right) = \int_{-\infty}^{\infty} \hat{f}(x)\delta\exp(\mathrm{i}xa)\,\mathrm{d}x$$

by the closedness of δ. $\square$

3.3.7 Theorem *Let* $a(=a^*)\in\mathscr{D}(\delta)$ *and let* $C^{(2)}([-\|a^-\|,\|a^+\|])$ *be the space of all twice continuously differentiable functions on* $[-\|a^-\|,\|a^+\|]$; *then* $f(a)\in\mathscr{D}(\delta)$ for $f\in C^{(2)}([-\|a^-\|,\|a^+\|])$.

Proof One can find a polynomial $p(x)$ on $\mathbb{R}$ such that $(f-p)(-\|a^-\|)=(f-p)(\|a^+\|)=0$ and $(f-p)'(-\|a^-\|)=(f-p)'(\|a^+\|)=0$. Put $g=f-p$ on $[-\|a^-\|,\|a^+\|]$, and $g(x)=0$, outside $[-\|a^-\|,\|a^+\|]$; then

$$\hat{g}(x)=\frac{1}{(2\pi)^{1/2}}\int_{-\infty}^{\infty}g(y)\exp(-\mathrm{i}xy)\,\mathrm{d}y=\frac{1}{(2\pi)^{1/2}}\int_{-\|a^-\|}^{\|a^+\|}g(y)\exp(-\mathrm{i}xy)\,\mathrm{d}y$$

$$=\frac{1}{(2\pi)^{1/2}}\left[g(y)\frac{\exp(-\mathrm{i}xy)}{-\mathrm{i}x}\right]_{-\|a^-\|}^{\|a^+\|}+\frac{1}{(2\pi)^{1/2}}\int_{-\|a^-\|}^{\|a^+\|}g'(y)\frac{\exp(-\mathrm{i}xy)}{\mathrm{i}x}\mathrm{d}y$$

$$=\frac{1}{(2\pi)^{1/2}\mathrm{i}x}\int_{-\|a^-\|}^{\|a^+\|}g'(y)\exp(-\mathrm{i}xy)\,\mathrm{d}y \qquad (x\neq 0).$$

Hence

$$\mathrm{i}x\hat{g}(x)=\frac{1}{(2\pi)^{1/2}}\int_{-\|a^-\|}^{\|a^+\|}g'(y)\exp(-\mathrm{i}xy)\,\mathrm{d}y.$$

Quite similarly,

$$(\mathrm{i}x)^2\hat{g}(x)=\frac{1}{(2\pi)^{1/2}}\int_{-\|a^-\|}^{\|a^+\|}g''(y)\exp(-\mathrm{i}xy)\,\mathrm{d}y.$$

Since $g''\in L^2(-\infty,\infty)$, $(\mathrm{i}x)^2\hat{g}(x)\in L^2(-\infty,\infty)$ and so

$$\int_{-\infty}^{\infty}|x\hat{g}(x)|\mathrm{d}x=\int_{-1}^{1}|x\hat{g}(x)|\,\mathrm{d}x+\int_{1}^{\infty}|x\hat{g}(x)|\,\mathrm{d}x+\int_{-\infty}^{-1}|x\hat{g}(x)|\,\mathrm{d}x$$

$$=\int_{-1}^{1}|x\hat{g}(x)|\mathrm{d}x+\int_{1}^{\infty}\left|\frac{1}{x}x^2\hat{g}(x)\right|\mathrm{d}x+\int_{-\infty}^{-1}\left|\frac{1}{x}x^2\hat{g}(x)\right|\mathrm{d}x$$

$$=\int_{-1}^{1}|x\hat{g}(x)|\,\mathrm{d}x+\left(\int_{1}^{\infty}\frac{1}{x^2}\mathrm{d}x\right)^{1/2}\left(\int_{1}^{\infty}|x^2\hat{g}(x)|^2\,\mathrm{d}x\right)^{1/2}$$

$$+\left(\int_{-\infty}^{-1}\frac{\mathrm{d}x}{x^2}\right)^{1/2}\left(\int_{-\infty}^{-1}|x^2\hat{g}(x)|^2\,\mathrm{d}x\right)^{1/2}<+\infty.$$

Hence $(f-p)(a)\in\mathscr{D}(\delta)$. Since $p(a)\in\mathscr{D}(\delta)$, $f(a)\in\mathscr{D}(\delta)$. □

Remark McIntosh [127] constructed an example of closed *-derivations δ in a C*-algebra in which there is an element $a(=a^*)\in\mathscr{D}(\delta)$ and a continuous diferentiable function f on $[-\|a^-\|,\|a^+\|]$ with $f(a)\notin\mathscr{D}(\delta)$.

Therefore, in Theorem 3.3.7, one cannot replace $C^{(2)}$ by $C^{(1)}$ in a non-commutative C^*-algebra though it is possible for a commutative C*-algebra (cf. Proposition 3.2.3).

There have been some developments in the spectral theory of non-self-adjoint operators in a Hilbert space (cf. [49], [100]). Let T be a bounded operator on a Hilbert space $\mathscr{H}$. Suppose that $\|\exp(itT)\| = O(|t|^k)$ as $|t| \to \infty$, where k is a non-negative number. Then it is known that $Sp(T) \subset \mathbb{R}$. Let f be a function belonging to $L^1(-\infty, \infty)$ such that $\int_{-\infty}^{\infty} |t|^k |\hat{f}(t)| \mathrm{d}t < +\infty$, where $\hat{f}$ is the Fourier transform of f. Then one can define

$$f(T) = \frac{1}{(2\pi)^{1/2}} \int_{-\infty}^{\infty} \hat{f}(t) \exp(itT) \, \mathrm{d}t$$

as a bounded operator on $\mathscr{H}$, because

$$\begin{aligned}\int_{-\infty}^{\infty} |\hat{f}(t)\exp(itT)| \, \mathrm{d}t &\leqslant \int_{-1}^{1} |\hat{f}(t)\exp(itT)| \mathrm{d}t + M_1 \int_{1}^{\infty} |\hat{f}(t)| t^k \, \mathrm{d}t \\ &\quad + M_2 \int_{-\infty}^{-1} |\hat{f}(t) t^k| \, \mathrm{d}t \\ &< +\infty \qquad (M_1, M_2 \text{ fixed positive numbers}).\end{aligned}$$

In particular if $\|\exp(itT)\| = O(|t|)$ as $|t| \to \infty$ and if $f \in C^{(2)}(\mathbb{R})$, then $f(T)$ is definable. If $\|\exp(itT)\| = O(1)$, then T is similar to a self-adjoint bounded operator by Nagy's theorem [131]; hence $f(T)$ is definable for $f \in C(\mathbb{R})$.

It has been an open question if $\|\exp(itT)\| = O(|t|)$ as $|t| \to \infty$ implies $f(T)$ is definable for $f \in C^{(1)}(\mathbb{R})$ (cf. [49], [100]). By McIntosh's example, one has a negative solution to this problem. On the other hand, if δ is a closed derivation in a commutative C*-algebra A and if $a(=a^*) \in \mathscr{D}(\delta)$, then $T = \begin{pmatrix} a & \delta(a) \\ 0 & a \end{pmatrix}$ satisfies the properties: $\|\exp(itT)\| = 0\ (|t|)$ as $|t| \to \infty$ and $f(T)$ is definable for $f \in C^{(1)}(\mathbb{R})$.

3.3.8 Problem *Can we characterize bounded operators S on a Hilbert space which have the properties:* $\|\exp(itS)\| = O(|t|)$ $(t \to \infty)$ *and* $f(S)$ *is definable for* $f \in C^{(1)}(\mathbb{R})$?

3.3.9 Problem *Let* δ *be a closed *-derivation in a C*-algebra A. Can we characterize* δ *which has the property: for* $a(=a^*) \in \mathscr{D}(\delta)$ *and* $f \in C^{(1)}(\mathbb{R})$, $f(a) \in \mathscr{D}(\delta)$?

3.3.10 Notes and remarks Proposition 3.3.5 is due to Powers [143]. In [143], he insisted that Theorem 3.3.7 was true for $C^{(1)}$. But, Robinson found an error in the proof; Bratteli and Robinson [39] corrected it in the form of

Theorem 3.3.7. Then McIntosh [127] constructed an example of closed *-derivation δ in a C*-algebra in which there is an element $a(=a^*)$ of $\mathscr{D}(\delta)$ and a continuously differentiable function f on $[-\|a^-\|, \|a^+\|]$ with $f(a)\notin\mathscr{D}(\delta)$. McIntosh's example also gives a negative solution to a longstanding problem in the spectral theory of non-self-adjoint bounded operators in a Hilbert space. This is one of the useful applications of the derivation theory to other branches.

References [39], [143], [166], [170].

3.4 Generators

In quantum physics, one of the most important problems is how to construct a time evolution from a given *-derivation. If one can show that δ is a pre-generator, then the problem is solved. There is a nice theory by Hille–Yoshida [78] and a powerful method of Nelson [134] for the problem. However for interacting continuous quantum systems, these methods cannot apply in most cases. The reader will discover this situation in Chapter 4. Therefore the generation problem is rather incomplete in interacting continuous quantum systems. On the other hand, general theories of generators are often useful for quantum lattice systems, although there are many important cases in which the so-called 'core problem' is left open as a very difficult problem.

In this section, we shall develop a general theory for generators.

Let δ be a closable *-derivation in a C*-algebra A. δ is said to be a pre-generator if its closure $\bar{\delta}$ is the (infinitesimal) generator of a strongly continuous one-parameter group of *-automorphisms on A. An important problem is the conditions under which δ is a pre-generator. It follows from semi-group theory [78, 188] that δ is a pre-generator if and only if $\|a+\lambda\delta(a)\|\geqslant\|a\|$ $(a\in\mathscr{D}(\delta))$ and the ranges $(\lambda 1+\delta)\mathscr{D}(\delta)$ are dense in A for all real $\lambda(\lambda\neq 0)$.

3.4.1 Proposition *Let δ be a pre-generator in A; then it is well-behaved.*

Proof Let $\bar{\delta}$ be the closure of δ and $\alpha_t=\exp(t\bar{\delta})$; then $\{\alpha_t \mid t\in\mathbb{R}\}$ is a strongly continuous one-parameter group of *-automorphisms on A. Suppose that $|f(a)|=\|f\|\,\|a\|$; then $|f(\alpha_t(a))|\leqslant\|f\|\,\|a\|$ $(t\in R)$; hence

$$\left\{\frac{\mathrm{d}}{\mathrm{d}t}f(\alpha_t(a))\right\}_{t=0}=f(\delta(a))=0 \qquad \text{for } a\ (=a^*)\in\mathscr{D}(\delta). \qquad \square$$

3.4.2 Definition *Let δ be a closable *-derivation in A, and let δ^* be its adjoint* *in* A^* – i.e., $\mathscr{D}(\delta^*) = \{f \in A^* | g(x) = f(\delta(x))$ *is bounded on* $A\}$ *and* $\delta^*(f) = g$. *Then* $\mathscr{D}(\delta^*)$ *is a* $\sigma(A^*, A)$*-dense linear subspace of* A^*. δ^* *is said to be well-behaved if for* $f(=f^*) \in \mathscr{D}(\delta^*)$, *there is an element* $x_f(=x_f^*) \in A^{**}$ *such that* $\|x_f\| = 1$, $|f(x_f)| = \|f\|$ *and* $(\delta^* f)(x_f) = 0$, *where* A^{**} *is the second dual of* A *(a* W^**-algebra).*

3.4.3 Proposition *Let* δ *be a closable *-derivation in* A. *Then* δ *is a pregenerator if and only if* δ *and* δ^* *are well-behaved.*

Proof Suppose that δ is a pre-generator and let $\alpha_t = \exp t\bar{\delta}$. Then $\|[1 + \lambda\delta]^{-1}\| \leqslant 1$ for real λ. By the general theory of linear operators on Banach spaces [98], $\{(1 + \lambda\delta)^{-1}\}^* = (1 + \lambda\delta^*)^{-1}$ and $\|\{(1 + \lambda\delta)^{-1}\}^*\| \leqslant 1$. Hence $\|(1 + \lambda\delta^*)f\| \geqslant \|f\|$ for $f \in \mathscr{D}(\delta^*)$. Let v be a real linear subspace of the self-adjoint portion of A^* generated by f and $\delta^* f$; then there is a linear functional F on v such that $\|F\| = 1$, $F(f) = \|f\|$ and $F(\delta^*(f)) = 0$. By the Hahn–Banach theorem, there is a self-adjoint linear functional $\tilde{F}$ on A^* such that $\|\tilde{F}\| = 1$, $\tilde{F}(f) = \|f\|$ and $\tilde{F}(\delta^*(f)) = 0$. Hence δ^* is well-behaved. Conversely suppose that δ^* and δ are well-behaved. Then $\|(1 + \lambda\delta)(a)\| \geqslant \|a\|$ for $a \in \mathscr{D}(\delta)$ and real λ (see (3.2.19).

If $(\lambda 1 + \delta)\mathscr{D}(\delta)$ is not dense in A for some real $\lambda \neq 0$ then there is an $f(=f^*) \in A^*$ with $f \neq 0$ such that $f((\lambda 1 + \delta)(a)) = 0$ for $a \in \mathscr{D}(\delta)$. Hence $f \in \mathscr{D}(\delta^*)$ and $(\lambda 1 + \delta^*)f = 0$. Since δ^* is well-behaved, there is an element x_f in A^{**} with $x_f^* = x_f$, $\|x_f\| = 1$ and $|f(x_f)| = \|f\|$ and $\delta^* f(x_f) = 0$. Hence $\|f + (\delta^*/\lambda)f\| \geqslant |(f + (\delta^*/\lambda)f)(x_f)| = \|f\| = 0$, a contradiction. □

3.4.4 Proposition *Let* δ *be a *-derivation in* $\mathscr{A}$. *Then* δ *is a pre-generator if and only if* δ *is well-behaved and* $(1 \pm \delta)\mathscr{D}(\delta)$ *are dense in* A.

Proof By 3.2.19, the well-behavedness of δ implies $\|(1 + \lambda\delta)(a)\| \geqslant \|a\|$ $(a \in \mathscr{D}(\delta), \lambda \in \mathbb{R})$. Since $(1 \pm \delta)\mathscr{D}(\delta)$ are dense, there exist $(1 \pm \bar{\delta})^{-1}$ and $\|(1 \pm \bar{\delta})^{-1}\| \leqslant 1$. In particular $(1 \pm \bar{\delta})\mathscr{D}(\bar{\delta}) = A$. Now suppose that $(\lambda_0 1 - \bar{\delta})\mathscr{D}(\bar{\delta}) = A$ for some real $\lambda_0 \neq 0$ and let $|\lambda_0 - \lambda|/|\lambda_0| < 1$; then we shall show that $(\lambda 1 - \bar{\delta})\mathscr{D}(\bar{\delta}) = A$. For this, it is enough to show that $(\lambda 1 - \bar{\delta})\mathscr{D}(\delta)$ is dense in A, for $\|(\lambda 1 - \bar{\delta})(a)\| \geqslant |\lambda| \|a\|$ $(a \in \mathscr{D}(\delta))$. Suppose that $(\lambda 1 - \delta)\mathscr{D}(\delta)$ is not dense in A; then there is an element x in A such that $\|x\| < |\lambda_0|/|\lambda_0 - \lambda|$ and $\|x - (\lambda 1 - \delta)(a)\| \geqslant 1$ for all $a \in \mathscr{D}(\delta)$. Put $(\lambda_0 1 - \bar{\delta})(b) = x$; then

$$1 \leqslant \|(\lambda_0 1 - \bar{\delta})(b) - (\lambda 1 - \bar{\delta})(b)\| = \|(\lambda_0 - \lambda)b\| = |\lambda_0 - \lambda| \, \|(\lambda_0 1 - \bar{\delta})^{-1}x\|$$

$$\leqslant |\lambda_0 - \lambda| \frac{1}{|\lambda_0|} \|x\| < \frac{|\lambda_0 - \lambda|}{|\lambda_0|} \frac{|\lambda_0|}{|\lambda_0 - \lambda|} = 1,$$

a contradiction. From this one can easily conclude that $(\lambda 1 - \bar{\delta})\mathscr{D}(\delta) = A$ for all real $\lambda(\neq 0)$. □

Let δ be a *-derivation in A. It is not so restrictive to assume that δ is well-behaved. However the density of $(1 \pm \delta)\mathscr{D}(\delta)$ is not apparent in many interesting cases. The real problem often involves proving the density from available conditions.

Analytic elements

Let $\mathscr{D}^{\infty}(\delta) = \bigcap_{n=1}^{\infty} \mathscr{D}(\delta^n)$. An element $a \in \mathscr{D}^{\infty}(\delta)$ is said to be analytic (with respect to δ) if there is a positive number γ (depending on a) such that $\sum_{n=0}^{\infty} (\| \delta^n(a) \| / n!) \gamma^n < +\infty$, where $\delta^0(a) = a$. Let $A(\delta)$ be the set of all analytic elements in $\mathscr{D}(\delta)$. Then $A(\delta)$ is a *-subalgebra of A.

3.4.5 Proposition *Let δ be a well-behaved *-derivation in A, and suppose that $A(\delta)$ is dense in A; then δ is a pre-generator.*

Proof Since δ is well-behaved, it is closable and

$$\|(1 + \lambda\bar{\delta})(a)\| \geqslant \|a\| \qquad \text{for } a \in \mathscr{D}(\delta) \text{ and } \lambda \in \mathbb{R}.$$

For $a \in A(\delta)$, there is the largest positive number $\gamma(a)$ (possibly $+\infty$) such that

$$\sum_{n=0}^{\infty} \frac{\| \delta^n(a) \|}{n!} \gamma^n < +\infty \qquad \text{for } 0 \leqslant \gamma < \gamma(a).$$

Define

$$\rho(z)(a) = \sum_{n=0}^{\infty} \frac{\delta^n(a)}{n!} z^n \qquad \text{for } z \in \mathbb{C} \text{ with } |z| < \gamma(a).$$

For $t \in R$ with $|t| < \gamma(a)$, let $a_n = (1 + (t/n)\delta)^n a$; then

$$\| a_n \| = \left\| \left\{ 1 + t\delta + \frac{1}{2!}\left(1 - \frac{1}{n}\right) t^2 \delta^2 + \cdots + \frac{1}{n!}\left(1 - \frac{1}{n}\right)\left(1 - \frac{2}{n}\right) \cdots \right.\right.$$
$$\left.\left. \times \left(1 - \frac{n-1}{n}\right) t^n \delta^n \right\} a \right\| \leqslant \sum_{n=0}^{\infty} \frac{\| \delta^m(a) \|}{m!} |t|^m < +\infty$$

$$\left\| a_n - \sum_{m=0}^{\infty} \frac{\delta^m(a)}{m!} t^m \right\| \leqslant \frac{1}{n} \sum_{m=2}^{n} \frac{\| \delta^m(a) \| \, |t|^m}{m!} + \sum_{m=n}^{\infty} \frac{\| \delta^m(a) \| \, |t|^m}{m!} \to 0 \qquad (n \to \infty).$$

Hence $a_n \to \rho(t)(a)$. Since

$$\left\| \left(1 + \frac{t\delta}{n}\right)^n a \right\| \geqslant \left\| \left(1 + \frac{t\delta}{n}\right)^{n-1} (a) \right\| \geqslant \cdots \geqslant \|a\|,$$

$\| \rho(t)(a) \| \geqslant \| a \|$. For $z \in \mathbb{C}$ with $|z| < \gamma(a)$,

$$\rho(z)(a) = \sum_{n=0}^{\infty} \frac{\delta^n(a)}{n!} z^n \text{ is analytic on } D = \{z \,|\, |z| < \gamma(a)\}.$$

By the closability of δ,

$$\delta\rho(z)(a) = \sum_{n=0}^{\infty} \frac{\delta^{n+1}(a)}{n!} z^n = \sum_{n=0}^{\infty} (n+1)\frac{\delta^{n+1}(a)}{(n+1)!} z^n = \frac{\mathrm{d}}{\mathrm{d}z}\rho(z)(a).$$

Analogously

$$\delta^m \rho(z)(a) = \frac{\mathrm{d}^m}{\mathrm{d}z^m}\rho(z)(a).$$

Hence $\gamma(\rho(z)(a)) \geqslant \gamma(a)$ for $z \in D$. For $|z| + |z_1| < \gamma(a)$,

$$\rho(z+z_1)(a) = \sum_{n=0}^{\infty} \frac{(\mathrm{d}^n/\mathrm{d}z^n)\rho(z)(a)}{n!} {z_1}^n = \sum_{n=0}^{\infty} \frac{\delta^n(\rho(z)(a))}{n!} {z_1}^n = \rho(z_1)\rho(z)(a).$$

Now suppose that $\|\rho(t_0)(a)\| > \|a\|$ for some real t_0 with $|t_0| < \gamma(a)/2$; then

$$\|a\| = \|\rho(t_0 - t_0)(a)\| = \|\rho(-t_0)\rho(t_0)(a)\| \geqslant \|\rho(t_0)(a)\| > \|a\|,$$

a contradiction; hence $\|\rho(t)(a)\| = \|a\|$ for real t with $|t| < \gamma(a)/2$. Since $\gamma(\rho(t)(a)) \geqslant \gamma(a)$ for $|t| < \gamma(a)$, $\rho(s)\rho(t)(a)$ is well-defined with $|s|$, $|t| < \gamma(a)$, and is real analytic with respect to s. Moreover $\|\rho(s)\rho(t)(a)\| = \|a\|$ if $|s|, |t| < \frac{1}{2}\gamma(a)$ and $\rho(s)\rho(t)(a) = \rho(s+t)(a)$ if $|s| + |t| < \gamma(a)$; hence $\|\rho(t)(a)\| = \|a\|$ for $|t| < \gamma(a)$. Continuing this process, we can define a real analytic vector-valued function $v(t)$ on $\mathbb{R}$ such that $\|v(t)\| = \|a\|$ and $v(t) = \rho(t)(a)$ with $|t| < \gamma(a)$. Now suppose that $(1-\delta)\mathscr{D}(\delta)$ is not dense in A; then there is a non-zero element f in A^* such that $f((1-\delta)\mathscr{D}(\delta)) = 0$, and so $f \in \mathscr{D}(\delta^*)$ and $\delta^* f = f$. Take an element a in $A(\delta)$ such that $f(a) \neq 0$; then

$$\exp(t)f(a) = \left(\sum_{n=0}^{\infty} \frac{\delta^{*n}}{n!} t^n f\right)(a) = f\left(\sum_{n=0}^{\infty} \frac{\delta^n(a)}{n!} t^n\right) = f(v(t)). \qquad \text{for } |t| < r(a).$$

Since $\exp(t)f(a)$ is real analytic on $\mathbb{R}$, $\exp(t)f(a) = f(v(t))$ for $t \in \mathbb{R}$. $|f(v(t))| \leqslant \|f\|$, a contradiction.

Similarly we have the density of $(1+\delta)\mathscr{D}(\delta)$. □

3.4.6 Proposition *Let δ be a *derivation in A and suppose that there exists a faithful state ϕ on A* (i.e. $\phi(axb) = 0$ *for all* $a, b \in A$ *implies* $x = 0$) *such that* $\phi(\delta(a)) = 0$ for $a \in \mathscr{D}(\delta)$. *Then if $A(\delta)$ is dense in A, then δ is a pregenerator.*

Proof Let $\{\pi_\phi, \mathscr{H}_\phi\}$ be the GNS construction of A via ϕ. Put $\delta(a)_\phi = Sa_\phi$ $(a \in \mathscr{D}(\delta))$; then

$$\begin{aligned}(Sa_\phi, b_\phi) &= \phi(b^*\delta(a)) = \phi(\delta(b^*a)) - \phi(\delta(b)^*a) \\ &= -(a_\phi, Sb_\phi) \qquad (a, b \in \mathscr{D}(\delta)).\end{aligned}$$

Hence $(\mathrm{i}S)^* \supset \mathrm{i}S$ and so $\mathrm{i}S$ is symmetric. It is easily seen that $\{A(\delta)\}_\phi$ are analytic for $\mathrm{i}S$, so that $\mathrm{i}S$ has a dense set of analytic elements. By Nelson's theorem, iS is essentially self-adjoint.

$$\pi_\phi(\delta(a))b_\phi = (\delta(a)b)_\phi = (\delta(ab) - a\delta(b))_\phi = S\pi_\phi(a)b_\phi - \pi_\phi(a)Sb_\phi$$
$$= -\mathrm{i}[\mathrm{i}S, \pi_\phi(a)]b_\phi.$$

Hence $-\mathrm{i}\overline{[\mathrm{i}S, \pi_\phi(a)]} = \pi_\phi(\delta(a))$ $(a \in \mathscr{D}(\delta))$.

Now we shall show that δ is closable. Let $a_n \to 0$ and $\delta(a_n) \to b$; then $\pi_\phi(a_n) \to 0$ and $\pi_\phi(\delta(a_n)) \to \pi_\phi(b)$. Therefore $-\mathrm{i}\overline{[\mathrm{i}S, 0]} = \pi_\phi(b)$; hence $b = 0$, so that δ is closable. Moreover

$$\exp(\mathrm{i}t\bar{S})\pi_\phi(a)\exp(-\mathrm{i}t\bar{S}) = \pi_\phi(\rho(t)(a)) \qquad \text{for } a \in A(\delta)$$

and t with $|t| < r(a)$. Hence

$$\|\rho(t)(a)\| = \|\pi_\phi(\rho(t)(a))\| = \|\pi_\phi(a)\| = \|a\|.$$

This implies that the restriction $\check{\delta}$ of δ to $A(\delta)$ is well-behaved and so $\check{\delta}$ is a pre-generator. If $\bar{\check{\delta}} \subsetneqq \bar{\delta}$, then there is a non-zero element a $(= a^*) \in \mathscr{D}(\delta)$ such that $\delta(a) = a$ or $-a$, for $(1 \pm \bar{\check{\delta}})\mathscr{D}(\bar{\check{\delta}}) = A$. Then $Sa_\phi = a_\phi$ or $(-a_\phi)$. This contradicts that iS is essentially self-adjoint. Hence $\bar{\check{\delta}} = \bar{\delta}$. □

Let δ be a generator. Then $\exp(t\delta) = \alpha_t$ ($\exp(t\delta)$ is the symbolic notation for the one-parameter group generated by an unbounded δ) is a strongly continuous one-parameter group of *-automorphisms on A. (Recall that δ is a *-derivation.) Let $A_1(\delta)$ be the set of all entire analytic elements a – i.e. $\sum_{n=0}^{\infty}(\|\delta^n(a)\|/n!)r^n < +\infty$ for all positive numbers r and $A_2(\delta)$ the set of all geometric elements b – i.e. $\|\delta^n(b)\| \leqslant M_b{}^n\|b\|$ $(n = 1, 2, 3, \ldots,)$, where M_b depends on b. Then $A(\delta) \supset A_1(\delta) \supset A_2(\delta)$, and each one is a dense *-subalgebra of A.

In fact, let f be a function on $(-\infty, \infty)$ such that $\int_{-\infty}^{\infty} f(t)\,\mathrm{d}t = 1$, the Fourier transform $\hat{f}$ is continuously differentiable and the support of $\hat{f}$ is included in $[-k, k]$. For $\lambda > 0$, let $a_\lambda = \int_{-\infty}^{\infty} f(t)\alpha_{\lambda t}(a)\,\mathrm{d}t$; then

$$\|a - a_\lambda\| \leqslant \int_{-\infty}^{\infty} |f(t)|\,\|\alpha_{\lambda t}(a) - a\|\,\mathrm{d}t \to 0 \qquad (\lambda \to 0).$$

$$\delta(a_\lambda) = \lim_{h\to 0}\frac{\alpha_h(a_\lambda) - a_\lambda}{h} = \lim_{h\to 0}\frac{1}{h}\left\{\int_{-\infty}^{\infty} f(t)\alpha_{\lambda t + h}(a)\,\mathrm{d}t - \int_{-\infty}^{\infty} f(t)\alpha_{\lambda t}(a)\,\mathrm{d}t\right\}$$
$$= \lim_{h\to 0}\left\{\int_{-\infty}^{\infty}\frac{f(t - h/\lambda) - f(t)}{h}\alpha_{\lambda t}(a)\,\mathrm{d}t\right\} = -\frac{1}{\lambda}\left\{\int_{-\infty}^{\infty} f^{(1)}(t)\alpha_{\lambda t}(a)\,\mathrm{d}t\right\}$$

Analogously,

$$\delta^n(a_\lambda) = \left(-\frac{1}{\lambda}\right)^n\int_{-\infty}^{\infty} f^{(n)}(t)\alpha_{\lambda t}(a)\,\mathrm{d}t.$$

Hence

$$\|\delta^n(a_\lambda)\| \leqslant \left(\frac{1}{\lambda}\right)^n \|f^{(n)}\|_1 \|a\|.$$

On the other hand

$$\begin{aligned}\|f^{(n)}\|_1 &= \int_{-\infty}^{\infty} |f^{(n)}(t)|\,\mathrm{d}t = \int_{-\infty}^{\infty} \left|\frac{1}{1+\mathrm{i}t}\right| |(1+\mathrm{i}t)f^{(n)}(t)|\,\mathrm{d}t \\ &\leqslant \left(\int_{-\infty}^{\infty} \frac{\mathrm{d}t}{1+t^2}\right)^{1/2} \left(\int_{-\infty}^{\infty} \left|\left(1-\frac{\mathrm{d}}{\mathrm{d}s}\right)s^n \hat{f}(s)\right|^2 \mathrm{d}s\right)^{1/2} \\ &\leqslant (\pi)^{1/2}(2k)^{1/2}(2k+1)^n \max(\|\hat{f}\|_\infty, \|\hat{f}^{(1)}\|_\infty).\end{aligned}$$

Hence $\|\delta^n(a_\lambda)\| \leqslant (1/\lambda)^n M^n \|a\|$, where M is a positive number. Put $\|a\| = \mu \|a_\lambda\|$; then $\|\delta^n(a_\lambda)\| \leqslant \mu(1/\lambda)^n M^n \|a_\lambda\|$.

Hence we have the following proposition.

3.4.7 Proposition *The set $A_2(\delta)$ of all geometric elements with respect to a generator δ is a dense *-subalgebra of A.*

3.4.8 Proposition *Let p be a projection in A and let ε be a positive number; then there is a projection q in $A(\delta)$ such that $\|p-q\| < \varepsilon$.*

Proof Put $\varepsilon' = \min(\frac{1}{4}, \varepsilon/8)$ and take a positive element a in $A(\delta)$ such that $\|a\| \leqslant 1$ and $\|p-a\| < \varepsilon'$; then

$$\begin{aligned}&\|a^2-a\| \\ &\quad\leqslant \|a^2-p^2\| + \|p^2-p\| + \|p-a\| \leqslant \|a^2-ap\| + \|ap-p^2\| + \|p-a\| \\ &\quad\leqslant \|a\|\,\|a-p\| + \|a-p\|\,\|p\| + \|p-a\| < 3\varepsilon'.\end{aligned}$$

Hence the spectrum $Sp(a)$ of a is contained in $[0, 1-(1-12\varepsilon')^{1/2}/2] \cup [1+(1-12\varepsilon')^{1/2}/2, 1]$. Let Γ be the circle in $\mathbb{C}$ with center 1 and radius $\frac{1}{2}$; define

$$q = \frac{1}{2\pi\mathrm{i}} \oint_\Gamma (\lambda 1 - a)^{-1}\,\mathrm{d}\lambda.$$

Let D be the commutative C^*-subalgebra of A generated by a; then $q \in D$ and $q(t) = 1$ for $Sp(a) \cap [1+(1-12\varepsilon')^{1/2}/2, 1]$ and $q(t) = 0$ for $t \in Sp(a) \cap [0, (1-(1-12\varepsilon')^{1/2}/2]$ and so q is a projection. Moreover $\|a-q\| \leqslant (1-(1-12\varepsilon')^{1/2})/2 < 6\varepsilon'$ and so $\|p-q\| \leqslant \|p-a\| + \|a-q\| < 7\varepsilon'$. Now we shall show $q \in A(\delta)$. For $\lambda \in \Gamma$,

$$\rho(z)(\lambda 1 - a) = \lambda 1 - \rho(z)(a) = \lambda 1 - \sum_{n=0}^{\infty} \frac{\delta^n(a)}{n!} z^n \qquad (|z| < \gamma(a))$$

$$= (\lambda 1 - a)\left\{1 - (\lambda 1 - a)^{-1} \sum_{n=1}^{\infty} \frac{\delta^n(a)}{n!} z^n\right\}.$$

There is a positive number $\alpha(<\gamma(a))$ such that

$$\sup_{|z|<\alpha} \left\| (\lambda 1 - a)^{-1} \left\{ \sum_{n=1}^{\infty} \frac{\delta^n(a)}{n!} z^n \right\} \right\| < 1 \qquad \text{for all } \lambda \in \Gamma.$$

Hence for $|z| < \alpha$,

$$\left\{1 - (\lambda 1 - a)^{-1} \sum_{n=1}^{\infty} \frac{\delta^n(a)}{n!} z^n\right\}^{-1} = \sum_{n=0}^{\infty} \left\{(\lambda 1 - a)^{-1} \sum_{n=1}^{\infty} \frac{\delta^{(n)}(a)}{n!} z^n\right\}^n.$$

This convergence is uniform on $(|z| < \alpha)$ and on Γ; hence it is analytic with respect to $z(|z| < \alpha)$. Consider $(1/2\pi i)\oint_\Gamma \{\lambda 1 - \rho(z)(a)\}^{-1} d\lambda$ (say $g(z)$); then

$$\frac{g(z) - g(z_0)}{z - z_0} = \frac{1}{2\pi i} \oint_\Gamma \frac{\{\lambda 1 - \rho(z)(a)\}^{-1} - \{\lambda 1 - \rho(z_0)(a)\}^{-1}}{z - z_0} d\lambda$$

with $|z|, |z_0| < \alpha$. Hence

$$\lim_{z \to z_0} \frac{g(z) - g(z_0)}{z - z_0}$$

$$= \lim_{z \to z_0} \frac{1}{2\pi i} \oint_\Gamma \frac{\{\lambda 1 - \rho(z)(a)\}^{-1}\{(\rho(z) - \rho(z_0))(a)\}\{\lambda 1 - \rho(z_0)(a)\}^{-1}}{z - z_0} d\lambda$$

$$= \frac{1}{2\pi i} \oint_\Gamma \left\{\lambda 1 - \rho(z_0)(a)\right\}^{-1} \left\{\frac{d}{dz}\rho(z)(a)\right\}_{z=z_0} \{\lambda 1 - \rho(z_0))(a)\}^{-1} d\lambda.$$

Hence $g(z)$ is analytic on $\{z \mid |z| < \alpha\}$ and so

$$\alpha_t(q) = \frac{1}{2\pi i} \oint_\Gamma \{\lambda 1 - \alpha_t(a)\}^{-1} d\lambda = \frac{1}{2\pi i} \oint_\Gamma \{\lambda 1 - \rho(t)(a)\}^{-1} d\lambda$$

is real analytic and so $q \in A(\delta)$. □

3.4.9 Proposition *Let δ be a closed *-derivation in A and p be a projection in A; then for $\varepsilon(>0)$, there is a projection q in $\mathscr{D}(\delta)$ such that $\|p - q\| < \varepsilon$.*

Proof In the proof of 3.4.8, take a $C^{(2)}(R)$-function F such that $F(t) = 0$ on $[0, 1 - (1 - 12\varepsilon')^{1/2}/2]$ and $F(t) = 1$ on $[1 + (1 - 12\varepsilon')^{1/2}/2, 1]$; then $f(a) \in \mathscr{D}(\delta)$ (3.3.7), $f(a)$ is a projection and $\|f(a) - p\| < \varepsilon$. □

The following problem seems to be important for the core problem of generators in UHF algebras (cf. 4.5.10).

3.4.10 Problem *In Proposition 3.4.8, can we replace $A(\delta)$ by $A_1(\delta)$ or $A_2(\delta)$?*

The proof of Proposition 3.4.8 cannot be applied to the problem. If $\rho(z)(a)$ does not change $Sp(a)$, then one can define

$$\frac{1}{2\pi i}\oint_{\gamma}\{\lambda 1-\rho(z)(a)\}^{-1}\,d\lambda \qquad \text{for } z \text{ with } |z|<\gamma(a).$$

Therefore we can solve the problem affirmatively. However $\rho(z)$ will change $Sp(a)$ generally, so that it is also an interesting problem under what conditions one can conclude that $\mathrm{Sp}(\rho(z)(a)) = Sp(a)$.

It is easily seen that $Sp(a) = Sp(\rho(z)(a))$ $(z\in\mathbb{C})$ if $\{\alpha_t | t\in R\}$ is a uniformly continuous one-parameter group of *-automorphisms on A.

Let $a\in A(\delta)$; then we can define

$$\rho(z)(a) = \sum_{n=0}^{\infty}\frac{\delta^n(a)}{n!}z^n \qquad (z\in\mathbb{C} \text{ with } |z|<\gamma(a)).$$

$\rho(z)(ab) = \rho(z)(a)\rho(z)(b)$ for $a, b\in A(\delta)$ if $|z| < \min(\gamma(a), \gamma(b))$. In particular, if $a\in A_1(\delta)$, then $\rho(z)(a)$ can be defined for all $z\in\mathbb{C}$, and $\{\rho(z) | z\in\mathbb{C}\}$ is a strongly continuous one-parameter group of automorphisms on $A_1(\delta)$. If $p\in A(\delta)$ is a projection, then $\rho(z)(p)$ is an idempotent, so that $Sp(\rho(z)(p)) = Sp(p)$. Generally the spectrum of $\rho(z)(a)$ will depend on z. In fact, let $C_0(R)$ be the C*-algebra of all continuous functions on the real line vanishing at infinity. Let $\delta = d/dt$; then $\exp(t\delta)f(x) = f(x+t)$ $(f\in C_0(R))$, where $x, t\in R$. Let $f(x) = 1/(1+x^2)$; then $f\in A(\delta)$ and $Sp(f)$ consists of non-negative real numbers.

On the other hand

$$\exp(z)(x) = \sum_{n=0}^{\infty}\frac{z^n f^{(n)}(x)}{n!} = \frac{1}{1+(x+z)^n}$$

is complex-valued. However if $A(\delta)$ contains many projections, the non-invariance of the spectrum under $\rho(z)$ is not clear. Later we shall discuss the problem in the case of UHF algebras.

3.4.11 Notes and remarks Proposition 3.4.7 is due to Bratteli and Robinson [40]. Proposition 3.4.8 is due to the author [166]. If Problem 3.4.10 has a positive solution, it may introduce a new technique into functional analysis. From this point of view, it would be an interesting problem.

References [41], [78], [170], [188].

3.5 Unbounded derivations in commutative C*-algebras

In the previous chapter, we have seen that any bounded derivation in a commutative C*-algebra is identically zero. On the other hand, there are many important non-trivial closable *-derivations in commutative C*-algebras. Partial differentiations in a differential manifold are typical closable *-derivations. We showed that a totally disconnected space has no non-trivial closable *-derivations (cf. 3.2.1). On the other hand, even a continuous manifold which is not differentiable may have non-trivial closable *-derivations. For example, a connected locally compact group is approximated by a Lie group, so that it has non-trivial closable *-derivations. Therefore the theory of derivations in a commutative C*-algebra can be developed even in a non-differential manifold. This shows the importance of the study of derivations in commutative C*-algebras. For example, the generalized Hilbert's fifth problem asks the following: Let Ω be a connected locally euclidean space and let G be a locally compact transformation group effectively acting on Ω. Then, can we conclude that G is a Lie group? This famous problem may be treated within the derivation theory in commutative C*-algebras, when G is connected. As yet, we do not have any definite study of the problem along this line. However it will certainly be one of the central interests in the future research program of derivation theory in commutative C*-algebras.

In this section, we shall discuss some general properties of unbounded closable *-derivations in commutative C*-algebras. We shall also describe a detailed analysis of derivations in $C([0,1])$. Recently there has been a significant development in this subject but we shall only discuss parts of it.

3.5.1 Definition *Let A be a semi-simple commutative Banach algebra with spectrum space Ω. Identify A, via the Gelfand transform, with a subalgebra of $C_0(\Omega)$. A is called a Šilov algebra if for each closed subset F of Ω and each $x \in \Omega \backslash F$, there is an $f \in A$ such that $f|F = 0$ and $f(x) = 1$.*

When we say that A is a Šilov subalgebra of $C_0(\Omega)$, we mean that the spectrum space of $A = \Omega$ and A is Šilov.

It is known that a Šilov algebra A satisfies the following properties ([130]):

(1) A contains partitions of unity subordinate to any finite open cover of a compact set $K \subseteq \Omega$. That is, if $\{U_i | i = 1, 2, \ldots, n\}$ is an open cover of K, then there exist functions f_i $(i = 1, 2, \ldots, n)$ in A such that f_i has compact support contained in U_i and $\sum_{i=1}^{n} f_i(x) = 1$ for $x \in K$. If A is self-adjoint, then the f_i can be chosen to satisfy $0 \leqslant f_i \leqslant 1$.

If X is compact, this implies that $1 \in A$.

(2) If f is a complex-valued function on Ω, we say that f is locally in A at $x \in \Omega$ (resp. f is locally in A at ∞) if there is a $g \in A$ such that $f = g$ in a neighborhood of x (resp. in a neighborhood of ∞). If U is a subet of

Ω, we say f is locally in A on U, if f is locally in A at each point of U. A consequence of the existence of partitions of unity is that if f is locally in A on a compact subset K, then there is a $g \in A$ such that $g|K = f|K$. If f is locally in A at each point of Ω and at ∞, then $f \in A$. This result is known as the local theorem for Šilov algebras.

(3) If f is a complex-valued function defined on an open subset U of Ω, the following conditions are equivalent:
  (i) f is locally in A on U.
  (ii) For each compact $K \subseteq U$, there is a $g \in A$ such that $g|K = f|K$
  (iii) Whenever $h \in A$ has compact support contained in U, $fh \in A$.
  If A is self-adjoint, another equivalent condition is:
  (iv) If $h \in A$, $0 \leqslant h \leqslant 1$, and h has compact support contained in U, then $hf \in A$.

(4) If $F \subseteq \Omega$, one defines $\ker(F) = \{f \in A \mid f|F = 0\}$. If $J \subseteq A$ is an ideal, one defines hull $(J) = \{x \in \Omega \mid f(x) = 0 \text{ for all } f \in J\}$. For a Šilov algebra A, hull $(\ker(F)) = F$ if and only if F is closed. For closed F, there is a minimal ideal $J(F)$ with hull F, $J(F) = \{f \in A \mid f = 0 \text{ in a neighborhood of } F\}$. The closure of $J(F)$ (denoted $\bar{J}(F)$) is the minimal closed ideal with hull F. An ideal whose hull is a singleton is called primary. For $x \in \Omega$, $\bar{J}(x)$ $(=\bar{J}(\{x\}))$ is called the minimum closed primary ideal at x.

(5) Let $\bar{\Omega}$ denote one point compactification of Ω. It is easily seen that A is a Šilov subalgebra of $C_0(\Omega)$ if and only if $A \oplus \mathbb{C}1$ is a Šilov subalgebra of $\bar{\Omega}$.

(6) It often happens that a commutative Banach algebra A is given explicitly as a subalgebra of $C(K)$ for some compact Hausdorff space K. Even if A satisfies the condition of Definition 3.5.1, it may not be true that the spectrum space of $A = K$. However it is well known that if A is self-adjoint, separates points, contains the constants, and contains the inverse of each strictly positive element, then the spectrum space of $A = K$.

Since any closable *-derivation in $C_0(\Omega)$ can be uniquely extended to a closable *-derivation in $C(\Omega \cup (\infty))$, for simplicity we shall discuss closable *-derivations on a compact space K.

3.5.2 Proposition *Let δ be a derivation in $C(K)$ and suppose that $\mathscr{D}(\delta)$ is a Šilov subalgebra of $C(K)$ under some norm $|||\cdot|||$; then $\|a\|_\delta \leqslant k|||a|||$ $(a \in \mathscr{D}(\delta))$, where k is a fixed positive number.*

Proof By Johnson's theorem (Theorem 3 in [83]) there is a finite family of mutually orthogonal idempotents $e_0, e_1, \ldots, e_n$ in $\mathscr{D}(\delta)$ such that for $p \in \operatorname{supp}(e_0)$, $a \to \delta(a)(p)$ $(a \in \mathscr{D}(\delta))$ is continuous with respect to $|||\cdot|||$, and $\mathscr{D}(\delta)e_i$

$(i=1,2,\ldots,n)$ has a unique maximal proper ideal and $\sum_{i=0}^{n} e_i = 1$. Since $\mathscr{D}(\delta)e_i$ $(i=1,2,\ldots,n)$ is semi-simple, it is one-dimensional and so the support of e_i $(i=1,2,\ldots,n)$ consists of one point p_i. Then $\delta(a)$ $(p_i)=\delta(a)(p_i)e_i(p_i)=\delta(ae_i)(p_i)=0$ $(\delta(e_i)=0)$; hence $a\rightarrow\delta(a)(p)=f_p(a)$ is continuous on $\mathscr{D}(\delta)$ with the norm $|||\cdot|||$ for each $p\in K$. Since $\{f_p|p\in K\}$ is $\sigma(\mathscr{D}(\delta)^*, \mathscr{D}(\delta))$-compact in $\mathscr{D}(\delta)^*$, where $\mathscr{D}(\delta)^*$ is the dual of $\mathscr{D}(\delta)$ with the norm $|||\cdot|||$, $\mathrm{Sup}_{p\in K}\|f_p\|<+\infty$. Since $a\rightarrow a(p)$ $(a\in\mathscr{D}(\delta))$ is a character of $\mathscr{D}(\delta)$, $\|a\|\leqslant|||a|||$ for $a\in\mathscr{D}(\delta)$. Hence

$$\|a\|_\delta=\left\|\begin{pmatrix} a & \delta(a)\\ 0 & a\end{pmatrix}\right\|=\sup_{p\in K}\left\|\begin{pmatrix} a(p) & \delta(a)(p)\\ 0 & a(p)\end{pmatrix}\right\|\leqslant k|||a||| \qquad (a\in\mathscr{D}(\delta)) \quad \square$$

Let δ be a closed *-derivation in $C(K)$. Then $\mathscr{D}(\delta)$ is a Banach *-algebra under the norm $\|\cdot\|_\delta$ and $\|a^*\|_\delta=\|a\|_\delta$ $(a\in\mathscr{D}(\delta))$.

3.5.3 Proposition *Let $\{\delta_\alpha|\alpha\in\Pi\}$ be a family of closed *-derivations in $C(K)$ and let $\mathscr{D}=\bigcap_{\alpha\in\Pi}\mathscr{D}(\delta_\alpha)$. For $a\in\mathscr{D}$, define $|||a|||=\sup_{\alpha\in\Pi}\|a\|_{\delta}$, and let $\mathscr{D}_0=\{a\in\mathscr{D}\,|\,|||a|||<+\infty\}$.*

If $\mathscr{D}_0$ is dense in $C(K)$, then $\mathscr{D}_0$ is a Šilov subalgebra of $C(K)$ under the norm $|||\cdot|||$.

Proof Let $\{a_n\}$ be a Cauchy sequence in $\mathscr{D}_0$ under $|||\cdot|||$; then it is Cauchy under $\|\cdot\|_{\delta_\alpha}$ so that there is an element b_α such that $\|a_n-b_\alpha\|\rightarrow 0$ and $\|\delta_\alpha(a_n)-\delta_\alpha(b_\alpha)\|\rightarrow 0$. Hence $b_\alpha=b_\beta=b$ for $\alpha,\beta\in\Pi$ and $b\in\mathscr{D}(\delta_\alpha)$ for each $\alpha\in\Pi$. For $\varepsilon>0$, there is a positive number $n(\varepsilon)$ such that $|||a_m-a_n|||=\sup_{\alpha\in\Pi}\|a_m-a_n\|_{\delta_\alpha}<\varepsilon$ for $m,n\geqslant n(\varepsilon)$, and so $\sup_{\alpha\in\Pi}\|a_m-b\|_{\delta_\alpha}=|||a_m-b|||\leqslant\varepsilon$ for $n\geqslant n(\varepsilon)$. This implies $|||b|||<+\infty$ and $a_m\rightarrow b$ in $\mathscr{D}_0$. Therefore $\mathscr{D}_0$ is a Banach *-algebra. $\mathscr{D}_0$ is self-adjoint, contains the constants and is dense in $C(K)$, so that it separates points in K.

Suppose that $a\in\mathscr{D}_0$ is strictly positive. Let $f\in C^{(1)}(R)$ satisfy $f(t)=t^{-1}$ for t in a neighborhood of $a(K)$. Then $a^{-1}=f(a)\in\mathscr{D}(\delta_\alpha)$ for each $\alpha\in\Pi$ (3.2.4) and $\delta_\alpha(a^{-1})=f'(a)\delta_\alpha(a)$; hence

$$\sup_{\alpha\in\Pi}\|\delta_\alpha(a^{-1})\|\leqslant\|f'(a)\|\sup_{\alpha\in\Pi}\|\delta_\alpha(a)\| \text{ and so } a^{-1}\in\mathscr{D}_0.$$

Hence K is the spectrum space of $\mathscr{D}_0$. Suppose that $F\subseteq K$ is closed and $x_0\in K\backslash F$. Since $\mathscr{D}_0$ is dense in $C(K)$, it contains a real-valued g such that $g(x_0)\geqslant 1$ and $|g(x)|\leqslant\frac{1}{2}(x\in F)$. Let $f\in C^{(1)}(R)$ satisfy $f(t)=0$ $(t\leqslant\frac{1}{2})$ and $f(t)=1$ $(t\geqslant 1)$. Then $f(g)\in\mathscr{D}(\delta_\alpha)$ for each $\alpha\in\Pi$ and $\|\delta_\alpha(f(g))\|\leqslant\|f'(g)\|\,\|\delta_\alpha(g)\|$. Hence $f(g)\in\mathscr{D}_0$. $\square$

Let δ be a closed *-derivation in $C(K)$ and suppose that for some positive integer n, $\mathscr{D}(\delta^n)$ is dense in $C(K)$; then $\mathscr{D}(\delta^n)$ is a dense *-subalgebra of $C(K)$. It is clear that $\mathscr{D}(\delta^m) \supset \mathscr{D}(\delta^n)$ $(m \leqslant n)$. For $a \in \mathscr{D}(\delta^n)$, define

$$\|a\|_{\delta^n} = \left\| \begin{pmatrix} a & \delta(a) & \frac{\delta^2(a)}{2!} & \cdots\cdots & \frac{\delta^n(a)}{n!} \\ 0 & a & \delta(a) & \cdots\cdots & \frac{\delta^{n-1}(a)}{(n-1)!} \\ 0 & 0 & a & \cdots\cdots & \\ & 0 & 0 & & \\ & & 0 & & \delta(a) \\ & & & & \\ 0 & 0 & 0 & \cdots\cdots & a \end{pmatrix} \right\| \quad \text{(say } \Phi(a))$$

Then $\mathscr{D}(\delta^n)$ becomes a normed *-algebra under the norm $\|\cdot\|_{\delta^n}$, for $a \to \Phi(a)$ is an isomorphism of $\mathscr{D}(\delta^n)$ into $C(K) \otimes B_{n+1}$, where B_{n+1} is the full matrix algebra of $(n+1) \times (n+1)$. One can easily see that $\Phi(f(a)) = f(\Phi(a))$ for $f \in C^\infty(\mathbb{R})$ – in particular, $f(a) \in \mathscr{D}(\delta^n)$ if $a \in \mathscr{D}(\delta^n)$ and $f \in C^\infty(\mathbb{R})$.

3.5.4 Proposition *Let $\{\delta_\alpha | \alpha \in \Pi\}$ be a family of closed *-derivations in $C(K)$, and let $\{n_\alpha | \alpha \in \Pi\}$ be a family of positive integers such that $\sup_{\alpha \in \Pi} n_\alpha < +\infty$. Let $\mathscr{D} = \bigcap_{\alpha \in \Pi} \mathscr{D}(\delta_\alpha^{n_\alpha})$. For $a \in \mathscr{D}$, define $|||a||| = \sup_{\alpha \in \Pi} \|a\|_{\delta_\alpha^{n_\alpha}}$, and let $\mathscr{D}_0 = \{a | \, |||a||| < +\infty, \, a \in \mathscr{D}\}$. If $\mathscr{D}_0$ is dense in $C(K)$, then $\mathscr{D}_0$ is a Šilov subalgebra of $C(K)$ under the norm $|||\cdot|||$.*

The proof is similar to the proof of Proposition 3.5.3.

3.5.5 Theorem *Let δ be a closed *-derivation in $C(K)$ and let δ_1 be a derivation in $C(K)$ with $\mathscr{D}(\delta_1) = \mathscr{D}(\delta)$; then there is a function ξ (not necessarily bounded) on K such that $\delta_1(f) = \xi \cdot \delta(f)$ $(f \in \mathscr{D}(\delta))$. In particular, δ_1 is closable.*

To prove the theorem, we shall provide some lemmas.

3.5.6 Lemma *Let δ be a closed *-derivation in $C(K)$. If $f \in \mathscr{D}(\delta)$ is a constant in a neighborhood of $p \in K$, then $\delta(f)(p) = 0$.*

Proof There is a constant λ such that $f - \lambda 1 = 0$ in an open neighborhood U of p. Take $g \in \mathscr{D}(\delta)$ such that $g(p) = 0$ and $g(K \backslash U) = 1$; then $(f - \lambda 1) = g(f - \lambda 1)$ and

$$\delta(f)(p) = \delta(f - \lambda 1)(p) = \delta(f - \lambda 1)(p) g(p) + (f - \lambda 1)(p) \delta(g)(p) = 0. \qquad \square$$

3.5.7 Lemma *For $p \in K$, let $I_p = \{f \in \mathscr{D}(\delta) | f(p) = (\delta f)(p) = 0\}$; then $\overline{J(p)} = I_p$.*

Proof By 3.5.6, $J(p) \subseteq I_p$. Since I_p is a closed ideal of $\mathscr{D}(\delta)$, $\overline{J(p)} \subseteq I_p$. Conversely suppose that $f(p) = (\delta f)(p) = 0$. Given $\varepsilon > 0$, let

$$U = \{q \in K \,|\, |f(q) - f(p)| + |(\delta f)(q) - (\delta f)(p)| < \varepsilon\}$$

and let $h \in \mathscr{D}(\delta)$ with $h = 1$ in a neighborhood of p, $h(K \backslash U) = 0$ and $0 \leqslant h \leqslant 1$. Let $g \in C^{(1)}(\mathbb{R})$ with $g(t) = t$ near 0, $0 \leqslant g' \leqslant 1$ and $\|g\| \leqslant \varepsilon\ (1 + \|\delta(h)\|)^{-1}$. Define $f_\varepsilon = h(g(\mathrm{Re}(f)) + ig(\mathrm{Im}(f)))$. Then $f = f_\varepsilon$ in a neighborhood of p. Hence $f - f_\varepsilon \in J(p)$. Moreover,

$$\|f - (f - f_\varepsilon)\|_\delta = \|f_\varepsilon\|_\delta \leqslant \|f_\varepsilon\| + \|\delta(f_\varepsilon)\|.$$

Since $\|f_\varepsilon\| \leqslant 2\varepsilon$ and

$$\begin{aligned}\|\delta(f_\varepsilon)\| \leqslant \|\delta(h)(g(\mathrm{Re}(f)) + ig(\mathrm{Im}(f)))\| + \|h(g'(\mathrm{Re}(f))\\ + ig'(\mathrm{Im}(f)))\delta(f)\| \leqslant 4\varepsilon,\end{aligned}$$

$\|f - (f - f_\varepsilon)\|_\delta \leqslant 6\varepsilon$ and so $\overline{J(p)} = I_p$. □

Proof of Theorem 3.5.5 Let

$$F_p = \left\{ a \in \mathscr{D}(\delta) \,\middle|\, \begin{pmatrix} a(p) & \delta_1(a)(p) \\ 0 & a(p) \end{pmatrix} = 0 \right\};$$

then F_p is a closed ideal of $\mathscr{D}(\delta)$, for $\|a\|_{\delta_1} \leqslant k\|a\|_\delta$ for $a \in \mathscr{D}(\delta)$. $\mathscr{D}(\delta)/F_p$ is at most a two-dimensional algebra, a unit element together with an element of square 0. Hence F_p is a primary closed ideal contained in $M_p = \{a \in \mathscr{D}(\delta) \,|\, a(p) = 0\}$ and so $J(p) \subseteq F_p$. Hence $\overline{J(p)} = I_p \subseteq F_p$.

Consequently

$$\left\| \begin{pmatrix} a(p) & \delta_1(a)(p) \\ 0 & a(p) \end{pmatrix} \right\| \leqslant k_p \left\| \begin{pmatrix} a(p) & \delta(a)(p) \\ 0 & a(p) \end{pmatrix} \right\|$$

($a \in \mathscr{D}(\delta)$ and $p \in K$), where k_p is a positive number depending on p.

$$\left\| \begin{pmatrix} a(p) - a(p)1 & \delta_1(a)(p) \\ 0 & a(p) - a(p)1 \end{pmatrix} \right\| \leqslant k_p \left\| \begin{pmatrix} a(p) - a(p)1 & \delta(a)(p) \\ 0 & a(p) - a(p)1 \end{pmatrix} \right\|$$

and so $|\delta_1(a)(p)| \leqslant k_p |\delta(a)(p)|$ ($a \in \mathscr{D}(\delta)$). Hence there is a number $\xi(p)$ such that $\delta_1(a)(p) = \xi(p)\delta(a)(p)$ ($a \in \mathscr{D}(\delta)$, $p \in K$).

Now define $G = \{p \in K \,|\, \exists a \in \mathscr{D}(\delta)\colon \delta_1(a)(p) \neq 0\}$; then G is an open set in K. It is easily seen that ξ is continuous on G. Also, one can define $\xi(p) = 0$ for $p \notin G$. Now we shall show the closability of δ_1. Suppose that $a_n \to 0$ and $\delta_1(a_n) \to b$; then $\xi(p)\delta(a_n)(p) \to b(p)$, ($p \in K$). Hence $\delta(a_n)(p) \to (1/\xi(p))b(p)$ ($p \in G$).

Let $h \in \mathscr{D}(\delta)$ with $\mathrm{supp}(h) \subset G$; then $a_n h \to 0$ and

$$\delta(a_n h)(p) = \delta(a_n)(p)h(p) + a_n(p)\delta(h)(p) \to \frac{1}{\xi(p)} b(p)h(p) \qquad (p \in K),$$

where $(1/\xi(p))b(p)h(p)$ defines 0 for $p \notin G$. Since $(1/\xi(p))b(p)h(p)$ is continuous on K, the closability of δ implies $(1/\xi(p))b(p)h(p)=0$ for $p \in G$; hence $b(p)=0$ for $p \in G$. On the other hand, $\xi(p)\delta(a_n)(p)=0$ $(p \notin G)$ implies $b(p)=0$ $(p \notin G)$. Hence $b=0$. □

In order to study properties of a *-derivation δ in $C(K)$ it is convenient to convert the definition of well-behaved elements of $\mathscr{D}(\delta)^s$ into a corresponding notion for points of K. Thus p in K is said to be well-behaved if $\delta(f)(p)=0$ whenever $f \in \mathscr{D}(\delta)^s$ and $|f(p)|=\|f\|$. We shall denote the set of all well-behaved points of K by K_δ.

3.5.8 Proposition *Let f be a real-valued function in $C(K)$ and let $\alpha_1=\sup\{|f(p)| \mid p \in K_\delta{}^0\}$ and $\alpha_2=\sup\{|f(p)| \mid p \in K \backslash K_\delta\}$ (where the supremum of the empty set is taken to be $-\infty$). Then for any $\varepsilon>0$ and $\beta<\frac{1}{2}(\alpha_1+\varepsilon+\min(\varepsilon-\alpha_2, 0))$, there is a function $g \in \mathscr{D}(\delta)^s$ with $\|f-g\|<\varepsilon$ such that $W(\delta)$ contains the closed ball in $\mathscr{D}(\delta)^s$ with center g and radius β.*

Proof If $K_\delta{}^0$ is empty, the statement is trivial, for $\beta<0$. If $K \backslash K_\delta$ is empty, $W(\delta)=\mathscr{D}(\delta)^s$ – in fact for any $g \in \mathscr{D}(\delta)^s$, there is a point p in K such that $|g(p)|=\|g\|$; hence $\delta(g)(p)=0$ and so $g \in W(\delta)$. Therefore we may assume that α_1 and α_2 are finite. It suffices also to assume that $f \in \mathscr{D}(\delta)$ and $\alpha_1, \alpha_2>0$. Choose real numbers ε_j $(1 \leqslant j \leqslant 6)$ such that $0<\varepsilon_j<1$,

$$\alpha_1\varepsilon_1+(1+\varepsilon)\varepsilon_3+(\alpha_1+\alpha_2)\varepsilon_4+\varepsilon_5+\alpha_2\varepsilon_6<\alpha_1+\varepsilon+\min(\varepsilon-\alpha_2, 0)-2\beta,$$

$$\|f\|\varepsilon_4<\varepsilon_5<\varepsilon \qquad \text{and} \qquad \varepsilon_2+\varepsilon\varepsilon_3<\alpha_2\varepsilon_6<\varepsilon.$$

Put $\varepsilon'=\min(\varepsilon\alpha_2{}^{-1}, 1)-\varepsilon_6$. There exists p_0 in $K_\delta{}^0$ such that $|f(p_0)|>\alpha_1(1-\varepsilon_1)$. We may suppose that $f(p_0) \geqslant 0$ (otherwise consider $-f$). There exists an open set V in K containing $K \backslash K_\delta{}^0$, but which does not have p_0 as a limit point, and which satisfies $|f(p)|<\alpha_2+\varepsilon_2$ for all p in V. Then there are functions g_1 and g_2 in $C(K)$ with $0 \leqslant g_j \leqslant 1$, $g_1(p_0)=1$, $g_1=0$ on V, $g_2=1$ on $K \backslash K_\delta$ and $g_2=0$ on $K \backslash V$. Since $\mathscr{D}(\delta)$ is dense, there exist g_3 and g_4 in $\mathscr{D}(\delta)$ with $0 \leqslant g_3 \leqslant 1$, $\|g_3-g_1\|<\varepsilon_3$, $0 \leqslant g_4 \leqslant 1$ and $\|g_4-g_2\|<\varepsilon_4$. Let $g=f(1-\varepsilon' g_4)+(\varepsilon-\varepsilon_5)g_3$. Then $g \in \mathscr{D}(\delta)$ and for p in $K \backslash V$,

$$|(g-f)(p)|<\varepsilon'\|f\|\varepsilon_4+\varepsilon-\varepsilon_5 \leqslant \varepsilon-(\varepsilon_5-\|f\|\varepsilon_4)<\varepsilon$$

while for p in V,

$$|(g-f)(p)|<(\alpha_2+\varepsilon_2)\varepsilon'+(\varepsilon-\varepsilon_5)\varepsilon_3<\varepsilon-\alpha_2\varepsilon_6+\varepsilon_2+\varepsilon\varepsilon_3<\varepsilon.$$

Thus $\|g-f\|<\varepsilon$. Also, for $p \in K \backslash K_\delta$,

$$\begin{aligned}|g(p)|+\beta &< \alpha_2(1-\varepsilon'(1-\varepsilon_4))+(\varepsilon-\varepsilon_5)\varepsilon_3+\beta \\ &< \alpha_2\varepsilon_6-\min(\varepsilon-\alpha_2, 0)+\alpha_2\varepsilon_4+\varepsilon\varepsilon_3+\beta\end{aligned}$$

$$< \alpha_1(1-\varepsilon_1) - \alpha_1\varepsilon_4 + \varepsilon - \varepsilon_3 - \varepsilon_5 - \beta$$
$$< \alpha_1(1-\varepsilon_1)(1-\varepsilon'\varepsilon_4) + (\varepsilon - \varepsilon_5)(1-\varepsilon_3) - \beta$$
$$< g(p_0) - \beta.$$

Hence if $h \in \mathscr{D}(\delta)^s$ and $\|g-h\| < \beta$, then $|h(p_1)| = \|h\|$ for some p_1 in K_δ, so $\delta(h)(p_1) = 0$ and $h \in W(\delta)$. □

3.5.9 Proposition *Let δ be a *-derivation in $C(K)$, $f \in \mathscr{D}(\delta)^s$, ε a positive number and suppose that $W(\delta)$ contains the open ball in $\mathscr{D}(\delta)^s$ with center f and radius ε. Then K_δ contains all points p of K with $|f(p)| > \|f\| - 2\varepsilon$.*

Proof Replacing f by $-f$ if necessary, we may assume that $f(p) \geqslant 0$. Adjusting f by a small function in $\mathscr{D}(\delta)^s$ non-zero at p, we may assume that $f(p) > 0$. There is a real $g \in C^{(1)}(R)$ such that $g(0) = 0$, $g(f(p)) = \|g(f)\|$ and $\|g(f) - f\| < \varepsilon$. Then $g(f) \in W(\delta)^0$, so $\delta(g(f))(p) = 0$ (3.2.16). Now suppose that $h(p) = \|h\|$ for some h in $\mathscr{D}(\delta)^s$. Then $\|f - g(f) - \lambda h\| < \varepsilon$ for small $\lambda > 0$, so $g(f) + \lambda h \in W(\delta)^0$. $\{g(f) + \lambda h\}(p) = \|g(f) + \lambda g\|$, so $\delta(g(f) + \lambda h)(p) = 0$ (3.2.16) Hence $\delta(h)(p) = 0$ and so $p \in K_\delta$. □

3.5.10 Theorem *A *-derivation δ in $C(K)$ is quasi-well-behaved* (*resp. well-behaved*) *if and only if $K_\delta{}^0$ is dense in K* (resp. $K_\delta = K$).

Proof Suppose that δ is quasi-well-behaved, but $K_\delta{}^0$ is not dense in K. Then there is an element g in $W(\delta)^0$ with $\|g\| < 1 + \varepsilon$, $|g(p)| \leqslant \frac{1}{2}$ for $p \in K_\delta{}^0$ and $g(p_0) = 1$ for some p_0 in K ($\varepsilon > 0$). By 3.5.8, and 3.5.9, $p_0 \in K_\delta{}^0$, a contradiction. Next suppose that $K_\delta{}^0$ is dense in K and f is a real function in $C(K)$. Then in the notation of Proposition 3.5.8, $\alpha_1 = \|f\| \geqslant \alpha_2$, so that Proposition 3.5.8 shows that for $0 < \beta < \varepsilon < \|f\|$, there exists g in $\mathscr{D}(\delta)^s$ such that $\|g - f\| < \varepsilon$ and $W(\delta)$ contains the closed ball in $\mathscr{D}(\delta)^s$ with center g and radius β. Hence $W(\delta)^0$ is dense in $\mathscr{D}(\delta)^s$ – i.e. δ is quasi well-behaved. In the proof of Proposition 3.5.8, we show that $K_\delta = K$ implies that δ is well-behaved. Conversely suppose that δ is well-behaved.

If $K_\delta \subsetneq K$, then there is an element $f \in \mathscr{D}(\delta)^s$ and a point $p_0 \in K \backslash K_\delta$ such that $f(p_0) = \|f\|$ and $\delta(f)(p_0) \neq 0$. On the other hand $W(\delta)^0 = \mathscr{D}(\delta)^s$, so $\delta(f)(p_0) = 0$ (3.2.16), a contradiction. □

Now we shall consider derivations in $C([0,1])$ with the unit interval $[0,1]$. Let $\delta_0 = \mathrm{d}/\mathrm{d}t$ with $\mathscr{D}(\delta_0) = C^{(1)}([0,1])$. Then δ_0 is a closed *-derivation. If δ_1 is a derivation in $C([0,1])$ with $\mathscr{D}(\delta_1) = C^{(1)}([0,1])$; then by 3.5.5., there is a function ξ on $[0,1]$ such that $\delta_1(f) = \xi \cdot \delta_0(f)$ ($f \in C^{(1)}[0,1]$). Let $f_0(t) = t$ ($t \in [0,1]$); then $\delta_0(f_0) = 1$. Hence $\delta_1(f_0) = \xi$ and so $\xi \in C([0,1])$. Therefore for any derivations δ in $C([0,1])$ with $\mathscr{D}(\delta) = C^{(1)}([0,1])$, there is a unique element λ in $C([0,1])$ such that $\delta(f)=\lambda\delta_0(f)$ $(f\in C^{(1)}([0,1]))$. Conversely for any $\lambda\in C([0,1])$ $f\to\lambda\delta_0(f)$ $(f\in C^{(1)}([0,1]))$ is a closable derivation in $C([0,1])$.

Let ρ be a homeomorphism on $[0,1]$; then ρ will define a *-automorphism on $C([0,1])$ (denoted by the same notation ρ) by $\rho(f)(t)=f(\rho(t))$ $(f\in C([0,1])$ and $t\in[0,1]$. Then for any derivation δ in $C([0,1])$, $\rho\delta\rho^{-1}$ is a derivation in $C([0,1])$ with domain $\mathscr{D}(\rho\delta\rho^{-1})=\rho(\mathscr{D}(\delta))$.

3.5.11 Proposition *Let δ be a closed *-derivation in $C([0,1])$ and suppose that $\mathscr{D}(\delta)$ contains a self-adjoint element h such that the C*-subalgebra of $C([0,1])$ generated by h is $C([0,1])$. Then there is a *automorphism ρ on $C([0,1])$ and a $\lambda\in C([0,1])^s$ such that $\rho^{-1}(C^{(1)}([0,1]))\subset\mathscr{D}(\delta)$ and $\rho\delta\rho^{-1}f=\lambda(\mathrm{d}/\mathrm{d}t)f$ $(f\in C^{(1)}([0,1]))$.*

Proof Let

$$k=\frac{\|h\|1+h}{\|\,\|h\|1+h\|};$$

then $k(t)\neq k(s)$ if $t\neq s$. Let $k(t_0)=\inf_{t\in[0,1]}$ and let

$$\eta=\frac{k-k_0(t_0)}{\|k-k_0(t_0)1\|};$$

then the spectrum of $\eta=[0,1]$ and $t\to\eta(t)$ is a homeomorphism on $[0,1]$. Moreover $\eta\in\mathscr{D}(\delta)$ and $\delta(f(\eta))=\delta(\eta)f'(\eta)$ for $f\in C^{(1)}([0,1])$. Consider the mapping $f(\eta)\to f$ on $C([0,1])$; then it is a *-automorphism ρ of $C([0,1])$.

Moreover

$$\rho\delta(f(\eta))=\rho\delta\rho^{-1}f\qquad\text{for } f\in C^{(1)}([0,1]).$$

Hence there is a unique real-valued continuous function λ on $[0,1]$ such that

$$\rho\delta\rho^{-1}f=\lambda\frac{\mathrm{d}}{\mathrm{d}t}f\qquad\text{for } f\in C^{(1)}([0,1])$$ □

Next we shall show a characterization of quasi-well-behaved *-derivations in $C([0,1])$. For this, we shall provide several lemmas.

3.5.12 Lemma *Let δ be a closed *-derivation in $C(K)$, and let F_1, F_2 be disjoint closed subsets of K. Then there is an element h in $\mathscr{D}(\delta)$ such that $0\leqslant h\leqslant 1$, $h=0$ on F_1 and $h=1$ on F_2.*

Proof Since $\mathscr{D}(\delta)$ is dense in $C(K)$, there is an element f in $\mathscr{D}(\delta)$ with $f\leqslant 0$

on F_1 and $f \geqslant 1$ on F_2. Let g be a C^1-function on R with $0 \leqslant g \leqslant 1$, $g = 0$ on $(-\infty, 0]$, $g = 1$ on $[1, \infty)$, and put $h = g(f)$. □

3.5.13 Lemma *Let δ be a closed *-derivation in $C(K)$, t_0 be a point of K_δ, and f be a self-adjoint element in $\mathscr{D}(\delta)$ with a local maximum at t_0* (i.e. $f(t_0) \geqslant f(t)$ *near* t_0). *Then $\delta(f)(t_0) = 0$.*

Proof Adding a constant, we may assume that $f(t_0) > 0$. By 3.5.12, there is an element h in $\mathscr{D}(\delta)$ with $0 \leqslant h \leqslant 1$, $h = 1$ near t_0 and such that $0 \leqslant f(t) \leqslant f(t_0)$ whenever $h(t) > 0$. Then fh attains its norm at t_0, so by 3.5.6 $\delta(hf)(t_0) = \delta(f)(t_0) = 0$. □

3.5.14 Lemma *Let δ be a closed *-derivation in $C(K)$ and F be a closed subset of K, and suppose that δ vanishes near F* (i.e. *there is an open set U such that $F \subset U$ and $\delta(f)(t) = 0$* for *$t \in U$ and all $f \in \mathscr{D}(\delta)$*). *Then any element f in $C(F)$ has an extension in $\mathscr{D}(\delta)$.*

Proof Let f_1 be any extension of f in $C(K)$, g_n be a sequence in $\mathscr{D}(\delta)$ converging uniformly to f_1 and h be an element in $\mathscr{D}(\delta)$ vanishing in U^c with $h = 1$ on F. Then $g_n h$ converges uniformly to $f_1 h$ and $\delta(g_n h) = 0$, so $f_1 h \in \mathscr{D}(\delta)$ for δ is closed. Thus $f_1 h = f$ on F. □

3.5.15 Lemma *Let δ be a closed *-derivation in $C(K)$ and let U_1 and U_2 be open subsets of K whose union is K. Suppose that f_1 and f_2 are elements in $\mathscr{D}(\delta)$ such that $f_1 = f_2$ on $U_1 \cap U_2$, and let f be the function on K defined by $f(t) = f_j(t)$ $(t \in U_j)$. Then $f \in \mathscr{D}(\delta)$ and $\delta(f) = \delta(f_j)$ on U_j.*

Proof Let h be an element in $\mathscr{D}(\delta)^s$ with $h = 0$ on $K \backslash U_1$ and $h = 1$ on $K \backslash U_2$. Then $f = hf_1 + (1 - h)f_2 \in \mathscr{D}(\delta)$, and $\delta(f) = \delta(f_j)$ on U_j by 3.5.6. □

Let δ be a closed *-derivation in $C(K)$. The set of points t where δ vanishes (i.e. $\delta(f)(t) = 0$ for all $f \in \mathscr{D}(\delta)$) will be denoted by N_δ, its complement in K by U_δ. Now we shall consider the case of $K = [0, 1]$. Let $P_\delta{}^+$ denote the set of points t in U_δ such that any element f in $\mathscr{D}(\delta)^s$ with $\delta(f)(t) > 0$ is strictly increasing near t. Put $P_\delta{}^- = P_{-\delta}{}^+$.

3.5.16 Lemma *The union of the sets $(P_\delta{}^+)^0$ and $(P_\delta{}^-)^0$ contains $K_\delta{}^0 \cap U_\delta$. If* 0 (*or* 1) *belongs to U_δ and is not a limit point of $[0, 1] \backslash K_\delta$, then* 0 (*or* 1) *belongs to $(P_\delta{}^+)^0$ or $(P_\delta{}^-)^0$.*

Proof Let t_0 be a point of $K_\delta{}^0 \cap U_\delta$, and suppose that t_0 does not belong to $P_\delta{}^+ \cup P_\delta{}^-$. Then there are two functions f_1 and f_2 in $\mathscr{D}(\delta)$ with

$\delta(f_j)(t_0) > 0$ $(j = 1, 2)$ but such that f_1 is not strictly decreasing and f_2 is not strictly increasing in any neighborhood of t_0. Let J be a compact interval which is a neighborhood of t_0 in K_δ and is so small that $\delta(f_j) > 0$ $(j = 1, 2)$ on J. Then by 3.5.13 f_j cannot have a local minimum or maximum at any point t in J.

Therefore f_1 is strictly increasing and f_2 is strictly decreasing. But then there are real numbers $\lambda_1 > 0$ and $\lambda_2 > 0$ such that $\lambda_1 f_1 + \lambda_2 f_2$ has the same value at the endpoints of J, and therefore has local minimum or maximum at some point t_1 of J. Then $\delta(\lambda_1 f_1 + \lambda_2 f_2)(t_1) = 0$, this contradicts $\delta(f_1)(t_1) > 0$, $\delta(f_2)(t_1) > 0$. Hence $t_0 \in P_\delta^+ \cup P_\delta^-$.

Suppose that $t_0 \in P_\delta^+$. Then there is a self-adjoint element f in $\mathscr{D}(\delta)$ and a neighborhood U of t_0 contained in K_δ^0 such that $\delta(f) > 0$ on U and f is strictly increasing on U. Then $U \cap P_\delta^- = (\phi)$, so by the first part of the proof, $U \subset P_\delta^+$.

Finally suppose that $0 \in U_\delta$ and is not a limit point of $[0, 1] \backslash K_\delta$. By the first part of the proof, there is a real number $\varepsilon > 0$ such that $(0, \varepsilon)$ is contained in $(P_\delta^+)^0 \cup (P_\delta^-)^0$. By connectedness, we can assume that $(0, \varepsilon) \subset (P_\delta^+)^0$. If $\delta(f)(0) > 0$ for some $f \in \mathscr{D}(\delta)$, then for all sufficiently small $t > 0$, $\delta(f)(t) > 0$ so f is strictly increasing near t. Hence f is strictly increasing near 0; hence $0 \in P_\delta^+$ and so $0 \in (P_\delta^+)^0$. □

3.5.17 Lemma *Let δ be a closed *-derivation in $C([0, 1])$ and t_0 be a point in $(P_\delta^+)^0$. Then there is an element f in $\mathscr{D}(\delta)^s$ such that f is monotone increasing on $[0, 1]$, strictly monotone near t_0, $\delta(f) \geqslant 0$, $\delta(f)(t_0) > 0$ and $\delta(f)$ vanishes outside $(P_\delta^+)^0$.*

Proof Take f_1 in $\mathscr{D}(\delta)^s$ with $\delta(f_1)(t_0) > 0$. There is a neighborhood $[t_1, t_2]$ of t_0 in $[0, 1]$ contained in $(P_\delta^+)^0$ on which $\delta(f_1)$ is positive and f_1 is strictly increasing. Let $t'_j = \frac{1}{2}(t_0 + t_j)$ $(j = 1, 2)$, and g be a monotone increasing function in $C^1(R)$ such that $g(s) = 0$ $(f_1(t_1) \leqslant s \leqslant f_1(t'_1))$, $g(s) = 1$ $(f_1(t'_2) \leqslant s \leqslant f_1(t_2))$ $g'(f_1(t_0)) > 0$ and put $f_2 = g(f_1)$. Then f_2 is increasing on $[t'_1, t'_2]$, constantly 0 on $[t_1, t'_1]$ and constantly 1 on $[t_2, t'_2]$, and $f_2 \in \mathscr{D}(\delta)^s$ and $\delta(f_2)(t_0) > 0$. By 3.5.12, there are elements h_1 and h_2 in $\mathscr{D}(\delta)^s$ such that $h_1 = 0$ on $[0, t_1]$, $h_1 = 1$ on $[t'_1, 1]$, $h_2 = 0$ on $[0, t'_2]$ and $h_2 = 1$ on $(t_2, 1]$. Put $f = (1 - h_2)h_1 f_2 + h_2$. Then $f \in \mathscr{D}(\delta)^s$ and

$$f(t) = \begin{cases} 0 & (0 \leqslant t \leqslant t'_1) \\ f_2(t) & (t'_1 < t \leqslant t'_2) \\ 1 & (t'_2 < t \leqslant 1). \end{cases}$$

Thus f is monotone increasing on $[0, 1]$, and by 3.5.6, $\delta(f)$ vanishes outside $[t'_1, t'_2)$, where $\delta(f)$ coincides with $\delta(f_2)$. Hence $\delta(f)(t_0) > 0$ so f is strictly increasing near t_0. □

3.5.18 Lemma *Let δ be a closed *derivation in $C([0,1])$. Then there is an increasing function f in $\mathscr{D}(\delta)^s$ such that $\delta(f)>0$ on $(P_\delta{}^+)^0$, $\delta(f)=0$ outside $(P_\delta{}^+)^0$, and $f(t_1)<f(t_2)$ for any t_1 and t_2 in $(P_\delta{}^+)^0$ with $t_1<t_2$.*

Proof For each t in $(P_\delta{}^+)^0$, by 3.5.17 there is an increasing function f_t in $\mathscr{D}(\delta)^s$ and a neighborhood U_t of t contained in $(P_\delta{}^+)^0$ such that $\delta(f_t)\geqslant 0$, $\delta(f_t)=0$ outside $(P_\delta{}^+)^0$, $\delta(f_t)>0$ in U_t and f_t is strictly increasing on U_t. Since $(P_\delta{}^+)^0$ is a separable metric space, there are points $\{t_n\}$ $(n=1,2,\ldots)$ such that $(P_\delta{}^+)^0=\bigcup_n U_{t_n}$. Let $f_N=\sum_{n=1}^N 2^{-n}(\|f_{t_n}\|+\|\delta(f_{t_n})\|)^{-1} f_{t_n}$. The $\{f_N\}$ converges uniformly to a function f and $\delta(f_N)$ to a function g. Since δ is closed, $f\in\mathscr{D}(\delta)$ and $\delta(f)=g$. □

3.5.19 Lemma *Let δ be a closed *-derivation in $C([0,1])$. Then there is an increasing function f in $\mathscr{D}(\delta)$ such that $\delta(f)=0$ and $f(t_1)<f(t_2)$ for any $t_1,t_2\in N_\delta{}^0$ with $t_1<t_2$.*

Proof One can easily deduce from 3.5.12, 3.5.14 and 3.5.15 that for each $t\in N_\delta{}^0$, there is an open interval U_t containing t and contained in $N_\delta{}^0$ and an increasing f_t in $\mathscr{D}(\delta)^s$ which is strictly increasing on U_t and satisfies $\delta(f_t)=0$. The remainder of the proof is as in Lemma 3.5.18. □

Now we shall show the following theorem.

3.5.20 Theorem *Let δ be a closed quasi-well-behaved *-derivation in $C([0,1])$. Then there is a function λ in $C([0,1])^s$ and a homeomorphism ρ on $[0,1]$ such that δ is an extension of $\lambda\rho^{-1}(\mathrm{d}/\mathrm{d}t)\rho$, where $\rho(\mathscr{D}(\delta))\supset C^1([0,1])$.*

Proof By 3.5.11, it is enough to show that $\mathscr{D}(\delta)$ contains a strictly monotone increasing function. By 3.5.17 and 3.5.18, there are increasing functions f_1,f_2 and f_3 in $\mathscr{D}(\delta)$ such that f_1 is strictly increasing on $(P_\delta{}^+)^0$ and $\delta(f_1)(t)\geqslant 0$ with strict inequality if and only if $t\in(P_\delta{}^+)^0$; f_2 is strictly increasing on $(P_\delta{}^-)^0$ and $\delta(f_2)(t)\leqslant 0$ with strict inequality if and only if $t\in(P_\delta{}^-)^0$; f_3 is strictly increasing on $N_\delta{}^0$ and $\delta(f_3)(t)=0$.

Let $f=f_1+f_2+f_3$. Then f is strictly increasing on $(P_\delta{}^+)^0\cup(P_\delta{}^-)^0\cup N_\delta{}^0$, which is dense in $[0,1]$ by 3.5.16, so f is strictly increasing on $[0,1]$. □

3.5.21 Corollary *Let δ be a closed *-derivation in $C([0,1])$. Then δ is quasi-well behaved if and only if $\mathscr{D}(\delta)$ contains a strictly monotone increasing function.*

This is an immediate corollary of 3.5.11 and 3.5.20. By Theorem 3.5.20, the study on quasi-well-behaved *-derivation in $C([0,1])$ can be essentially reduced to the extension problem of *-derivations with the form of $\lambda(\mathrm{d}/\mathrm{d}t)$. First of all we shall discuss the extension problem of the *-derivation d/dt.

3.5.22 Lemma *Let δ be a closed *-derivation in $C(K)$. For any open subset $U \subseteq K$, define $\delta_{\bar{U}}(f|_{\bar{U}}) = \delta(f)|\bar{U}$ for $f \in \mathscr{D}(\delta)$; then $\delta_{\bar{U}}$ is a closable *-derivation in $C(\bar{U})$.*

Proof If $f \in \mathscr{D}(\delta)$ and $f|\bar{U} = 0$, then by 3.5.6 $\delta(f)|U = 0$. By continuity, $\delta(f)|\bar{U} = 0$. Let $\{f_n\}$ be a sequence in $\mathscr{D}(\delta)$ and g an element of $C(K)$ such that $f_n \to 0$ and $\delta(f_n) \to g$ uniformly in $\bar{U}$. Given $t \in U$, choose $e \in \mathscr{D}(\delta)$ such that $e = 1$ near x and $\operatorname{supp}(e) \subseteq U$. Then $f_n e \to 0$ and $\delta(f_n e) = f_n \delta(e) + e\delta(f_n) \to eg$ uniformly on K. Since δ is closed, $eg = 0$. Therefore $g(t) = 0$. Hence $g|\bar{U} = 0$.

□

3.5.23 Definition *A real-valued function $f \in C([0, 1])$ is said to be a generalized Cantor function (abbreviated GCF) if f is monotone on $[0, 1]$, but not strictly monotone on any subinterval of $[0, 1]$.*

Every real constant function is a GCF. The familiar Cantor middle third function is a GCF. One can show that every non-constant GCF resembles the usual Cantor function as follows: let f be a non-constant GCF, and let ϕ be the usual Cantor function. Then there exist

(1) a homeomorphism h of $[0, 1]$ onto Range (f);
(2) numbers α and β with $0 \leqslant \alpha < \beta \leqslant 1$, and
(3) a function g: $[0, 1] \mapsto [0, 1]$ satisfying
  (i) $g(t) = 0$ $(0 \leqslant t \leqslant \alpha)$,
  (ii) $g|_{[\alpha,\beta]}$ is a homeomorphism onto $[0, 1]$, and
  (iii) $g(t) = 1$ $(\beta \leqslant t \leqslant 1)$ such that $f = h \circ \phi \circ g$.

3.5.24 Proposition *Let δ be a closed *-derivation in $C([0, 1])$ such that δ extends d/dt. Then there is a GCF Φ such that $\operatorname{Ker}(\delta) =$ the C*-subalgebra of $C([0, 1])$ generated by Φ and 1, and $\mathscr{D}(\delta) = C^1([0, 1]) + \operatorname{Ker}(\delta)$.*

Proof Since δ is closed, $\operatorname{Ker}(\delta)$ is a C*-subalgebra of $C([0, 1])$, and therefore there is a compact Hausdorff space X and a continuous mapping ϕ of $[0, 1]$ onto X such that $\operatorname{Ker}(\delta) = \{f \circ \phi \mid f \in C(X)\}$.

Suppose that ϕ is one-to-one on an open subset U. Then take an open subset V such that $\bar{V} \subset U$; then $\operatorname{Ker}(\delta)|\bar{V} = C(\bar{V})$; hence $\delta_{\bar{V}} \equiv 0$, a contradiction. Now suppose that $s < t$ and $\phi(s) = \phi(t)$. Let $g \in \operatorname{Ker}(\delta)$ satisfy $g(s) = g(t) = 0$. For $f \in C^1([0, 1])$, define

$$h(u) = -f(u)g(u) + \int_0^u f'(v)g(v)\,\mathrm{d}v.$$

Then

$$\delta(h) = -f'g - f\delta(g) + f'g = 0;$$

hence $h \in \operatorname{Ker}(\delta)$. Therefore

$$0 = h(t) - h(s) = \int_s^t f'(v)g(v)\,\mathrm{d}v.$$

Since g is orthogonal to every continuous function on $[s, t]$, $g = 0$ on $[s, t]$. It follows that every function in $\operatorname{Ker}(\delta)$ is constant on $[s, t]$ and each set of constant $\operatorname{Ker}(\delta)$ is connected. Define $\Phi(t) = \inf\{s \mid \phi(s) = \phi(t), s \in [0, 1]\}$; then Φ is a GCF. If $\phi(s) \neq \phi(t)$ and $s < t$, then $\Phi(s) \neq \Phi(t)$; therefore the C*-subalgebra of $C([0, 1])$ generated by Φ and 1 is $\operatorname{Ker}(\delta)$. □

3.5.25 Proposition *Let Φ be a non-constant GCF and let $C(\Phi)$ be the C*-subalgebra of $C([0, 1])$ generated by Φ and 1. Define $\delta(f + g) = (\mathrm{d}/\mathrm{d}t)f$ ($f \in C^1([0, 1])$ and $g \in C(\Phi)$); then δ is a closed *-derivation which is a proper extension of* $\mathrm{d}/\mathrm{d}t$.

Proof Since $C(\Phi) \cap C^1([0, 1]) = \mathbb{C}1$ and $(\mathrm{d}/\mathrm{d}t)1 = 0$, δ is a well-defined linear mapping. To show that $C^1([0, 1]) + C(\Phi)$ is a *-subalgebra of $C([0, 1])$, it suffices to show that for $f \in C^1([0, 1])$ and $g \in C(\Phi)$. fg has a decomposition $fg = h + k$, where $h \in C^1([0, 1])$ and $k \in C(\Phi)$.

Define

$$h(t) = \int_0^t f'(u)g(u)\,\mathrm{d}u \qquad \text{for } t \in [0, 1]$$

and let $k = fg - h$. Clearly $h \in C^1([0, 1])$. If $s < t$ and $g(s) = g(t)$, then g is a constant on $[s, t]$.

$$\begin{aligned} k(t) - k(s) &= f(t)g(t) - f(s)g(s) - \int_s^t f'(u)g(u)\,\mathrm{d}u \\ &= f(t)g(t) - f(s)g(s) - g(s)\int_s^t f'(u)\,\mathrm{d}u \\ &= g(s)(f(t) - f(s)) - g(s)(f(t) - f(s)) = 0; \end{aligned}$$

hence $k \in C(\Phi)$.

$$\begin{aligned} \delta((f_1 + g_1)(f_2 + g_2)) &= \delta(f_1 f_2 + f_1 g_2 + f_2 g_1 + g_1 g_2) \\ &= \delta(f_1 f_2) + \delta(f_1 g_2) + \delta(f_2 g_1), \end{aligned}$$

where $f_1, f_2 \in C^{(1)}([0, 1])$ and $g_1, g_2 \in C(\Phi)$. Let

$$h_1(t) = \int_0^t f'_1(u)g_2(u)\,\mathrm{d}u, \qquad h_2(t) = \int_0^t f'_2(u)g_1(u)\,\mathrm{d}u;$$

then

$$\delta(f_1 g_2) = \delta\left(\int_0^t f'_1(u) g_2(u)\,\mathrm{d}u\right) = f'_1 g_2$$

and $\delta(f_2 g_1) = f'_2 g_1$. Hence

$$\delta((f_1 + g_1)(f_2 + g_2)) = \delta(f_1 f_2) + f'_1/g_2 + f'_2 g_1$$
$$= f'_1 f_2 + f_1 f'_2 + f'_1 g_2 + f'_2 g_1.$$

On the other hand,

$$\delta(f_1 + g_1)(f_2 + g_2) + (f_1 + g_1)\delta(f_2 + g_2) = f'_1(f_2 + g_2) + (f_1 + g_1) f'_2.$$

Therefore δ is a *-derivation. Finally we shall show the closedness of δ. If $f_n + g_n \to 0$ and $f'_n \to h$, then $\inf_{\lambda \in \mathbb{C}} \| f_n - \lambda 1 \| \to 0$. Let $\inf_{\lambda \in \mathbb{C}} \| f_n - \lambda 1 \| = \| f_n - \lambda_n 1 \|$; then $\delta(f_n - \lambda_n 1) = f'_n \to h$. Hence $h = 0$. □

Next we shall discuss the extension problem of $\lambda(\mathrm{d}/\mathrm{d}t)$. Let δ be a closed *-derivation in $C([0, 1])$ with $\mathscr{D}(\delta) \supset C^1([0, 1])$. Let f_0 be the identity function on $[0, 1]$ such that $f_0(t) = t$ and let $\lambda = \delta(f_0)$. By 3.5.5,

$$\delta(f) = \lambda(\mathrm{d}/\mathrm{d}t) f \quad (f \in C^1([0, 1])).$$

If δ is a closed *-derivation with $N_\delta = \phi$, then λ is never zero; $\lambda^{-1}\delta$ is a closed *-derivation with the same domain and kernel as δ, and $\lambda^{-1}\delta$ extends $\mathrm{d}/\mathrm{d}t$. Therefore by 3.5.23 and 3.5.24, we can see the structure of δ. Now let $\lambda \in C([0, 1])^s$ be arbitrary. Let $Z_\lambda = \lambda^{-1}(\{0\})$ and let $U_\lambda = [0, 1] \backslash Z_\lambda$. Let V be an open subset of U_λ and let L be the algebra of functions in $C([0, 1])$ which are locally in $\mathscr{D}(\delta)$ on V. For $t \in V$, choose $e \in \mathscr{D}(\delta)$ such that $e = 1$ near t, $\mathrm{supp}(e) \subset V$ and $0 \leqslant e \leqslant 1$. Then for any $f \in L$, $ef \in \mathscr{D}(\delta)$, and the number $\delta(ef)(t)$ depends only on f and t, not on the choice of e. Define $(\delta f)(t) = \delta(ef)(t)$. For $f \in L$, define

$$(\delta_0 f)(t) = \begin{cases} \lambda(t)(\delta f)(t) & (t \in V) \\ 0 & (t \in [0, 1] \backslash V). \end{cases}$$

Let $\mathscr{D}(\delta_0) = \{f \in L: \delta_0(f) \in C([0, 1])\}$. If $f \in \mathscr{D}(\delta)$ and $\mathrm{supp}(f) \subseteq V$, then $f \in \mathscr{D}(\delta_0)$ and $\delta_0(f) = \lambda \circ \delta(f)$; therefore $\mathscr{D}(\delta_0)$ at least separates points of V. One can easily see that δ_0 is closed. To express that δ_0 has been obtained in this way from δ, λ and V, we write $\delta_0 = \delta_0(\delta, \lambda, V)$. Consider the special case $V = U_\lambda$. In this case, $\mathscr{D}(\delta) \subseteq \mathscr{D}(\delta_0)$ and $\delta_0 f = \lambda \delta f$ $(f \in \mathscr{D}(\delta))$. Hence $\delta_0(\delta, \lambda, V)$ is the closed extension of $\lambda\delta$.

We can also define for $f \in L$

$$(\tilde{\delta} f)(t) = \begin{cases} \lambda(t)\delta f(t) & (t \in V) \\ 0 & (t \in Int([0, 1] \backslash V). \end{cases}$$

The natural domain $\mathscr{D}(\tilde{\delta})$ is the set of $f \in L$ such that $\tilde{\delta} f$ has a (necessarily unique) continuous extension to $[0, 1]$. $\tilde{\delta}$ is also a closed *-derivation in $C([0, 1])$ (with a not necessarily dense domain). To express that $\tilde{\delta}$ has been obtained from δ, λ and V in this way, we write $\tilde{\delta} = \tilde{\delta}(\delta, \lambda, V)$. Then we have the following theorem.

3.5.26 Theorem ([9]) *Let δ_1 be a closed *-derivation in $C([0,1])$ extending $\lambda(\mathrm{d}/\mathrm{d}t)$. Then, there is a unique closed *-derivation δ in $C([0,1])$ such that*

(1) *δ extends* $\mathrm{d}/\mathrm{d}t$;
(2) *$\tilde{\delta}(\delta, \lambda, U_\lambda)$ extends δ_1;*
(3) *δ is minimum among all closed *-derivations satisfying* (1) *and* (2);
(4) *δ_1 extends $\lambda \cdot \delta$.*

Finally we shall note that there are many closed non-quasi well-behaved *-derivations in $C([0,1])$, for the study of which we shall refer to papers in [112], [113], [114].

3.5.27 Notes and remarks In 1977, the author [170] initiated the study of unbounded *-derivations in commutative C*-algebras and raised many problems on them. Those problems have been extensively studied by Batty, Goodman, Kurose, Takenouchi, Watatani and others. Since these studies, Batty, Bratteli, Elliott, Goodman, Jorgensen, Kurose, Robinson, Tomiyama and others have widened the study in various directions and have obtained many interesting results.

Now the subject has become one of the main branches in derivation theory. The reader will find these works in [10], [11], [12], [19], [20], [21], [22], [24], [26], [27], [112], [113], [114], [116], [119], [136], [153], [154], [155], [186], [187] etc.

Theorems 3.5.10 and 3.5.20 are due to Batty [8], [9]. Propositions 3.5.24 and 3.5.25 are due to Goodman [63]. The first examples of non-quasi well-behaved closed *-derivations were constructed by Kurose [112].

3.6 Transformation groups and unbounded derivations

Let Ω be a Hausdorff space and G a topological group each element of which is a homeomorphism of Ω onto itself:

(1) $$f(g{:}x) = g(x) = x' \in \Omega \qquad \text{for } g \in G,\ x \in \Omega.$$

The pair (G, Ω) is called a transformation group if for every pair g_1, g_2 of elements and every $x \in \Omega$.

(2) $$g_1(g_2(x)) = (g_1 g_2)(x)$$

and if $x' = g(x) = f(g{:}x)$ is continuous simultaneously in $x \in \Omega$ and $g \in G$.

If the identity e in G is the only element in G which leaves all of Ω fixed, then G is called effective.

Let E_n (n, a positive integer) denote a euclidean n-space, with real coordinates $x_1, x_2, \ldots, x_n$. The term 'locally euclidean' is used to describe a topological space E of fixed dimension n each point of which has a neighborhood that is homeomorphic to an open subset in E_n. If a locally euclidean space is connected it is called a manifold. A manifold is said to be a differential manifold and to have a differential structure of class C^r ($r \geqslant 1$) if there is a covering family of coordinate neighborhoods given in such a way that where any two of the neighborhoods overlap the coordinate transformation in both directions is given by n functions with continuous partial derivatives of order r. In the same way a manifold is said to be a (real) analytic manifold and to have a (real) analytic structure if there is a covering family of coordinate neighborhoods given in such a way that where any two overlap the coordinate transformation in both directions is given by n functions which are real analytic. The definition of a complex analytic manifold and structure is similar to the above.

In transformation group theory, the following question is important.

Question *If a locally compact group acts effectively on a manifold Ω, then is G necessarily a Lie group*?

If Ω is a differential manifold, and for each g, $f(g; x)$ is differentiable, then the answer is yes (cf. [128]). For the general case, it is an open question. The application of the unbounded derivation theory of commutative C*-algebra to this outstanding question in transformation group theory was proposed by the author in lectures given at the Kingston conference in 1980 (cf. [175]). In this section we shall discuss related matters.

Now we shall define transformation groups as dynamical systems in commutative C*-algebras. Let Ω be a locally compact Hausdorff space and let $C_0(\Omega)$ be the C*-algebra of all complex-valued continuous functions on Ω vanishing at infinity. Let $\mathrm{Aut}(C_0(\Omega))$ be the group of all *-automorphisms on $C_0(\Omega)$ and let $\mathrm{Hom}(\Omega)$ be the group of all homeomorphisms of Ω onto itself. For $\rho \in \mathrm{Aut}(C_0(\Omega))$, there is a unique homeomorphism ξ of Ω onto itself such that $\rho(h)(x) = h(\xi(x))$ for $x \in \Omega$ and $h \in C_0(\Omega)$. Conversely for $\xi \in \mathrm{Hom}(\Omega)$, define $\rho(h)(x) = h(\xi(x))$ for $x \in \Omega$ and $h \in C_0(\Omega)$; then ρ is a *-automorphism on $C_0(\Omega)$. Put $\phi(\rho) = \xi$; then ϕ is a one-to-one mapping of $\mathrm{Aut}(C_0(\Omega))$ onto $\mathrm{Hom}(\Omega)$.

Moreover

$$(\rho_1\rho_2)(h)(x) = \rho_1(\rho_2(h))(x) = \rho_2(h)(\phi(\rho_1)(x)) = h(\phi(\rho_2)\phi(\rho_1)(x));$$

hence, $\phi(\rho_1\rho_2) = \phi(\rho_2)\phi(\rho_1)$. Therefore the mapping $\rho \to \phi(\rho^{-1})$ of

$\mathrm{Aut}(C_0(\Omega))$ onto $\mathrm{Hom}(\Omega)$ is a group isomorphism. Now let (G,Ω) be an effective transformation group on a locally compact space Ω. For simplicity, we shall assume that G is a locally compact group satisfying the first countability axiom. Suppose that $g_n \to e$ $(n\to\infty)$; then $g_n(x)\to x$ for $x\in\Omega$ and so $h(g_n(x))\to h(x)$ for $h\in C_0(\Omega)$ and $x\in\Omega$. From this one can easily conclude that $\rho_{g_n}(h)\to h$ in $\sigma(C_0(\Omega), C_0(\Omega)^*)$, where $\phi(\rho_{g_n})=g_n$ and $C_0(\Omega)^*$ is the dual Banach space of $C_0(\Omega)$. Therefore the dynamical system $\{C_0(\Omega), G, \alpha\}$ defined by $\alpha_g{}^{-1}(h)=\phi^{-1}(g)$ is a $\sigma(C_0(\Omega), C_0(\Omega)^*)$ – continuous dynamical system, so that by the well-known theorem of the representation theory, $\{C_0(\Omega), G, \alpha\}$ is a (strongly continuous) C*-dynamical system. Conversely let $\{C_0(\Omega), G, \alpha\}$ be any C*-dynamical system with a locally compact group G. Then the mapping $(g,x)\to\phi(\alpha_g{}^{-1})(x)$ is simultaneously continuous. Therefore the question stated above is equivalent to the question, when G satisfies the first countability axiom, as follows. Let $\{C_0(\Omega), G, \alpha\}$ be a faithful C*-dynamical system (i.e. $\alpha_g=1$ implies $g=e$) with a connected locally euclidean space Ω; then can we conclude that G is a Lie group?

Now we shall assume that G is a connected locally compact metric group. Then by the structure theorem of Iwasawa ([128]), G has maximal compact subgroups, and all such subgroups are connected and are mutually conjugate (i.e. only one conjugate class). Now let K denote one of them. Then G contains subgroups $H_1, H_2, \ldots, H_r$ all isomorphic to the vector group V_1 such that every element $g\in G$ can be decomposed uniquely and continuously in the form $g=h_1h_2\cdots h_rk$, $h_i\in H_i$, $k\in K$. In particular the space of G is the product of the compact space of K and that $H_1\times H_2\times\cdots\times H_r$ which is homeomorphic to the r-dimensional euclidean space E_r.

Therefore if K is locally euclidean, then G is a Lie group by the Gleason and Montgomery-Zippin Theorem ([128]). Therefore it is sufficient to study the case of a connected compact group if G is connected. It is known that if a compact metric group acts effectively on a connected locally euclidean space and if all the orbits are locally connected, then G is a Lie group ([128]). On the other hand, if G is a compact connected separable metric abelian group, then there is a one-parameter group T which is dense in G (cf. p. 254, Lemma in [128]). Therefore it becomes increasingly important to study the C*-dynamical system $\{C_0(\Omega), \mathbb{R}, \alpha\}$, where Ω is a connected locally euclidean space. In the previous section, we showed that for any C*-dynamical system $\{C[0,1], \mathbb{R}, \alpha\}$, there is a homeomorphism ρ of $[0,1]$ such that $\delta_{\rho\alpha\rho^{-1}}$ extends λD for some $\lambda\in C[0,1]$, where D denotes differentiation defined on $[0,1]$. It is easily seen that similar remarks apply to dynamical systems on $C_0(\mathbb{R})$, where D is the generator for the flow of translations, and λ is a function on $\mathbb{R}$. In fact, we can, by choosing ρ appropriately, arrange that $\rho T\rho^{-1}$ is one of the flow $T_U{}^{\varepsilon}$ described in the following examples, where T is the flow corresponding to α (cf. page 26; [179]).

3.6.1 Examples For each open interval I in R, define a flow T_I on I as follows:

$$T_{(a,b)}(x,t) = \frac{b(x-a)\exp(b-a)t + a(b-x)}{b-x+(x-a)\exp(b-a)t},$$

$$T_{(a,\infty)}(x,t) = a + (x-a)\exp t,$$

$$T_{(-\infty,b)}(x,t) = b + (x-b)\exp(-t)$$

$$T_R(x,t) = x+t.$$

Now let U be an open subset of $\mathbb{R}$, C_U be the set of all connected components of U, and ε be a function of C_U into $\{-1,1\}$. Define

$$T_U^{\varepsilon}(x,t) = \begin{cases} T_I(x,\varepsilon(I)t) & (x \in I \in C_U), \\ x & (x \in \mathbb{R} \setminus U). \end{cases}$$

Then T_U^{ε} is a flow on $\mathbb{R}$, and its generator is the closure of $\lambda_U^{\varepsilon} D \mid C_0^{\infty}(\mathbb{R})$, where

$$\lambda_U^{\varepsilon}(x) = \begin{cases} \varepsilon((a,b))(x-a)(b-x) & (x \in (a,b) \in C_U), \\ \varepsilon((a,\infty))(x-a) & (x \in (a,\infty) \in C_U), \\ \varepsilon((-\infty,b))(b-x) & (x \in (-\infty,b) \in C_U), \\ \varepsilon(R) & (\text{if } U = \mathbb{R}), \\ 0 & (x \in \mathbb{R} \setminus U). \end{cases}$$

This fact may suggest the possibility to choose a nicely behaved homeomorphism ρ on general spaces.

Now let $\{C_0(\Omega), \mathbb{R}, \alpha\}$ be a C*-dynamical system with a connected locally euclidean space Ω such that the closure of $\alpha(\mathbb{R})$ in $\mathrm{Aut}(C_0(\Omega))$ with respect to the strong operator topology is compact. Let K = the closure of $\alpha(\mathbb{R})$; then K is a compact group. Let $\alpha_t = \exp(t\delta)$; then δ is a well-behaved closed *-derivation in $C_0(\Omega)$. For $p \in \Omega$, let $U(p)$ be a neighborhood of p such that $U(p)$ is homeomorphic to an open n-cell and its closure is homeomorphic to the closed n-cell. Then the restriction $\delta \mid \overline{U(p)}$ of δ is a quasi-well-behaved *-derivation in $C(\overline{U(p)})$.

Let δ_p be the closure of $\delta \mid \overline{U(p)}$; then it is a quasi-well-behaved *-derivation. Let $\mathscr{D}(\delta_p)$ be the domain of δ_p.

3.6.2 Problem *Can we take n self-adjoint elements in* $\mathscr{D}(\delta_p)$ *such that* $C(\overline{U(p)}) = C(h_1) \otimes C(h_2) \otimes \cdots \otimes C(h_n)$, *where* $C(h_j)$ $(j = 1, 2, \ldots, m)$ *is a C*-subalgebra of* $C(\overline{U(p)})$ *generated by* h_j *and* 1, *and* $\otimes$ *is the C*-tensor product of* $\{C(h_j) \mid j = 1, 2, \ldots, n\}$?

Suppose that this problem is affirmatively solved. Then by replacing h_j by

$$h_j - \left(\inf_{q \in \overline{U(p)}} h_j(q)\right)1 \Big/ \left\| h_j - \left(\inf_{q \in \overline{U(p)}} h_j(q)\right)1 \right\|,$$

we may assume that $0 \leqslant h_j \leqslant 1$. Let $\overline{U(p)} = \{x = (x_1, x_2, \ldots, x_n) \in E^n | 0 \leqslant x_j \leqslant 1$ $(j = 1, 2, \ldots, n)\}$ and $f_j(x) = x_j$ $(j = 1, 2, \ldots, n)$. Then a mapping $h_j \to f_j$ can be uniquely extended to a *-automorphism ρ of $C(\overline{U(p)})$, so that there is a homeomorphism θ on $\overline{U(p)}$ such that $\rho(g)(x) = g(\theta(x))$ for $g \in C(\overline{U(p)})$. Then $\rho\delta_p\rho^{-1} f_j = \rho\delta_p(h_j)$ $(j = 1, 2, \ldots, n)$. Since $\rho\delta_p\rho^{-1}$ is a closed *-derivation, $\mathscr{D}(\rho\delta_p p^{-1}) \supset \bigcap_{j=1}^n \mathscr{D}(\partial/\partial x_j)$.

By a discussion similar to that in the previous section, we can show that

$$\rho\delta_p\rho^{-1} = \sum_{j=1}^n \lambda_j \frac{\partial}{\partial x_j} \qquad \text{on } \bigcap_{j=1}^n \mathscr{D}\left(\frac{\partial}{\partial x_j}\right).$$

Since

$$\rho\delta\rho^{-1}(f_j) = \rho(\delta_p(h_j)) = \sum_{i=1}^n \lambda_i \frac{\partial f_j}{\partial x_i} = \lambda_j,$$

we conclude that

$$\lambda_j \in C(\overline{U(p)}) \qquad (j = 1, 2, \ldots, n).$$

Therefore it becomes important to study *-derivations with the form $\sum_{j=1}^n \lambda_j(\partial/\partial x_j)$.

3.6.3 Lemma *Suppose that the solution to Problem 3.6.2 is affirmative. Then if $x \in U(p)$ and $\hat{\alpha}_t\theta(x) = \theta(x)$ for $t \in \mathbb{R}$, then $\lambda_j(x) = 0$ $(j = 1, 2, \ldots, n)$, where $\hat{\alpha}_t$ is the flow defined by $\alpha_t = \exp(t\delta)(t \in \mathbb{R})$.*

Proof For $h \in \mathscr{D}(\delta)$ and $y \in U(p)$,

$$\frac{\mathrm{d}}{\mathrm{d}t} h(\hat{\alpha}_t\theta(y)) = (\delta(h))(\hat{\alpha}_t\theta(y)).$$

Take a sequence $\{h_{m,j}\}$ such that $h_{m,j} \in \mathscr{D}(\delta)$ and $\| h_{m,j} | \overline{U(p)} - h_j \| \to 0$, $\| \delta(h_{m,j}) | \overline{U(p)} - \delta_p(h_j) \| \to 0$; then for a sufficiently small $|t|$ with $\hat{\alpha}_t\theta(y) \in U(p)$,

$$\delta(h_{m,j})(\hat{\alpha}_t\theta(y)) \to \delta_p(h_j)(\hat{\alpha}_t\theta(y)).$$

Hence,

$$\begin{aligned}\int_0^t \delta(h_{m,j})(\hat{\alpha}_p\theta(y))\mathrm{d}s &= \int_0^t \frac{\mathrm{d}}{\mathrm{d}s} h_{m,j}(\hat{\alpha}_s\theta(y))\mathrm{d}s \\ &= h_{m,j}(\hat{\alpha}_t\theta(y)) - h_{m,j}(\theta(y)) \to h_j(\hat{\alpha}_t\theta(y)) - h_j(\theta(y)).\end{aligned}$$

On the other hand,

$$\int_0^t \delta(h_{m,j})(\hat{\alpha}_s\theta(y))\mathrm{d}s \to \int_0^t \delta_p(h_j)(\hat{\alpha}_s\theta(y))\mathrm{d}s.$$

Hence,

$$\int_0^t \delta_p(h_j)(\hat{\alpha}_s\theta(y))\mathrm{d}s = h_j(\hat{\alpha}_t\theta(y)) - h_j(\theta(y)),$$

and so

$$\frac{\mathrm{d}}{\mathrm{d}t} h_j(\hat{\alpha}_t\theta(y)) = \delta_p(h_j)(\hat{\alpha}_t\theta(y)).$$

If $\hat{\alpha}_t\theta(x) = \theta(x)(t \in R)$, then $\delta_p(h_j)(\hat{\alpha}_t\theta(y)) = 0$.

$$\begin{aligned}\delta_p(h_j)(\hat{\alpha}_t\theta(x)) &= \rho\delta_p(h_j)(\theta^{-1}\hat{\alpha}_t\theta(x)) = \rho\delta_p(h_j)(x)\\ &= \rho\delta_p\rho^{-1}f_j(x) = \lambda_j(x);\end{aligned}$$

hence $\lambda_j(x) = 0$ $(j = 1, 2, \ldots, n)$. □

3.6.4 Lemma *Let* $F = \{x \in U(p) \mid \lambda_j(x) = 0, j = 1, 2, \ldots, n\}$ *and let* F^0 *be the interior of* F. *If* $x \in F^0$, *then* $\hat{\alpha}_t\theta(x) = \theta(x)(t \in \mathbb{R})$.

Proof There is a positive number $\varepsilon > 0$ and a neighborhood $V(\theta(x))$ of $\theta(x)$ such that $\hat{\alpha}_t\theta(x) \subset V(\theta(x)) \subset U(p)$ for t with $|t| < \varepsilon$. Then

$$\int_0^t \delta_p(h_j)(\hat{\alpha}_s\theta(x))\mathrm{d}s = h_j(\hat{\alpha}_t\theta(x)) - h_j(\theta(x)) \qquad \text{for } t \text{ with } |t| < \varepsilon.$$

Since

$$\begin{aligned}\delta_p(h_j)(\hat{\alpha}_s\theta(x)) &= \rho\delta_p\rho^{-1}f_j(\theta^{-1}\hat{\alpha}_s\theta(x))\\ &= \sum_{i=1}^n \lambda_i(\theta^{-1}\hat{\alpha}_s\theta(x))\frac{\partial f_j}{\partial x_i}(\theta^{-1}\hat{\alpha}_s\theta(x))\\ &= \lambda_j(\theta^{-1}\hat{\alpha}_s\theta(x)).\end{aligned}$$

without loss of generality, we may assume that $\theta^{-1}\hat{\alpha}_s\theta(x) \in \theta^{-1}V(\theta(x)) \subset F^0$. Hence

$$\lambda_j(\theta^{-1}\hat{\alpha}_s\theta(x)) = 0 \qquad \text{for } s \text{ with } |s| < \varepsilon,$$

and so

$$h_j(\hat{\alpha}_t\theta(x)) - h_j(\theta(x)) = 0 \qquad \text{for all } j.$$

Since $\{h_1, h_2, \ldots, h_n, 1\}$ generates $C(\overline{U(p)})$, $\hat{\alpha}_t\theta(x)) = \theta(x)$ for t with $|t| < \varepsilon$; hence $\hat{\alpha}_t\theta(x) = \theta(x)$ for all $t \in \mathbb{R}$. □

3.6.5 Lemma *Let $\mathscr{D}_0$ be a *-subalgebra of $\mathscr{D}(\delta_p)$ generated by $h_1, h_2, \ldots, h_n$ and* 1. *If $\mathscr{D}_0$ is dense in $\mathscr{D}(\delta_p)$ under the norm $\|\cdot\|_{\delta_p}$, and if $\lambda_j(x) = 0$ for some $x \in U(p)$ $(j = 1, 2, \ldots, n)$, then $\hat{\alpha}_t\theta(x) = \theta(x)(t \in \mathbb{R})$.*

Proof

$$\frac{d}{dt} h_j(\hat{\alpha}_t\theta(x))|_{t=0} = \lambda_j(\theta^{-1}\hat{\alpha}_t\theta(x))|_{t=0};$$

hence

$$\frac{d}{dt} h_j(\hat{\alpha}_t\theta(x))|_{t=0} = \lambda_j(x) = 0.$$

Since $\mathscr{D}_0$ is dense in $\mathscr{D}(\delta_p)$ with the norm $\|\cdot\|_{\delta_p}$,

$$\frac{d}{dt} g(\hat{\alpha}_t\theta(x))|_{t=0} = 0 \qquad \text{for } g \in \mathscr{D}(\delta)$$

and so $(\delta g)(\theta(x)) = 0$ for $g \in \mathscr{D}(\delta)$. Hence for $h \in A_2(\delta)$,

$$(\delta^n h)(\theta(x)) = 0 \qquad (n = 1, 2, \ldots)$$

and so

$$\exp(t\delta)h(\theta(x)) = \alpha_t(h)(\theta(x)) = h(\hat{\alpha}_t\theta(x)) = h(\theta(x)).$$

Since $A_2(\delta)$ is dense in $C_0(\Omega)$,

$$\hat{\alpha}_t\theta(x) = \theta(x) \qquad \text{for } t \in \mathbb{R}. \qquad \square$$

While in this section, we have not given any definite result for transformation groups. However the reader may understand the importance of studying Problem 3.6.2 and, more generally, derivations with the form $\sum_{j=1}^{n} \lambda_j(\partial/\partial x_j)$.

References [12], [128], [175].

4

C*-dynamical systems

4.0 Introduction

In this chapter, we shall discuss a unified axiomatic treatment of quantum lattice systems and quasi-free dynamics in Fermion field theory within the framework of C*-dynamical systems. In these systems, time evolution, equilibrium states (KMS states), ground states, stability under bounded perturbations and phase transitions are important physical notions. Here we shall present an abstract treatment of these notions within the theory of C*-dynamical systems. About time evolution, we shall emphasize the approximate innerness property for the corresponding time automorphism group in a given C*-dynamical system. In fact, the approximate innerness assures the existence of a ground state, and the existence of a KMS state at each inverse temperature under the assumption of 'the existence of a tracial state', which is always satisfied in quantum lattice systems and in canonical anticommutation relation algebras.

We shall also discuss in detail one of the most important open problems (the Powers–Sakai conjecture) in C*-dynamical systems. In each section we shall explain relations between C*-dynamical systems and physical systems.

For simplicity we shall assume that every C*-algebra has an identity (unless otherwise stated).

4.1 Approximately inner C*-dynamics

A quantum lattice system consists of a set of particles confined to a lattice and interacting at distance. There are two physical interpretations of these models. One is a lattice gas and the other is a spin system. The lattice gas views each point of the lattice as a possible site for a finite number of N-particles, i.e. each point of the configuration space can be empty or occupied by $1, 2, \ldots, N$-particles. These particles interact with each other, and this leads to time evolution in which we envisages the particles jumping from lattice site to lattice site. The spin system assumes that every lattice site is permanently occupied by a particle but the particles have various internal degrees of freedom, i.e. the particles could have an intrinsic spin with several possible orientations. The interaction between particles follows from a coupling of the internal degrees of freedom and it results in an evolution in which the spin orientations are constantly changing. In spite of these two different interpretations, they can be treated mathematically as one object. For many problems, it suffices to assume that the lattice L is a countable set of points. For each point $x \in L$, a finite-dimensional full matrix algebra $\mathscr{A}_x$ is given as the totality of physical observables associated with the particles at the point x of L. The C*-infinite tensor product $\mathscr{A} = \bigotimes_{x\in L}\mathscr{A}_x$ then corresponds to the observables of the entire quantum lattice system. Let Λ be a non-empty finite subset of L and let $\mathscr{A}_\Lambda = \bigotimes_{x\in\Lambda}\mathscr{A}_x$, i.e. $\mathscr{A}_\Lambda$ represents the physical observables associated with the particles at the points of Λ. Then $\mathscr{A}_\Lambda$ is canonically embedded into $\mathscr{A}$ such that $\mathscr{A}_\Lambda = \bigotimes_{x\in\Lambda}\mathscr{A}_x \otimes \bigotimes_{y\in L\setminus\Lambda} 1_y$, where 1_y is the identity of $\mathscr{A}_y$.

The C*-algebra $\mathscr{A}$ defines the kinematics of the quantum lattice system, i.e. the instantaneous observables. An interaction Φ is defined as a function from non-empty finite subsets X of L into self-adjoint elements $\Phi(X)$ of $\mathscr{A}$ with $\Phi(X)\in\mathscr{A}_X$. Each $\Phi(X)$ represents the energy of interactions of the set of all particles in the finite subset X. Define $H = \sum_{X\subset L}\Phi(X)$, where X varies over all non-empty finite subsets of L; then mathematically H has no real meaning, because it diverges in most cases. However it may still be considered physically to be the total energy corresponding to the given quantum lattice system. Let $\mathscr{D}(\delta) = \bigcup_{\Lambda\subset L}\mathscr{A}_\Lambda$, where Λ moves on all non-empty finite subsets of L.

For $A\in\mathscr{A}_\Lambda$, define $\delta(A) = \mathrm{i}\sum_{X\subset L}[\Phi(X), A]$; then if $X\cap\Lambda = \phi$, $[\Phi(X), A] = 0$ and so $\delta(A) = \mathrm{i}\sum_{X\cap\Lambda\neq\phi}[\Phi(X), A]$. Moreover if $\Lambda_1\subset\Lambda$ and $A\in\mathscr{A}_{\Lambda_1}$, then

$$\mathrm{i}\sum_{X\cap\Lambda\neq\phi}[\Phi(X), A] = \mathrm{i}\sum_{X\cap\Lambda_1\neq\phi}[\Phi(X), A].$$

Therefore if $\sum_{X\cap\Lambda\neq\phi}\Phi(X)$ converges uniformly in $\mathscr{A}$ for each finite subset Λ, then δ will define a *-derivation in $\mathscr{A}$ with its domain $\mathscr{D}(\delta)$. This is the case for almost all interesting models. An interaction is said to be of finite range if there is no interaction between distant particles. We can make this definition precise only for $L = \mathbb{Z}^n$, where $\mathbb{Z}$ is the group of all integers. The interaction Φ is then defined to have finite range if there exists a finite subset Λ_Φ such that $\Phi(X) = 0$ whenever $X - X \not\subset \Lambda_\Phi$.

If $C(\subset\mathscr{A})$ is a commutative C*-subalgebra describing a classical lattice system and $\Phi(X)\in C_X\subset\mathscr{A}_X$ for all non-empty finite subsets of L, then Φ is called a classical interaction, where C_X is a commutative C*-subalgebra of $\mathscr{A}_X$ and $C_X \subset C$.

The above considerations of quantum lattice systems can be generalized as follows. Given a C*-algebra $\mathscr{A}$, and a *-derivation δ in $\mathscr{A}$ satisfying the following conditions:

(1) There is an increasing sequence $\{\mathscr{A}_n\}$ of full matrix algebras such that $1 \in \mathscr{A}_1 \subset \mathscr{A}_2 \subset \cdots \subset \mathscr{A}_n \subset \cdots$ and $\mathscr{A}$ is the uniform closure of $\bigcup_{n=1}^{\infty} \mathscr{A}_n$.
(2) The domain $\mathscr{D}(\delta)$ of δ is $\bigcup_{n=1}^{\infty} \mathscr{A}_n$.

Namely, $\mathscr{A}$ is a UHF algebra and δ is a normal *-derivation in $\mathscr{A}$ (cf. §4.5). If $\delta(\mathscr{D}(\delta)) \subset \mathscr{D}(\delta)$, then δ is said to be of *finite range.* We can define commutative normal *-derivations as a generalization of classical lattice systems (cf. §4.6) and we shall discuss them in later sections.

In almost all interesting cases of quantum lattice systems, the closure $\bar{\delta}$ of δ is an infinitesimal generator. In fact, in a quantum lattice system with $L = \mathbb{Z}^n$, the interaction function Φ is said to be translation-invariant if $\Phi(X + z) = \Phi(X)$ for each finite subset $X \subset \mathbb{Z}^n$ and each $z \in \mathbb{Z}^n$. If a quantum lattice system has a translation-invariant, finite range interaction, then $A(\delta) = \mathscr{D}(\delta)$ and so δ is a pre-generator (cf. [42]). In particular, Examples (4) and (5) in §3.1 are pre-generators, even in quantum lattice systems with infinite range interaction, and there are many important models in which *-derivations δ are pre-generators (cf. [42]).

If the *-derivation δ associated with a quantum lattice system is a pre-generator, then $\{\exp(t\bar{\delta}) \mid t \in \mathbb{R}\}$ is the time evolution corresponding to the system. Then one can conclude that the time automorphism group $\{\exp(t\bar{\delta}) \mid t \in \mathbb{R}\}$ is approximately inner (cf. §4.6). In addition in the quasi-free dynamics of Fermion field theory, one can easily see that the time automorphism group is approximately inner. Therefore it is important to study approximately inner C*-dynamics.

Let A be a C*-algebra and $t \to \alpha_t$ be a strongly continuous one-parameter group of *-automorphisms on A. The system $\{A, \alpha\}$ is then said to be a C*-dynamics. Let $\alpha_t = \exp(t\delta)$; then δ is a well-behaved closed *-derivation.

In mathematical physics, it is important to study the strong convergence of the one-parameter groups of *-automorphisms.

4.1.1 Definition *Let $\alpha_{n,t}: t \mapsto \alpha_{n,t}$ $(n = 1, 2, \ldots)$ and $t \mapsto \alpha_t$ be a family of strongly continuous one-parameter groups of *-automorphisms on a C*-algebra A. α is said to be a strong limit of $\{\alpha_n\}$ (denoted by $\alpha = \text{strong}\lim_n \alpha_n$ or $\alpha_t = \text{strong}\lim \alpha_{n,t}$) if $\|\alpha_{n,t}(a) - \alpha_t(a)\| \to 0$ uniformly on every compact subset of* $\mathbb{R}$ for each fixed $a \in A$. *(By using Baire's category theorem, one can easily* *see that* $\|\alpha_{n,t}(a)-\alpha_t(a)\|\to 0$ *(simple convergence) for each* $a\in A$ *implies the uniform convergence* $\|\alpha_{n,t}(a)-\alpha_t(a)\|\to 0$ *on every compact subset of* $\mathbb{R}$.*)*

4.1.2 Proposition *Let* $\alpha_{n,t}=\exp(t\delta_n)$ *and* $\alpha_t=\exp(t\delta)$; *then* $\alpha=\text{strong}\lim\alpha_n$ *if and only if* $(1-\delta_n)^{-1}\to(1-\delta)^{-1}$ *strongly in* $B(A)$, *where* $B(A)$ *is the algebra of all bounded operators on* A.

Proof By the Kato–Trotter theorem ([98]) in semi-group theory, $(1-\delta_n)^{-1}\to(1-\delta)^{-1}$ (strongly) is equivalent to $\|\alpha_{n,t}(a)-\alpha_t(a)\|\to 0$ $(n\to\infty)$ for $t\geqslant 0$. For $t<0$, $\|\alpha_{n,t}(a)-\alpha_t(a)\|=\|\alpha_{n,t}\{\alpha_{-t}-\alpha_{n,-t}\}\alpha_t(a)\|=\|(\alpha_{-t}-\alpha_{n,-t})(\alpha_t(a))\|\to 0$ $(n\to\infty)$. Hence $\alpha=\text{strong}\lim\alpha_n$. □

Now suppose that $\{\alpha_n|n=1,2,\ldots\}$ is a sequence of uniformly continuous one-parameter groups of *-automorphisms on a C*-algebra A and $\alpha=\text{strong}\lim_n\alpha_n$. Let $\alpha_{n,t}=\exp(t\delta_n)$; then δ_n is a bounded *-derivation on A. Let $\{\Pi,\mathscr{H}\}$ be any *-representation of A on a Hilbert space $\mathscr{H}$; then by Theorem 2.5.4, there is a sequence (h_n) of self-adjoints in the weak closure of $\Pi(A)$ such that $\Pi(\delta_n(a))=\mathrm{i}[h_n,\Pi(a)]$ $(a\in A)$. Hence

$$\Pi(\alpha_{n,t}(a))=\exp(tih_n)\Pi(a)\exp(-tih_n)\qquad(a\in A).$$

In this sense, we shall define the following:

4.1.3 Definition *A C*-dynamics* $\{A,\alpha\}$ *is said to be weakly approximately inner if there exists a sequence* $\{\alpha_n\}$ *of uniformly continuous one-parameter groups of *-automorphisms on* A *such that* $\alpha=\text{strong}\lim\alpha_n$ – *i.e. there is a sequence of bounded *-derivations* $\{\delta_n\}$ *on* A *such that* $(1-\delta_n)^{-1}\to(1-\delta)^{-1}$ *(strongly), where* $\alpha_t=\exp(t\delta)$.

A C*-dynamics appearing in mathematical physics usually satisfies a stronger property than the weak approximate innerness:

4.1.4 Definition *A C*-dynamics* $\{A,\alpha\}$ *is said to be approximately inner, if there is a sequence* $\{\alpha_n\}$ *of uniformly continuous one-parameter groups of inner *-automorphisms on* A *such that* $\alpha=\text{strong}\lim\alpha_n$ – *i.e. there is a sequence* (h_n) *of self-adjoint elements in* A *such that* $(1-\delta_{ih_n})^{-1}\to(1-\delta)^{-1}$ *strongly, where* $\alpha_t=\exp(t\delta)$ *and* $\delta_{ih_n}(x)=\mathrm{i}[h_n,x]$ $(x\in A)$. *If* A *is a simple C*-algebra with identity (often enough for C*-physics), then any bounded derivation is inner (2.5.8), so that a weakly approximately inner dynamics is approximately inner in this case.*

In mathematical physics, we are often concerned with a C*-algebra A containing an identity and an increasing sequence $\{A_n\}$ of C*-subalgebras such that $1\in A_n$ and the uniform closure of $\bigcup_{n=1}^{\infty}A_n$ is A. In addition, we are given a *-derivation δ in A satisfying the following conditions:

(1) $\mathscr{D}(\delta) = \bigcup_{n=1}^{\infty} A_n$;
(2) there is a sequence of self-adjoint elements $\{h_n\}$ in $\mathscr{A}$ such that $\delta(a) = \mathrm{i}[h_n, a]$ $(a \in A_n)$ $(n = 1, 2, \ldots,)$.

4.1.5 Definition *We shall call such a *-derivation a general normal *-derivation in A* (later we shall define normal *-derivations in a UHF algebra more restrictively).

A general normal *-derivation is approximately inner, so that it is well-behaved (cf. 3.2.13, 3.2.14).

4.1.6 Proposition *Suppose that a *-derivation in A is approximately inner – i.e. there is a sequence (h_n) of self-adjoint elements in A such that* $\lim_n \delta_{\mathrm{i}h_n}(a) = \delta(a)$ *for each* $a \in \mathscr{D}(\delta)$. *If* $(1 \pm \delta)\mathscr{D}(\delta)$ *is dense in A, then the closure* $\bar{\delta}$ *of* δ *is a generator and* $\exp(t\bar{\delta}) =$ *strong* $\lim \exp(t\delta_{\mathrm{i}h_n})$ *– in particular,* $\{A, \exp(t\bar{\delta})(t \in R)\}$ *is approximately inner.*

Proof Since δ is well-behaved, the densities of $(1 \pm \delta)\mathscr{D}(\delta)$ imply that $\bar{\delta}$ is a generator (cf. 3.4.4).

$$\begin{aligned}
&\|(1 \pm \delta_{\mathrm{i}h_n})^{-1}(1 \pm \bar{\delta})(a) - (1 \pm \bar{\delta})^{-1}(1 \pm \bar{\delta})(a)\| \\
&\quad = \|(1 \pm \delta_{\mathrm{i}h_n})^{-1}(1 \pm \bar{\delta})(a) - (1 \pm \delta_{\mathrm{i}h_n})^{-1}(1 \pm \delta_{\mathrm{i}h_n})(a)\| \\
&\quad \leqslant \|(1 \pm \bar{\delta})(a) - (1 \pm \delta_{\mathrm{i}h_n})(a)\| \to 0 \qquad (n \to \infty)(a \in \mathscr{D}(\delta)).
\end{aligned}$$

Since $\|(1 \pm \delta_{\mathrm{i}h_n})^{-1}\| \leqslant 1$ and $(1 \pm \delta)\mathscr{D}(\delta)$ are dense in A, $(1 \pm \delta_{\mathrm{i}h_n})^{-1} \to (1 \pm \bar{\delta})^{-1}$ (strongly). □

4.1.7 Proposition *Let δ be a *-derivation in A and suppose that there is a sequence $\{\delta_n\}$ of bounded *-derivations in A such that* $\lim_n \delta_n(a) = \delta(a)$ *for* $a \in \mathscr{D}(\delta)$. *Then if* $(1 \pm \delta)\mathscr{D}(\delta)$ *is dense, then δ is a pre-generator and* $\{A, \exp(t\bar{\delta})\}$ *is weakly approximately inner.*

The proof is the same as the proof of Proposition 4.1.6.

4.1.8 Definition *Let δ be a general normal *-derivation in A such that* $\delta(a) = \mathrm{i}[h_n, a]$ $(a \in A_n)$ $(n = 1, 2, \ldots)$. *δ is said to have bounded surface energy if there is a sequence $\{k_n\}$ of self-adjoint elements in A such that $k_n \in A_n$ and* $\|k_n - h_n\| = O(1)$ $(n = 1, 2, \ldots)$

One-dimensional quantum lattice systems with finite range interactions have bounded surface energy.

4.1.9 Proposition *If a general normal *-derivation δ in A has bounded surface energy, then δ is a pre-generator and* $\exp(t\bar{\delta}) =$ strong $\lim_n \exp(\delta_{\mathrm{i}h_n})$.

Proof Suppose that $\|h_n - k_n\| \leqslant M$ $(n = 1, 2, \ldots,)$. Suppose that $(10M - \delta)\mathscr{D}(\delta)$ is not dense in A; then there is an $f(=f^*) \in \mathscr{D}(\delta^*)$ such that $\delta^* f = 10Mf$ and $\|f\| = 1$. Since $\bigcup_{n=1}^{\infty} A_n$ is dense in A, there is an n_0 and an element $a_0(=a_0^*)$ in A_{n_0} such that $\|a_0\| = 1$ and $|f(a_0)| \geqslant \frac{1}{2}$. For $a \in A_n$,

$$(\delta^* f)(a) = f(\delta(a)) = f(\mathrm{i}[h_n, a])$$

and so

$$|\delta^* f(a) - f(\mathrm{i}[k_n, a])| = |f(\mathrm{i}[h_n - k_n, a])| \leqslant 2M \|a\|.$$

Since $\delta^* f = 10Mf$,

$$|10Mf(a) - f(\mathrm{i}[k_n, a])| = |f((10M1 - \delta \mathrm{i}k_n)(a))| \leqslant 2M \|a\|.$$

Hence

$$\sup_{\substack{\|a\| \leqslant 1 \\ a \in A_n}} f((10M1 - \delta \mathrm{i}k_n)(a)) \leqslant 2M.$$

On the other hand,

$$\|(10M1 - \delta \mathrm{i}k_n)(a)\| \geqslant 10M \|a\|.$$

Since $(10M1 - \delta \mathrm{i}k_n)(A_n) = A_n$, there is an element $b(=b^*)$ in A_{n_0} such that $(10M1 - \delta \mathrm{i}k_{n_0})(b) = a_0$. Then $1 = \|a_0\| \geqslant 10M \|b\|$ and so $1/\|b\| \geqslant 10M$. Put $c = b/\|b\|$; then $(10M1 - \delta \mathrm{i}k_{n_0})(c) = a_0/\|b\|$ and $|f((10M1 - \delta \mathrm{i}k_{n_0})(c))| = |f(a_0/\|b\|)| \geqslant \frac{1}{2}/\|b\| \geqslant 5M$, a contradiction. Hence $(10M1 - \delta)\mathscr{D}(\delta)$ is dense and analogously we have the density of $(10M1 + \delta)\mathscr{D}(\delta)$. Hence δ is a pre-generator. Since $\delta_{\mathrm{i}h_n}(a) \to \delta(a)$ $(a \in \mathscr{D}(\delta))$, $\exp(t\delta) = \text{strong lim} \exp(t\delta_{\mathrm{i}h_n})$ □

4.1.10 Definition *A general normal *-derivation δ in A is said to be commutative if we can choose the sequence (h_n) such that $h_m h_n = h_n h_m$ $(m, n = 1, 2, \ldots)$.*

Any *-derivation arising from classical lattice systems is commutative.

4.1.11 Proposition *If a general normal *-derivation δ in A is commutative, then it has an extension δ_1 such that δ_1 is a generator and $\exp(t\delta_1)a = \exp(t\delta \mathrm{i}h_n)a$ $(a \in B_n)$ $(n = 1, 2, 3, \ldots)$, where B_n is the C*-subalgebra of A generated by A_n and $h_1, h_2, \ldots, h_n$, and $\delta_1(a) = \mathrm{i}[h_n, a]$ $(a \in B_n)$. In particular, $\exp(t\delta_1) = \text{strong lim}_n \delta_{\mathrm{i}h_n}$.*

Proof Let B_n be the C*-subalgebra of A generated by A_n and $h_1, h_2, \ldots, h_n$. Then $[\mathrm{i}h_m, h_n] = 0$ and $[\mathrm{i}h_m, A_n] = [\mathrm{i}h_n, A_n] \subset B_n$ for $m \geqslant n$. Therefore B_n is invariant under $\delta \mathrm{i}h_m$ $(m \geqslant n)$. Moreover $\mathrm{i}[h_m, a] = \mathrm{i}[h_n, a]$ $(a \in A_n)$; hence $\delta \mathrm{i}h_m = \delta \mathrm{i}h_n$ on B_n. Therefore there is a unique strongly continuous one-parameter group α of *-automorphisms on A such that $\alpha_t(a) = \lim_m \exp(t\delta_{ih_m})(a) = \exp(t\delta_{ih_n})(a)$ $(a \in B_n;\ n = 1, 2, \ldots,)$. Let $\alpha_t = \exp(t\delta_1)$; then, clearly $\delta \subseteq \delta_1$. □

4.1.12 Notes and remarks Proposition 4.1.9 is due to Kishimoto [103]. Proposition 4.1.11, due to Sakai [167], is applicable to classical lattice systems. The fact that $\exp(t\delta_1)(a) = \exp(t\delta_{ih_n})(a)$ for $a \in B_n$ in Proposition 4.1.11 was first realized by the author. Previously, mathematical physicists have always used $\exp(t\delta) = \text{strong}\lim_n \exp t(\delta_{ih_n})$ only. The exactness of the expression may be useful for the study of classical lattice systems. The notion of approximate innerness was introduced by Powers and Sakai [145].

4.2 Ground states

A ground state is a zero-temperature state. Physically it corresponds to the positive energy state associated with a given dynamical group. In this section, we shall introduce it into a C*-dynamics, and show its abstract characterization. We shall also demonstrate the various properties of a ground state. One of the main results in this section is that an approximately inner C*-dynamics always has a ground state (Theorem 4.2.5).

4.2.1 Definition *Let* $\{A, \alpha\}$ *be a C*-dynamics, and let* δ *be the generator of* α. *A state* ϕ *on* A *is said to be a ground state for* $\{A, \alpha\}$ *if* $-i\phi(a^*\delta(a)) \geqslant 0$ *for* $a \in \mathscr{D}(\delta)$.

Let $\{A, \exp(t\delta_{ih})\ (t \in \mathbb{R})\}$ be a C*-dynamics such that h is a self-adjoint element in A; then there is a real number λ such that $h - \lambda 1 \geqslant 0$ and h is not invertible. Take a state ϕ on A such that $\phi(h - \lambda 1) = 0$; then ϕ is a ground state for $\{A, \exp(t\delta_{ih})\ (t \in \mathbb{R})\}$. In fact, let $k = h - \lambda 1$; then $\delta_{ih} = \delta_{ik}$. For $a \in A$,

$$\begin{aligned} -i\phi(a^*\delta_{ik}(a)) &= -i\phi(a^*[ik, a]) \\ &= -i\phi(a^*(ika - iak)) \\ &= \phi(a^*ka - a^*ak). \end{aligned}$$

Since $|\phi(a^*ak)| = |\phi(a^*ak^{1/2}k^{1/2})| \leqslant \phi(a^*aka^*a)^{1/2}\phi(k)^{1/2} = 0$,

$$-i\phi(a^*\delta_{ik}(a)) = \phi(a^*ka) \geqslant 0.$$

In the following, we shall develop a general theory of ground states.

4.2.2 Proposition *A ground state* ϕ *for* α *is invariant under* α *– i.e.* $\phi(\alpha_t(a)) = \phi(a)$ $(t \in \mathbb{R}, a \in A)$.

Proof

$$-\mathrm{i}\phi((a+\lambda 1)^*\delta(a+\lambda 1)) = -\mathrm{i}\phi((a^*+\bar{\lambda}1)\delta(a)) = -\mathrm{i}\phi(a^*\delta(a)) - \mathrm{i}\bar{\lambda}\phi(\delta(a))$$
$$\geqslant 0 \qquad \text{for } a\in\mathscr{D}(\delta) \text{ and } \lambda\in\mathbb{C}.$$

Therefore, $-\mathrm{i}\lambda\phi(\delta(a))$ is real for $\lambda\in\mathbb{C}$ and so $\phi(\delta(a))=0$; hence $\phi(\alpha_t(a))=\phi(a)$ for $a\in A_2(\delta)$. Since $A_2(\delta)$ is dense, $\phi(\alpha_t(b))=\phi(b)$ for $b\in A$. □

Let ϕ be a ground state for $\{A,\alpha\}$ and let $\{\pi_\phi, \mathscr{H}_\phi\}$ be the GNS representation of A constructed via ϕ. Put $U_\phi(t)a_\phi = (\alpha_t(a))_\phi$ $(a\in A)$; then

$$\|U_\phi(t)a_\phi\|^2 = \phi(\alpha_t(a)^*\alpha_t(a)) = \phi(\alpha_t(a^*a)) = \phi(a^*a) = \|a_\phi\|^2.$$

$U_\phi(t)$ can be uniquely extended to a unitary operator (denoted by $U_\phi(t)$ again) on $\mathscr{H}_\phi$.

$$U_\phi(t_1+t_2)a_\phi = (\alpha_{t_1+t_2}(a))\phi = \{\alpha_{t_1}(\alpha_{t_2}(a))\}_\phi$$
$$= \{U_\phi(t_1)U_\phi(t_2)\}a_\phi \qquad (t_1,t_2\in\mathbb{R}).$$
$$\|U_\phi(t)a_\phi - a_\phi\|^2 = \phi((\alpha_t(a)-a)^*(\alpha_t(a)-a))$$
$$= \phi(a^*a - \alpha_t(a^*)a - a^*\alpha_t(a) + a^*a) \to 0 \qquad (t\to 0).$$

Hence $t\mapsto U_\phi(t)$ is a strongly continuous one-parameter group of unitary operators in $\mathscr{H}_\phi$. Moreover,

$$U_\phi(t)\pi_\phi(a)U_\phi(-t)b_\phi = (\alpha_t(a\alpha_{-t}(b)))_\phi = (\alpha_t(a)b)_\phi$$
$$= \pi_\phi(\alpha_t(a))b_\phi \qquad (a,b\in A).$$

Hence

$$U_\phi(t)\pi_\phi(a)U_\phi(-t) = \pi_\phi(\alpha_t(a)) \qquad (a\in A).$$

Namely, $\{\pi_\phi, U_\phi, \mathscr{H}_\phi\}$ is a covariant representation of the system $\{A,\alpha\}$. By Stone's theorem, there is a self-adjoint operator H_ϕ in $\mathscr{H}_\phi$ such that $U_\phi(t)=\exp(\mathrm{i}tH_\phi)$.

For $a\in\mathscr{D}(\delta)$,

$$\pi_\phi(\delta(a)) = \lim_{t\to 0}\frac{\pi_\phi(\alpha_t(a)) - \pi_\phi(a)}{t}$$
$$= \lim_{t\to 0}\frac{U_\phi(t)\pi_\phi(a)U_\phi(-t) - \pi_\phi(a)}{t} = \mathrm{i}\overline{[H_\phi, \pi_\phi(a)]},$$

where $\overline{[\ ,\ ]}$ is the closure of $[\ ,\ ]$.

$$-\mathrm{i}\phi(a^*\delta(a)) = -\mathrm{i}(\pi_\phi(\delta(a))1_\phi, \pi_\phi(a)1_\phi)$$
$$= -\mathrm{i}(\mathrm{i}\overline{[H_\phi,\pi_\phi(a)]}1_\phi, 1_\phi) = (H_\phi\pi_\phi(a)1_\phi - \pi_\phi(a)H_\phi 1_\phi, \pi_\phi(a)1_\phi).$$

Since $U_\phi(t)1_\phi = 1_\phi$, $H_\phi 1_\phi = 0$, and so $-\mathrm{i}\phi(a^*\delta(a)) = (H_\phi\pi_\phi(a)1_\phi, \pi_\phi(a)1_\phi)$. Since $(1\pm\delta)\mathscr{D}(\delta) = A$, $(1\pm H_\phi)\mathscr{D}(\delta)_\phi = A_\phi$ so that the closure of the restriction of H_ϕ to A_ϕ is H_ϕ; hence $(H_\phi\xi,\xi)\geqslant 0$ for $\xi\in\mathscr{D}(H_\phi)$ and so $H_\phi\geqslant 0$. Therefore $U(t)\in\pi_\phi(\mathscr{A})''$ (cf. 2.4.1). Hence we have:

4.2.3 Proposition *Let ϕ be a ground state for a C*-dynamics $\{A,\alpha\}$ and let $\{\pi_\phi, U_\phi, \mathscr{H}_\phi\}$ be the covariant representation of $\{A,\alpha\}$ constructed via ϕ. Let $U_\phi(t)=\exp(\mathrm{i}tH_\phi)$ be the Stone representation of U_ϕ. Then $H_\phi\geqslant 0$, $H_\phi 1_\phi=0$ and $U_\phi(t)\in\pi_\phi(A)''$ for $t\in R$.*

Let B' be a non-zero element in $\pi_\phi(A)'$ and define

$$\phi_{B'}(x)=\frac{(\pi_\phi(a)B'1_\phi, B'1_\phi)}{(B'1_\phi, B'1_\phi)} \qquad (a\in A).$$

Since $H_\phi\eta\pi_\phi(A)''$, $H_\phi B'\supset B'H_\phi$. Hence

$$\begin{aligned}\pi_\phi(a^*\delta(a))B'1_\phi &= \pi_\phi(a^*)\,\mathrm{i}\overline{[H_\phi,\pi_\phi(a)]}B'1_\phi\\ &=\pi_\phi(a^*)B'\,\mathrm{i}\overline{[H_\phi,\pi_\phi(a)]}1_\phi\\ &=\pi_\phi(a^*)B'\,\mathrm{i}H_\phi\pi_\phi(a)1_\phi=\mathrm{i}\pi_\phi(a^*)H_\phi B'\pi_\phi(a)1_\phi\\ &=\mathrm{i}\pi_\phi(a^*)H_\phi\pi_\phi(a)B'1_\phi.\end{aligned}$$

Therefore

$$-\mathrm{i}\phi_{B'}(a^*\delta(a))=\frac{(\pi_\phi(a^*)H_\phi\pi_\phi(a)B'1_\phi, B'1_\phi)}{(B'1_\phi, B'1_\phi)}\geqslant 0\,(a\in\mathscr{D}(\delta)).$$

So, $\phi_{B'}$ is also a ground state for $\{A,\alpha\}$.

Let $\sigma^g(\alpha)$ be the set of all ground states for $\{A,\alpha\}$; then $\sigma^g(\alpha)$ is a compact convex subset of the state space $\mathscr{S}$ of A. By the above considerations, a $\psi\in\mathscr{S}$ with $\psi\leqslant\lambda\phi$ for some $\phi\in\sigma^g(\alpha)$ and $\lambda>0$ is again a ground state, so that $\sigma^g(\alpha)$ is a face in $\mathscr{S}$.

If ϕ is an extreme point in $\sigma^g(\alpha)$, then $\pi_\phi(A)''=B(\mathscr{H}_\phi)$ and so ϕ is pure. Hence we have:

4.2.4 Proposition *Let $\sigma^g(\alpha)$ be the compact convex set of all ground states for a C*-dynamics $\{A,\alpha\}$; then $\sigma^g(\alpha)$ is a face of the state space of A. Moreover if $\sigma^g(\alpha)\neq\emptyset$; then any extreme point of $\sigma^g(\alpha)$ is a pure state.*

4.2.5 Theorem *Let $\{A,\alpha\}$ be a weakly approximately inner C*-dynamics; then $\sigma^g(\alpha)$ is not empty.*

Proof Let $\{\pi,\mathscr{H}\}$ be a faithful *-representation of A on a Hilbert space $\mathscr{H}$. Since $\{A,\alpha\}$ is weakly approximately inner, there is a sequence of bounded *-derivations (δ_n) on A such that $(1\pm\delta_n)^{-1}\to(1\pm\delta)^{-1}$ strongly in $B(A)$, where $\alpha_t=\exp(t\delta)$. Then there is a sequence (h_n) of self-adjoint elements in the weak closure of $\pi(A)$ such that $\pi(\delta_n(x)) = \mathrm{i}[h_n, \pi(x)]$ $(x \in A)$. Now we shall identify A with $\pi(A)$. Let B be the C*-subalgebra of $B(\mathscr{H})$ generated by A and h_n $(n = 1, 2, \ldots)$. Without loss of generality, we may assume that h_n is positive and not invertible. Let ϕ_n be a state on B such that $\phi_n(h_n) = 0$ and let ϕ be an accumulation point of (ϕ_n) in the state space of B. Without loss of generality, we may assume that $\phi_n \to \phi$ in $\sigma(B^*, B)$. Let $C_c^\infty(\mathbb{R})$ be the space of all infinitely differentiable functions with compact supports on $\mathbb{R}$. For $f \in C_c^\infty(\mathbb{R})$, let

$$a_n = \int f(t) \exp(t\delta_{ihn}(a)) \, \mathrm{d}t$$

and

$$a_n^* = \int \overline{f(t)} \exp(t\delta_{\mathrm{i}h_n}(a^*)) \, \mathrm{d}t \qquad (a \in A);$$

then

$$\begin{aligned} -\mathrm{i}\phi_n(a_n^* \delta_n(a_n)) &= -\mathrm{i}\phi_n\left(\left(\int \overline{f(t)} \exp(t\delta_{\mathrm{i}h_n}(a^*) \, \mathrm{d}t\right)\left(-\int f'(t) \exp(t\delta_{\mathrm{i}h_n})(a) \, \mathrm{d}t\right)\right) \\ &= -\mathrm{i}\phi_n(a_n^* \, \mathrm{i}[h_n, a_n]) = -\mathrm{i}\phi_n(a_n^* \, \mathrm{i}(h_n a_n - a_n h_n)) \\ &= \phi_n(a_n^* h_n a_n) \geqslant 0. \end{aligned}$$

Since $(1 \pm \delta_n)^{-1} \to (1 \pm \delta)^{-1}$ (strongly), $a_n \to \int f(t) \exp(t\delta)(a) \, \mathrm{d}t$ and

$$\begin{aligned} \delta_n(a_n) &= -\int f'(t) \exp(t\delta_{\mathrm{i}h_n})(a) \, \mathrm{d}t \to -\int f'(t) \exp(t\delta)(a) \, \mathrm{d}t \\ &= \delta\left(\int f(t) \exp(t\delta)(a)\right). \end{aligned}$$

Hence

$$\begin{aligned} &\left| -\mathrm{i}\phi\left(\left(\int f(t) \exp(t\delta)(a) \, \mathrm{d}t\right)^* \delta\left(\int f(t) \exp(t\delta)(a) \, \mathrm{d}t\right)\right) - \{-\mathrm{i}\phi_n(a_n \delta_n(a_n))\} \right| \\ &\leqslant \left| (\phi_n - \phi)\left(\left(\int f(t) \exp(t\delta)(a) \, \mathrm{d}t\right)^* \left(\int f(t) \exp(t\delta)(a) \, \mathrm{d}t\right)\right) \right| \\ &\quad + \left\| a_n^* \delta_n(a_n) - \left(\int f(t) \exp(t\delta)(a) \, \mathrm{d}t\right)^* \delta\left(\int f(t) \exp(t\delta)(a) \, \mathrm{d}t\right) \right\| \\ &\to 0 \qquad (n \to \infty). \end{aligned}$$

Therefore

$$-\mathrm{i}\phi\left(\left(\int f(t) \exp(t\delta)(a) \, \mathrm{d}t\right)^* \delta\left(\int f(t) \exp(t\delta)(a) \, \mathrm{d}t\right)\right) \geqslant 0.$$

Take $(f_j) \subset C_c^{\infty}(\mathbb{R})$ such that $\{f_j\}$ is a suitable approximate identity in $L^1(\mathbb{R})$; then $\int f_j(t)\exp(t\delta)(a)\,dt \to a$ and

$$\delta\left(\int f_j(t)\exp(t\delta)(a)\,dt\right) = \int f_j(t)\exp(t\delta)(\delta(a))\,dt \to \delta(a) \qquad \text{for } a \in \mathscr{D}(\delta);$$

hence

$$-i\phi(a^*\delta(a)) \geqslant 0 \qquad (a \in \mathscr{D}(\delta)). \qquad \square$$

If a C*-dynamics $\{A, \alpha\}$ has a ground state ϕ, then there is a strongly continuous one-parameter unitary group $t \mapsto U_\phi(t)$ on $\mathscr{H}_\phi$ such that $U_\phi(t) \in \pi_\phi(A)''$ and $\pi_\phi(\alpha_t(a)) = U_\phi(t)\pi_\phi(a)U_\phi(-t)$ $(a \in A, t \in \mathbb{R})$. Therefore if A is commutative, then $\pi_\phi(\alpha_t(a)) = \pi_\phi(a)$ $(a \in A)$. Moreover let $\mathscr{I}_\phi = \{a \mid \phi(a^*a) = 0, a \in A\}$; then $\mathscr{I}_\phi$ is a closed two-sided ideal of A, $\alpha_t(I_\phi) \subset I_\phi$ and α_t is the identity on $A/\mathscr{I}_\phi$. Therefore a C*-dynamics $\{A, \alpha\}$ with a commutative A does not have a ground state unless A has a closed two-sided ideal $\mathscr{I}$ such that $\alpha_t(\mathscr{I}) \subset \mathscr{I}$ and α_t is the identity on $A/\mathscr{I}$. For example, let $A = C(T)$ (T is a one-dimensional torus group) and $\delta = d/dt$; then $\{A, \alpha_t = \exp(t\delta)(t \in \mathbb{R})\}$ has no ground state.

On the other hand, it is not trivial to construct a C*-dynamics $\{A, \alpha\}$ with a simple C*-algebra A which does not have a ground state. This is understandable, because all the C*-dynamics $\{A, \alpha\}$ appearing in quantum physics have ground states. The first such example was constructed by Lance and Niknam ([117]). Nowadays we know that many C*-dynamics $\{A, \alpha\}$ with a simple (even separable) C*-algebra A which have no ground state exist (cf. pp. 499–501 in [91]).

4.2.6 Definition *A ground state for a C*-dynamics* $\{A, \alpha\}$ *is said to be a physical ground state if the representation* $\{\pi_\phi, U_\phi, \mathscr{H}_\phi\}$ *constructed via* ϕ *satisfies the following condition:* $K_\phi = \{\xi \in \mathscr{H}_\phi \mid H_\phi \xi = 0\}$ *is one-dimensional, where* $U_\phi(t) = \exp(tH_\phi)$.

The vacuum state in the field theory is a physical ground state. It is the state of zero energy and momentum in quantum field theory. A physical ground state must be a pure state, for $\pi_\phi(A)'H_\phi 1_\phi = H_\phi \pi_\phi(A)' 1_\phi$ and so $[\pi_\phi(A)'1_\phi]$ is one-dimensional, and so $\pi_\phi(A)' = \mathbb{C}1_{\mathscr{H}_\phi}$, where $1_{\mathscr{H}_\phi}$ is the identity operator on $\mathscr{H}_\phi$. But a pure ground state is not necessarily a physical ground state (even in the finite-dimensional case).

Let $\{A, \alpha\}$ be a C*-dynamics, and let $\alpha_t = \exp(t\delta)(t \in \mathbb{R})$. Let $\sigma(\alpha)$ be the set of all invariant states ϕ – i.e. $\phi(\delta(a)) = 0$ $(a \in \mathscr{D}(\delta))$. $\sigma(\alpha)$ is a compact convex subset of the state space of A. Since $\mathbb{R}$ is amenable, $\sigma(\alpha) \neq (0)$. An extreme point ϕ in $\sigma(\alpha)$ is said to be ergodic; then one can easily see that

(1) $\phi \in \sigma(\alpha)$ is ergodic $\Leftrightarrow$;
(2) $\{\pi_\phi(A), U_\phi(\mathbb{R})\}' = \mathrm{Cl}_{\mathscr{H}_\phi} \Leftarrow$;
(3) $\dim K_\phi = 1$ (cf. [165]).

A C*-dynamics $\{A, \alpha\}$ is said to be α-abelian if for any $\phi \in \sigma(\alpha)$, $P_\phi \pi_\phi(A) P_\phi$ is a family of mutually commuting operators, where P_ϕ is the orthogonal projection of $\mathscr{H}_\phi$ onto K_ϕ.

If a dynamics $\{A, \alpha\}$ is α-abelian, then the above three conditions are equivalent (cf. [165]). Hence we have:

4.2.7 Proposition *Suppose that a C*-dynamics $\{A, \alpha\}$ is α-abelian; then any pure ground state (i.e., any extreme point in $\sigma^g(\alpha)$) for $\{A, \alpha\}$ is a physical ground state.*

A dynamics $\{A, \alpha\}$ is said to be asymptotically abelian if there is a sequence of real numbers (t_n) such that $\lim_n [\alpha_{t_n}(a), b] = 0$ $(a, b \in A)$. There are many important examples in quantum physics which are asymptotically abelian (for example, consider the anticommutation relation algebra $\mathscr{A}(L^2(\mathbb{R}))$, let $H = \mathrm{i}(\mathrm{d}/\mathrm{d}t)$ in $L^2(\mathbb{R})$; then the dynamics $\{\mathscr{A}(L^2(\mathbb{R})), \exp(t\bar{\delta}_{\mathrm{i}H}\}$ is asymptotically abelian, where $\delta_{\mathrm{i}H}$ is the quasi-free derivation in $\mathscr{A}(L^2(\mathbb{R}))$ defined by H).

On the other hand, there are also many important dynamics in quantum physics, which are not asymptotically abelian (for example, Ising models). It is known that if a dynamics $\{A, \alpha\}$ is asymptotically abelian, then it is α-abelian (cf. [118]); hence we have:

4.2.8 Proposition *Suppose that a C*-dynamics $\{A, \alpha\}$ is asymptotically abelian; then any pure ground state for $\{A, \alpha\}$ is a physical ground state.*

If a dynamics $\{A, \alpha\}$ is α-abelian, then $\sigma(\alpha)$ is a Choquet simplex (cf. [118]). Since $\sigma^g(\alpha) \subset \sigma(\alpha)$, $\sigma^g(\alpha)$ is also a Choquet simplex, so that any ground state can be *uniquely* expressed as an integral of pure ground states (physical ground states). Various models appearing in quantum field theory do not satisfy α-abelianness. Nevertheless, in all of these models, there exist physical ground states and moreover it is very likely that all pure ground states are physical. Therefore the following problems are interesting.

4.2.9 Problems

(1) Characterize those approximately inner dynamics that possess a physical ground state;
(2) characterize those C*-dynamics which possess a physical ground state;
(3) characterize those C*-dynamics in which all pure ground states are physical.

If a C*-dynamics has a unique ground state, then it must be a physical ground state. If a C*-dynamics has two different physical ground states, they are centrally orthogonal. Gross [70] gave a method for proving the existence of physical ground states which is applicable to a wide variety of quantum field theoretic models.

4.2.10 Definition *Let $\{A, \alpha\}$ be a C*-dynamics with at least one ground state. It is said to have a phase transition for ground states if it has at least two ground states.*

There are non-trivial models in quantum field theory; some have phase transitions for ground states and others do not.

4.2.11 Problem *Characterize those C*-dynamics which possess a unique ground state.*

We have seen that a ground state ϕ for a C*-dynamics $\{A, \alpha\}$ is invariant under α and $H_\phi \geqslant 0$, $H_\phi \eta \pi(A)''$.

On the other hand, by Borchers's theorem (2.4.3), we can extend some results on ground states to a class of more general states. Let ϕ be a state on A and let $\{\pi_\phi, \mathscr{H}_\phi\}$ be the GNS construction of A via ϕ. Suppose that there is a strongly continuous one-parameter unitary group $t \mapsto U_\phi(t)$ on $\mathscr{H}_\phi$ as follows.

$$\pi_\phi(\alpha_t(a)) = U_\phi(t)\pi_\phi(a)U_\phi(-t) \qquad (a \in A).$$

Let $U_\phi(t) = \exp(\mathrm{i}tH_\phi)$ be the Stone representation. Moreover, assume that $H_\phi \geqslant 0$; then there is a strongly continuous one-parameter unitary group $t \mapsto V(t) \in \pi_\phi(A)''$ and

$$V(t)\pi_\phi(a)V(-t) = U_\phi(t)\pi_\phi(a)U_\phi(-t) \qquad (a \in A).$$

4.2.12 Definition *We shall call a state ϕ as described above a quasi-ground state for $\{A, \alpha\}$.*

4.2.13 Proposition *Suppose that a C*-dynamics $\{A, \alpha\}$ has a quasi-ground state; then it has a ground state.*

Proof Let ϕ be a quasi-ground state for $\{A, \alpha\}$. Without loss of generality, we may assume that H_ϕ is not invertible.

Let

$$H_\phi = \int_0^\infty \lambda \, \mathrm{d}E_\lambda$$

and

$$E(n^{-1}) = \int_0^{n^{-1}} dE_\lambda \qquad \text{for a positive integer } n;$$

then $E(n^{-1}) \neq 0$. Take an element ξ_n in $\mathscr{H}_\phi$ such that $E(n^{-1})\xi_n = \xi_n$ and $\|\xi_n\| = 1$. Since

$$\lim_{t\to 0} \frac{U_\phi(t)\pi_\phi(a)U_\phi(-t) - \pi_\phi(a)}{t} = \pi_\phi(\delta(a)) \qquad \text{(uniformly)}$$

and

$$\lim_{t\to 0} U_\phi(t)\frac{\pi_\phi(a)U_\phi(-t) - \pi_\phi(a)}{t}\xi$$

$$= \lim_{t\to 0} \left[U_\phi(t)\left\{\frac{\pi_\phi(a)U_\phi(-t) - \pi_\phi(a)}{t} + i\pi_\phi(a)H_\phi\right\} - iU_\phi(t)\pi_\phi(a)H_\phi\right]\xi$$

$$= -i\pi_\phi(a)H_\phi\xi \qquad (\xi \in \mathscr{D}(H_\phi)),$$

$$\lim_{t\to 0} \frac{U_\phi(t)\pi_\phi(a) - \pi_\phi(a)}{t}\xi = iH_\phi\pi_\phi(a)\xi;$$

hence $\pi_\phi(A)\mathscr{D}(H_\phi) \subset \mathscr{D}(H_\phi)$. Put $\phi_n(a) = (\pi_\phi(a)\xi_n, \xi_n)$ $(a \in A)$ and let ϕ_0 be an accumulation point of (ϕ_n) in the state space of A. Then

$$\begin{aligned} -i\phi_n(a^*\delta(a)) &= ((H_\phi\pi_\phi(a) - \pi_\phi(a)H_\phi)\xi_n, \pi_\phi(a)\xi_n) \\ &= (H_\phi\pi_\phi(a)\xi_n, \pi_\phi(a)\xi_n) - \left(\pi_\phi(a)\int_0^{n^{-1}} \lambda\, dE_\lambda\xi_n, \pi_\phi(a)\xi_n\right), \end{aligned}$$

and

$$\left|\left(\pi_\phi(a)\int_0^{n^{-1}} \lambda\, dE_\lambda\xi_n, \pi_\phi(a)\xi_n\right)\right| \leqslant \|\pi_\phi(a)\|^2\, \|n^{-1}\xi_n\| \to 0.$$

Hence $-i\phi_0(a^*\delta(a)) \geqslant 0$ for $a \in \mathscr{D}(\delta)$. □

4.2.14 Notes and remarks Definition 4.2.1 is due to Powers and Sakai [145], Theorem 4.2.5 to Powers and Sakai [145].

References [145], [170].

4.3 KMS states

The KMS condition was first noted by Kubo [198] in 1957, and subsequently by Martin and Schwinger [199] in 1959 for finite-volume Gibbs states. It was proposed as a criterion for equilibrium by Haag, Hugenholtz and

Winnink [72] in 1967. Now the KMS condition gives every evidence of being the correct abstract formulation of the condition for the equilibrium states. In this section, we shall show various nice properties of KMS states. We shall also review the relationship between KMS states and the Tomita–Takesaki theory in von Neumann algebras. One of the main results in this section is that an approximately inner C*-dynamics with a tracial state has a KMS state at each inverse temperature.

4.3.1 Definition *Let $\{A, \alpha\}$ be a C*-dynamics. For a real number β, a state ϕ_β on A is said to be a KMS state for $\{A, \alpha\}$ at inverse temperature β if for $a, b \in A$, there is a bounded continuous function $F_{a,b}$ on the strip $S_\beta = \{z \in \mathbb{C} \mid 0 \leqslant \mathrm{Im}(z) \leqslant \beta\}$ for $\beta \geqslant 0$ (resp. $0 \geqslant \mathrm{Im}(z) \geqslant \beta$ for $\beta < 0$) in the complex plane which is analytic on $0 < \mathrm{Im}(z) < \beta$ (resp. $0 > \mathrm{Im}(z) > \beta$) so that $F_{a,b}(t) = \phi_\beta(a\alpha_t(b))$ and $F_{a,b}(t + \mathrm{i}\beta) = \phi_\beta(\alpha_t(b)a)$ for $t \in \mathbb{R}$.*

From the definition of KMS states, one can easily see that a KMS state ϕ_0 for $\{A, \alpha\}$ at zero inverse temperature is a tracial state on A. Moreover if A has a tracial state τ and $\alpha_t = \exp(t\delta_{\mathrm{i}h})(t \in \mathbb{R})$ with an element h in A, then for each $\beta \in \mathbb{R}$, $\phi_\beta(a) = \tau((a\exp(-\beta h))/\tau(\exp(-\beta h))(a \in A)$ is a KMS state for $\{A, \exp(t\delta_{\mathrm{i}h})\}(t \in \mathbb{R})$ at β. In fact, for $a, b \in A$, let $F_{a,b}(z) = \phi_\beta(a\exp(z\delta_{\mathrm{i}h})(b))$; then $F_{a,b}$ is entire analytic on $\mathbb{C}$. Moreover,

$$F_{a,b}(t) = \phi_\beta(a\alpha_t(b))$$

and

$$\begin{aligned} F_{a,b}(t + \mathrm{i}\beta) &= \phi_\beta(a(\exp(t\delta_{\mathrm{i}h})\exp(\mathrm{i}\beta\delta_{\mathrm{i}h}))(b) \\ &= \frac{\tau(a\exp(-\beta h)\exp(tih)b\,exp(-tih)\exp(\beta h)\exp(-\beta h))}{\tau(\exp(-\beta h))} \\ &= \frac{\tau(a\exp(-\beta h)\alpha_t(b))}{\tau(\exp(-\beta h))} = \frac{\tau(\alpha_t(b)a\exp(-\beta h))}{\tau(\exp(-\beta h))} = \phi_\beta(\alpha_t(b)a). \end{aligned}$$

Suppose that A has a unique tracial state τ and ϕ_β is a KMS state for $\{A, \exp(t\delta_{\mathrm{i}h}(t \in \mathbb{R}))\}$; then

$$\phi_\beta(a) = \frac{\tau(a\exp(-\beta h))}{\tau(\exp(-\beta h))} \qquad (a \in A).$$

In fact, $\phi_\beta(a\exp(z\delta_{\mathrm{i}h})(b))$ is entire analytic and bounded on the strip S_β, for

$$|\phi_\beta(a\exp(z\delta_{\mathrm{i}h})b)| \leqslant \|a\| \, \|\exp(\mathrm{i}\beta\delta_{\mathrm{i}h})b\| \leqslant \|a\| \, \|b\| \exp|\beta| \, \|h\|.$$

Since

$$\phi_\beta(a\exp(t\delta_{\mathrm{i}h})b) = F_{a,b}(t), \qquad F_{a,b}(z) = \phi_\beta(a\exp(z\delta_{\mathrm{i}h})b) \qquad (z \in S_\beta).$$

Hence,

$$\begin{aligned}\phi_\beta(\alpha_t(b)a) &= \phi_\beta(\exp(ith)b\exp(-ith)a)\\ &= \phi_\beta(a\exp(-\beta h)\exp(ith)b\exp(-ith)\exp(\beta h))\end{aligned}$$

and so

$$\phi_\beta(ba) = \phi_\beta(a\exp(-\beta h)b\exp(\beta h)).$$

Now put

$$\tau_0(a) = \frac{\phi_\beta(a\exp(\beta h))}{\phi_\beta(\exp(\beta h))}(a \in A);$$

then

$$\begin{aligned}\tau_0(ba) &= \frac{\phi_\beta(ba\exp(\beta h))}{\phi_\beta(\exp(\beta h))} = \frac{\phi_\beta(a\exp(\beta h)\exp(-\beta h)b\exp(\beta h))}{\phi_\beta(\exp(\beta h))}\\ &= \frac{\phi_\beta(ab\exp(\beta h))}{\phi_\beta(\exp(\beta h))} = \tau_0(ab).\end{aligned}$$

Since $\phi_\beta(\exp(\beta h/2)a\exp(\beta h/2)) = \phi_\beta(\exp(\beta h/2)\exp(-\beta h)\exp(\beta h/2)a\exp(\beta h)) = \phi_\beta(a\exp(\beta h)), \tau_0$ is positive, and since $\tau_0(1) = 1, \tau_0$ is a tracial state. Hence $\tau_0 = \tau$, and so

$$\phi_\beta(a) = \tau(a\exp(-\beta h))\phi_\beta(\exp(\beta h)) \qquad \text{and}$$
$$\tau(\exp(-\beta h)) = \phi_\beta(1)/\phi_\beta(\exp(\beta h)).$$

Hence

$$\phi_\beta(a) = \tau(a\exp(-\beta h))/\tau(\exp(-\beta h)) \qquad (a \in A).$$

In the following, we shall develop a general theory of KMS states.

4.3.2 Proposition *A KMS state ϕ_β for $\{A, \alpha\}$ is invariant under α.*

Proof By the theory of harmonic functions (1.13), there exist kernel functions $K_1(t,z), K_2(t,z) (z \in S_\beta{}^0$ and $t \in \mathbb{R})$ such that

$$K_1 \geqslant 0, K_2 \geqslant 0, \int_{-\infty}^{\infty} K_1(t,z)\mathrm{d}t + \int_{-\infty}^{\infty} K_2(t,z)\mathrm{d}t = 1$$

and

$$F_{a,b}(z) = \int_{-\infty}^{\infty} K_1(t,z)\phi_\beta(a\alpha_t(b))\mathrm{d}t + \int_{-\infty}^{\infty} K_2(t,z)\phi_\beta(\alpha_t(b)a)\mathrm{d}t.$$

Put $a = 1$ and $b = b^*$; then $F_{a,b}(z)$ is real-valued and analytic on the strip; hence $F_{a,b}(z) =$ constant on the strip. □

4.3.3 Proposition *Let ϕ_β be a KMS state for $\{A,\alpha\}$ at β and let $I=\{x\in A \mid \phi_\beta(x^*x)=0\}$; then I is a closed two-sided ideal of A (i.e. $\phi_\beta(x^*x)=0$ implies $\phi_\beta(xx^*)=0$).*

Proof $\phi_\beta(x^*x)=0$ implies $\phi_\beta(x^*\alpha_t(x))=0$ $(t\in\mathbb{R})$ (Schwartz's inequality). Define $F_{x^*,x}(z)=\overline{F_{x^*,x}(\bar{z})}$ for $(z\in -\beta<\operatorname{Im}(z)<0$ or $0<\operatorname{Im}(z)<-\beta)$. Since $F_{x^*,x}(t)$ is real $t\in\mathbb{R}$, by Schwartz's reflection theorem, $F_{x^*,x}$ is an analytic function on $(-\beta<\operatorname{Im}(z)<\beta$ or $\beta<\operatorname{Im}(z)<-\beta)$ and $F_{x^*,x}(t)=0$ $(t\in\mathbb{R})$; hence $F_{x^*,x}(z)=0$ on the strip. □

4.3.4 Proposition *A state ϕ on A is a KMS state for $\{A,\alpha\}$ at β if and only if $\phi(a\alpha_{i\beta}(b))=\phi(ba)$ $(a,b\in A_2(\delta))$, where $\exp(t\delta)=\alpha_t$, $A_2(\delta)$ is the set of all geometric elements with respect to δ and $\alpha_{i\beta}(a)=\sum_{n=0}^{\infty}(\delta^n(a)/n!)(i\beta)^n$.*

Proof Suppose that ϕ is a KMS state for $\{A,\alpha\}$ at β. Then $F_{a,b}(t)=\phi(a\alpha_t(b))$ and $F_{a,b}(t+i\beta)=\phi(\alpha_t(b)a)$. Since $b\in A_2(\delta)$, $\alpha_z(b)=\sum_{n=0}^{\infty}(\delta^n(b)/n!)z^n$ is well-defined on the whole complex plane and analytic on it; hence $F_{a,b}(z)=\phi(a\alpha_z(b))(z\in S_\beta{}^0)$. Hence $F_{a,b}(i\beta)=\phi(ba)=\phi(a\alpha_{i\beta}(b))$. Conversely suppose that $\phi(ba)=\phi(a\alpha_{i\beta}(b))$ for $a,b\in A_2(\delta)$. Then

$$\phi(\alpha_t(b)a)=\phi(a\alpha_{i\beta}\alpha_t(b))=\phi(a\alpha_{t+i\beta}(b)),\qquad \text{for } \alpha_t(A_2(\delta))\subset A_2(\delta).$$

Hence

$$\phi(\alpha_t(b)a)=\phi(a\alpha_{t+i\beta}(b)).$$

Put $F_{a,b}(z)=\phi(a\alpha_z(b))$; then $F_{a,b}(t)=\phi(a\alpha_t(b))$ and $F_{a,b}(t+i\beta)=\phi(\alpha_t(b)a)$. For $x,y\in A$, there are sequences $(x_n),(y_n)$ in $A_2(\delta)$ such that $\|x_n-x\|\to 0$ and $\|y_n-y\|\to 0$. Then $\phi(x_n\alpha_t(y_n))\to\phi(x\alpha_t(y))$ (uniformly on $\mathbb{R}$) and $\phi(\alpha_t(y_n)x_n)\to\phi(\alpha_t(y)x)$ (uniformly on $\mathbb{R}$).

$$\begin{aligned}|F_{x_n,y_n}(z)-F_{x_m,y_m}(z))| &\leqslant \sup_{t\in\mathbb{R}}|\phi(x_n\alpha_t(y_n))-\phi(x_m\alpha_t(y_m))|\\ &\quad+\sup_{t\in\mathbb{R}}|\phi(\alpha_t(y_n)x_n)-\phi(\alpha_t(y_m)x_m)|\\ &\leqslant 2(\|x_n-x_m\|\,\|y_n\|+\|x_m\|\,\|y_n-y_m\|)\to 0\end{aligned}$$

Hence $\{F_{x_n,y_n}\}$ converges uniformly on the strip. Let $F_{x,y}$ be the limit; then it is analytic in the interior $S_\beta{}^0$ and bounded, continuous on the strip S_β. Moreover $F_{x,y}(t)=\phi(x\alpha_t(y))$ and $F_{x,y}(t+i\beta)=\phi(\alpha_t(y)x)$. □

4.3.5 Proposition *Let $\{\phi_{\beta_n}\}$ be a KMS state for $\{A,\alpha\}$ at β_n and $\beta_n\to\beta$; then any accumulation point ϕ of $\{\phi_{\beta_n}\}$ in the state space is a KMS state for $\{A,\alpha\}$ at β (if $\beta=\infty$, ϕ is a ground state; if $\beta=-\infty$, then ϕ is a ceiling state* – i.e. $i\phi(a^*\delta(a))\geqslant 0$ *for* $a\in\mathscr{D}(\delta)$*).*

Proof Let ϕ be an accumulation point of $\{\phi_{\beta_n}\}$. Without loss of generality, we may assume that $\phi_{\beta_n} \to \phi$. If $-\infty < \beta < +\infty$, $\alpha_{i\beta_n}(b) \to \alpha_{i\beta}(b)$ for $b \in A_2(\delta)$ and so

$$|\phi(a\alpha_{i\beta}(b)) - \phi_{\beta_n}(a\alpha_{i\beta_n}(b))| \leqslant |(\phi - \phi_{\beta_n})(a\alpha_{i\beta}(b))| + |\phi_{\beta_n}(a(\alpha_{i\beta}(b) - \alpha_{i\beta_n}(b))| \to 0 \qquad \text{and} \qquad |\phi(ba) - \phi_{\beta_n}(ba)| \to 0.$$

Hence

$$\phi(a\alpha_{i\beta}(b)) = \phi(ba)(a, b \in A_2(\delta)).$$

If $\beta = +\infty$, then

$$F_{a,b,n}(t) = \phi_{\beta_n}(a\alpha_t(b))$$
$$F_{a,b,n}(t + i\beta_n) = \phi_{\beta_n}(\alpha_t(b)a) \qquad (a, b \in A).$$

and

$$|F_{a,b,n}(z)| \leqq \|a\| \, \|b\| \qquad \text{on } S_{\beta_n}{}^0.$$

For $a, b \in \mathscr{D}(\delta)$, $F'_{a,b,n}(t) = \phi_{\beta_n}(a\alpha_t(\delta(b)))$ and so $|F'_{a,b,n}(t)| \leqslant \|a\| \, \|\delta(b)\|$. Therefore from the theory of analytic functions, $\{F_{a,b,n}\}$ has a subsequence $\{F_{a,b,n_j}\}$ such that $\{F_{a,b,n_j}\}$ converges uniformly to a bounded holomorphic function $F_{a,b}$ on the upper half-plane on every compact subset of the upper half-plane, and $F_{a,b}$ is continuous on $\{\mathrm{Im}(z) \geqslant 0\}$. Moreover, $F_{a,b}(t) = \phi(a\alpha_t(b))$. Since a KMS state is invariant under α, ϕ is invariant under α.

Let $\{\pi_\phi, U_\phi, \mathscr{H}_\phi\}$ be the covariant representation of $\{A, \alpha\}$ via ϕ and let $U_\phi(t) = \exp(itH_\phi)$.

$$F_{a,b}(z) = (a_\phi{}^*, \exp(izH_\phi)b_\phi) = (a_\phi{}^*, \exp(itH_\phi)\exp(-yH_\phi)b_\phi),$$

where $z = t + iy$ for $a, b \in A_1(\delta)$. It follows that $H_\phi \geqslant 0$.

For $\beta = -\infty$, the proof is similar. □

Now let ϕ be a KMS state for $\{A, \alpha\}$ at β and let $\{\pi_\phi, U_\phi, \mathscr{H}_\phi\}$ be the covariant representation of $\{A, \alpha\}$ constructed via ϕ.

Let $U_\phi(t) = \exp(itH_\phi)$ and let $A_1(H_\phi)$ be the set of all entire analytic vectors, with respect to H_ϕ in $\mathscr{H}_\phi$ – i.e. $A_1(H_\phi) = \{\xi \in \mathscr{H}_\phi | \sum_{n=0}^{\infty}(\|H_\phi{}^n\xi\|/n!)r^n < +\infty$ for all positive numbers $r\}$.

Define

$$U_\phi(z)\xi = \sum_{n=0}^{\infty} \frac{z^n}{n!}(iH_\phi)^n\xi \qquad \text{for } \xi \in A_1(H_\phi);$$

then for $a \in A_1(\delta)(\alpha_t = \exp(t\delta))$, we have $U_\phi(t)a_\phi = (\alpha_t(a))_\phi$, and $z \to U_\phi(z)a_\phi$ and $z \to (\alpha_z(a))_\phi$ are entire analytic; hence $U_\phi(z)a_\phi = (\alpha_z(a))_\phi$ $(z \in \mathbb{C})$. Let

$\mathcal{M} = \pi_0(A)''$; then the mapping $x \to U_\phi(t)xU_\phi(-t)$ (say $\tilde{\alpha}_t(x)$) of $\mathcal{M}$ is a *-automorphism and $t \to \tilde{\alpha}_t$ is a σ-weakly continuous one-parameter group of *-automorphisms on $\mathcal{M}$. By using Kaplansky's density theorem and the discussions in the proof of 4.3.4, one can easily show that for $x, y \in \mathcal{M}$, there is a bounded continuous function $\tilde{F}_{x,y}$ on the strip S_β which is analytic on $S_\beta{}^0$ so that $\tilde{F}_{x,y}(t) = \tilde{\phi}(x\tilde{\alpha}_t(y))$ and $\tilde{F}_{x,y}(t + i\beta) = \tilde{\phi}(\tilde{\alpha}_t(y)x)$, where $\tilde{\phi}(x) = (x1_\phi, 1_\phi)$ $(x \in \mathcal{M})$. Therefore by a similar discussion in the proof of 4.3.3, $\tilde{\phi}(x^*x) = 0$ implies $\tilde{\phi}(xx^*) = 0$ and so $x = 0$; hence 1_ϕ is a cyclic and separating vector for $\mathcal{M}$ – i.e. $[\mathcal{M}1_\phi] = [\mathcal{M}'1_\phi] = \mathcal{H}_\phi$.

Define conjugate linear operators P, Q in $\mathcal{H}_\phi$ as follows.

$$Pa1_\phi = a^*1_\phi (a \in \mathcal{M}) \qquad \text{and} \qquad Qa'1_\phi = (a')^*1_\phi (a' \in \mathcal{M}').$$

Then

$$(P(a)1_\phi, a'1_\phi) = (a^*1_\phi, a'1_\phi) = (a'^*1_\phi, a1_\phi) = \overline{(a1_\phi, a'^*1_\phi)}.$$

Hence $P^*(a')1_\phi = a'^*1_\phi = Q(a'1_\phi)$. Since P^* has a dense domain, P has the least closed extension $\bar{P}(= P^{**})$. Analogously $Q^* \supset P$ and so Q has the least closed extension $\bar{Q}(= Q^{**})$.

4.3.6 Lemma *$\bar{Q} = P^*$, and $\bar{P} = Q^*$.*

Proof For $\xi \in \mathcal{D}(P^*)$, $(Pa1_\phi, \xi) = \overline{(a1_\phi, P^*\xi)}$ $(a \in \mathcal{M})$. Since $\xi \in [\mathcal{M}'1_\phi]$, by the BT theorem (cf. (2.7.14) in [165]), there exists a bounded operator b' and a closed operator t' such that $b't'1_\phi = \xi$, $b' \in \mathcal{M}'$ and $t' \eta \mathcal{M}'$. Thus, $(Pa1_\phi, \xi) = (a^*1_\phi, b't'1_\phi) = (b'^*1_\phi, t'a1_\phi)$. Hence $(t'a1_\phi, b'^*1_\phi) = (a1_\phi, P^*\xi)$ $(a \in \mathcal{M})$. Let s' be the restriction of t' to $\mathcal{M}1_\phi$, and let $\bar{s}'$ be the least closed extension of s'. Then for every unitary $u \in \mathcal{M}$, $u^*s'u = s'$ and so $u^*\bar{s}'u = \bar{s}$. Hence $\bar{s}'\eta\mathcal{M}'$. Moreover $(\bar{s}')^*b'^*1_\phi = P^*\xi$. Let $\bar{s}' = v'h'$ be the polar decomposition of $\bar{s}'$. Then by the uniqueness of polar decomposition, $v' \in \mathcal{M}'$ and $h'\eta\mathcal{M}'$. Let $h' = \int_0^\infty \lambda de'_\lambda$ be the spectral decomposition of h' and set $h_n = \int_0^n \lambda de'_\lambda$. Then

$$(\bar{s}')^*b'^*1_\phi = h'v'^*b'^*1_\phi = \lim_n h'_n v'^*b'^*1_\phi$$

and

$$b'\bar{s}'1_\phi = b'v'h'1_\phi = \lim_n b'v'h'_n 1_\phi.$$

Since $Q(b'v'h'_n 1_\phi) = h'_n v'^*b'^*1_\phi \to P^*\xi, \bar{Q} = P^*$. Also $\bar{P} = P^{**}$, so that $\bar{P} = Q^*$. □

For $a \in A_1(\delta)$,

$$\begin{aligned}(U_\phi(i\beta)a_\phi, a_\phi) &= \phi(a^*\alpha_{i\beta}(a)) = \phi(aa^*) = (a_\phi{}^*, a_\phi{}^*) \\ &= (Pa_\phi, Pa_\phi) = (P^*\bar{P}a_\phi, a_\phi).\end{aligned}$$

Hence $U(\mathrm{i}\beta)a_\phi = P^*\bar{P}a_\phi$ $(a\in A_2(\delta))$.

$$\|\delta^n\alpha_{\mathrm{i}\beta}(a)\| = \|\alpha_{\mathrm{i}\beta}(\delta^n(a))\| = \left\| \sum_{m=0}^{\infty} \frac{(\mathrm{i}\beta)^m}{m!}\delta^m(\delta^n(a))\right\|$$

$$\leqslant \sum_{m=0}^{\infty} \frac{|\beta|^m}{m!} M_a{}^{m+n}\|a\| = M_a{}^n\|a\|\exp(|\beta|M_a)$$

$$\leqslant (\exp(|\beta|M_a)M_a{}^n\|a\| \qquad (n=1,2,3,\ldots).$$

Hence $\alpha_{\mathrm{i}\beta}(a)\in A_2(\delta)$ and so $\alpha_{\mathrm{i}\beta}(A_2(\delta))\subset A_2(\delta)$. $\{A_2(\delta)\}_\phi$ is invariant under $U_\phi(\mathrm{i}\beta)$ and it consists of analytic vectors for $\{U_\phi(\mathrm{i}\beta)\}$. Hence by Nelson's theorem, the symmetric operator $U_\phi(\mathrm{i}\beta)|\{A_2(\delta)\}_\phi$ is essentially self-adjoint and so

$$\overline{U_\phi(\mathrm{i}\beta)|\{A_2(\delta)\}_\phi} = \overline{U_\phi(\mathrm{i}\beta)} = \exp(-\beta H_\phi)$$

and $\overline{U_\phi(\mathrm{i}\beta)} = P^*\bar{P}$. Since $(\exp -\beta H_\phi)^{\mathrm{i}t} = \exp(-\beta \mathrm{i}tH_\phi)$, $\overline{U_\phi(\mathrm{i}\beta)}^{\mathrm{i}t} = U_\phi(-\beta t)$.

Now let $\bar{P} = J(P^*\bar{P})^{1/2} = J\overline{U_\phi(\mathrm{i}\beta)}^{1/2} = J\exp((-\beta/2)H_\phi) = J\overline{U_\phi(\mathrm{i}\beta/2)}$ be the polar decomposition of $\bar{P}$.

Since $\alpha(\mathrm{i}\beta/2)(A_2(\delta)) = A_2(\delta)$, Range $(U_\phi(\mathrm{i}\beta/2))\supset\{A_2(\delta)\}_\phi$. Range $(\bar{P})\supset \mathscr{M}1_\phi$. Thus, J is a conjugate linear isometry of $\mathscr{H}_\phi$ onto $\mathscr{H}_\phi$. Since $J = P\overline{U_\phi(-\mathrm{i}\beta/2)}$, for $a\in A_2(\delta)$

$$J^2a_\phi = PU_\phi\left(-\frac{\mathrm{i}\beta}{2}\right)PU_\phi\left(-\frac{\mathrm{i}\beta}{2}\right)a_\phi = PU_\phi\left(-\frac{\mathrm{i}\beta}{2}\right)P\left(\alpha - \frac{\mathrm{i}\beta}{2}(a)\right)_\phi$$

$$= PU_\phi\left(-\frac{\mathrm{i}\beta}{2}\right)\left(\alpha\frac{\mathrm{i}\beta}{2}(a^*)\right)_\phi = P(a^*)_\phi = a_\phi;$$

hence $J^2 = 1$ and so $J^* = J$. Since

$$P^2 = J\overline{U_\phi\left(\frac{\mathrm{i}\beta}{2}\right)}J\overline{U_\phi\left(\frac{\mathrm{i}\beta}{2}\right)} = 1 \qquad \text{on } \{A_2(\delta)\}_\phi,$$

$$J\overline{U_\phi\left(\frac{\mathrm{i}\beta}{2}\right)}J = \overline{U_\phi\left(-\frac{\mathrm{i}\beta}{2}\right)} \qquad \text{on } \{A_2(\delta)\}_\phi.$$

By the self-adjointness of $J\overline{U_\phi(\mathrm{i}\beta/2)}J$ and $\overline{U_\phi(-\mathrm{i}\beta/2)}$, $J\overline{U_\phi(\mathrm{i}\beta/2)}J = \overline{U_\phi(-\mathrm{i}\beta/2)}$. Hence

$$\bar{Q} = P^* = \overline{U_\phi\left(\frac{\mathrm{i}\beta}{2}\right)}J^* = \overline{U_\phi\left(\frac{\mathrm{i}\beta}{2}\right)}J = JJ\overline{U_\phi\left(\frac{\mathrm{i}\beta}{2}\right)}J = J\overline{U_\phi\left(-\frac{\mathrm{i}\beta}{2}\right)}.$$

4.3.7 Lemma $J\mathscr{M}J\subset\mathscr{M}'$.

Proof For $a, b, c \in A_2(\delta)$,

$$J\pi_\phi(a)J\pi_\phi(b)c_\phi = PU_\phi\left(-\frac{i\beta}{2}\right)\pi_\phi(a)PU_\phi\left(-\frac{i\beta}{2}\right)\pi_\phi(b)c_\phi$$
$$= PU_\phi\left(-\frac{i\beta}{2}\right)\pi_\phi(a)\left(\left(\alpha-\frac{i\beta}{2}(bc)\right)^*\right)_\phi$$
$$= PU_\phi\left(-\frac{i\beta}{2}\right)\left(a\alpha\frac{i\beta}{2}(c^*b^*)\right)_\phi$$
$$= \left(\alpha-\frac{i\beta}{2}(a)(c^*b^*)\right)^*_\phi$$
$$= \left(bc\alpha\frac{i\beta}{2}(a^*)\right)_\phi.$$
$$\pi_\phi(b)J\pi_\phi(a)Jc_\phi = \pi_\phi(b)PU_\phi\left(-\frac{\beta i}{2}\right)\pi_\phi(a)PU_\phi\left(-\frac{\beta}{2}i\right)c_\phi$$
$$= \pi_\phi(b)PU_\phi\left(-\frac{\beta i}{2}\right)\left(a\left(\alpha-\frac{\beta i}{2}(c)\right)^*\right)_\phi$$
$$= \pi_\phi(b)\left(\alpha-\frac{\beta}{2}i(a)c^*\right)^*_\phi$$
$$= \left(bc\alpha\frac{\beta}{2}i(a^*)\right)_\phi.$$

Hence $J\pi_\phi(b)J\pi_\phi(a) = \pi_\phi(a)J\pi_\phi(b)J$ and so $J\mathcal{M}J \subset \mathcal{M}'$. □

4.3.8 Lemma $J\mathcal{M}J = \mathcal{M}'$.

Proof Since

$$\alpha-\frac{i\beta}{2}(A_2(\delta)) = A_2(\delta), U_\phi\left(-\frac{i\beta}{2}\right)\{A_2(\delta)\}_\phi = \{A_2(\delta)\}_\phi.$$

Since $\{A_2(\delta)\}_\phi$ are analytic vectors for $U_\phi(-(i\beta/2))$, $\overline{U_\phi(-(i\beta/2))|\{A_2(\delta)\}_\phi} = \overline{U_\phi(-(i\beta/2))}$.

Hence for $\xi \in \mathcal{D}(\overline{U_\phi(-(i\beta/2))})$, there is a sequence $\{a_n\}$ in $A_2(\delta)$ such that $\pi_\phi(a_n)1_\phi \to \xi$ and $U_\phi(-(i\beta/2))\pi_\phi(a_n)1_\phi \to \overline{U_\phi(-(i\beta/2))}\xi$ and so $JU_\phi(-(i\beta/2))\pi_\phi(a_n)1_\phi \to J\overline{U_\phi(-(i\beta/2))}\xi$.

For $a' \in \mathcal{M}'$, put $\xi = a'1_\phi$ then $a'1_\phi \in \mathcal{D}(\bar{Q}) = \mathcal{D}(\overline{U_\phi(-(i\beta/2))})$. Therefore $\pi_\phi(a_n)1_\phi \to a'1_\phi$ and

$$JU_\phi(-(\mathrm{i}\beta/2))\pi_\phi(a_n)1_\phi = J\pi_\phi(\alpha-(\mathrm{i}\beta/2))(a_n)J1_\phi \to \bar{Q}a'1_\phi = a'^*1_\phi.$$

$$\begin{aligned}\pi_\phi(a_n)1_\phi &= J\cdot J(a_n)_\phi = JPU(-\beta\mathrm{i}/2)(a_n)_\phi = J(\alpha-(\beta\mathrm{i}/2)(a_n))_\phi{}^* \\ &= J(\alpha(\beta\mathrm{i}/2)(a_n{}^*))_\phi = J\pi_\phi(\alpha(\beta\mathrm{i}/2)(a_n{}^*))1_\phi \\ &= J\pi_\phi(\alpha(\beta\mathrm{i}/2)(a_n{}^*))J1_\phi \to a'1_\phi.\end{aligned}$$

$J\pi_\phi(\alpha(\beta\mathrm{i}/2)(a_n{}^*))J \in J\mathcal{M}J$ and $(J\pi_\phi(\alpha(\beta\mathrm{i}/2)(a_n{}^*))J)^* = J\pi_\phi(\alpha-(\beta\mathrm{i}/2)(a_n))J$.

Hence there is a sequence $\{b'_n\}$ of elements in $J\mathcal{M}J$ such that $b'_n1_\phi \to a'1_\phi$ and $b'^*_n1_\phi \to a'^*1_\phi$. For $x' \in \mathcal{M}'$ and $a, b \in A$

$$\begin{aligned}(Jx'Ja'\pi_\phi(a)1_\phi, \pi_\phi(b)1_\phi) &= (Jx'J\pi_\phi(a)a'1_\phi, \pi_\phi(b)1_\phi) \\ &= \lim(Jx'J\pi_\phi(a)b'_n1_\phi, \pi_\phi(b)1_\phi) \\ &= \lim(b'_nJx'J\pi_\phi(a)1_\phi, \pi_\phi(b)1_\phi) \qquad (b'_nJx'J = Jx'Jb'_n) \\ &= \lim(Jx'J\pi_\phi(a)1_\phi, b'^*_n\pi_\phi(b)1_\phi) \\ &= \lim(Jx'J\pi_\phi(a)1_\phi, \pi_\phi(b)b'^*_n1_\phi) \\ &= (Jx'J\pi_\phi(a)1_\phi, \pi_\phi(b)a'^*1_\phi) \\ &= (Jx'J\pi_\phi(a)1_\phi, a'^*\pi_\phi(b)1_\phi) \\ &= (a'Jx'J\pi_\phi(a)1_\phi, \pi_\phi(b)1_\phi).\end{aligned}$$

Hence $Jx'Ja' = a'Jx'J$ and so $J\mathcal{M}'J \subset (\mathcal{M}')' = \mathcal{M}$. By Lemma 4.3.7, $J\mathcal{M}J \subset \mathcal{M}'$ and so $\mathcal{M} \subset J\mathcal{M}'J$. Finally we have $J\mathcal{M}'J = \mathcal{M}$ and $J\mathcal{M}J = \mathcal{M}'$.

□

We have proved the following theorem.

4.3.9 Theorem *Let ϕ be a KMS state for $\{A, \alpha\}$* at β and let $\{\pi_\phi, U_\phi, \mathcal{H}_\phi\}$ *be the covariant representation of $\{A, \alpha\}$ constructed via ϕ. Let $U_\phi(t) = \exp(\mathrm{i}tH_\phi)$ be the Stone representation of U_ϕ and let $Px1_\phi = x^*1_\phi$ $(x \in \mathcal{M})$, $Qx'1_\phi = x'^*1_\phi$ $(x \in \mathcal{M}')$, where $\mathcal{M} = \pi_\phi(A)''$, and let $\bar{P} = J(P^*\bar{P})^{1/2}$ be the polar decomposition of $\bar{P}$. Then we have: $(P^*\bar{P})^{1/2} = \overline{U_\phi(\beta\mathrm{i}/2)} = \exp((-\beta/2)H_\phi)$; $J = J^*$; $J^2 = 1$; $J\overline{U_\phi(\mathrm{i}\beta/2)}J = \overline{U_\phi(-\mathrm{i}\beta/2)} = \exp((\beta/2)H_\phi)$; $\bar{Q} = J\overline{U_\phi(-\beta\mathrm{i}/2)}$; $(P^*\bar{P})^{-\mathrm{i}t}x(P^*\bar{P})^{\mathrm{i}t} = U_\phi(\beta t)xU_\phi(-\beta t)$ $(t \in R; x \in \mathcal{M})$; $J\mathcal{M}J = \mathcal{M}'$ and $J\mathcal{M}'J = \mathcal{M}$.*

Remark $P^*\bar{P}$ is the modular operator Δ and J is the modular involution J in the theory of Tomita–Takesaki (cf. [184]).

4.3.10 Proposition *Let ϕ be a KMS state for $\{A, \alpha\}$ at β. Then ϕ is an extreme point in the set $K_\beta(\alpha)$ of all KMS states for $\{A, \alpha\}$ at β if and only if it is a factorial state.*

Proof Consider $\{\pi_\phi, U_\phi, \mathcal{H}_\phi\}$. Let Z be the center of $\pi_\phi(A)''$ and let p be a non-zero projection in Z; then

$$(\pi_\phi(a)U_\phi(\mathrm{i}\beta)\pi_\phi(b)U_\phi(-\mathrm{i}\beta)p1_\phi, 1_\phi) = (\pi_\phi(a)pU_\phi(\mathrm{i}\beta)\pi_\phi(b)U_\phi(-\mathrm{i}\beta)1_\phi, 1_\phi)$$
$$= (\pi_\phi(b)\pi_\phi(a)p1_\phi, 1_\phi) \qquad \text{(KMS condition) for } a, b \in A_2(\delta).$$

Hence $\phi_1(a) = (\pi_\phi(a)p1_\phi, 1_\phi)/(p1_\phi, 1_\phi)$ $(a \in A)$ is a KMS state for $\{A, \alpha\}$ at β. If ϕ is extreme in $K_\beta(\alpha)$, then $Z = \{\lambda 1 \mid \lambda \in \mathbb{C}\}$; hence ϕ is factorial.

Conversely suppose that ϕ is not extreme; then there is a KMS state ψ at β such that $\psi \leqslant \lambda\phi$ for some positive λ. Hence there is a positive element h' in $\pi_\phi(A)'$ such that $\psi(a) = (\pi_\phi(a)h'1_\phi, 1_\phi)(a \in A)$. For $a, b \in A_2(\delta)$,

$$\begin{aligned}(\pi_\phi(a)U_\phi(\mathrm{i}\beta)\pi_\phi(b)U_\phi(-\mathrm{i}\beta)1_\phi, h'1_\phi) &= (\pi_\phi(a)P^*\bar{P}\pi_\phi(b)1_\phi, h'1_\phi)\\ &= (\pi_\phi(a)P^*\pi_\phi(b^*)1_\phi, h'1_\phi)\\ &= (h'P^*\pi_\phi(b^*)1_\phi, \pi_\phi(a^*)1_\phi).\end{aligned}$$

On the other hand,

$$\begin{aligned}(\pi_\phi(a)U_\phi(\mathrm{i}\beta)\pi_\phi(b)U_\phi(-\mathrm{i}\beta)1_\phi, h'1_\phi) &= (\pi_\phi(b)\pi_\phi(a)1_\phi, h'1_\phi)\\ &= (P\pi_\phi(a^*)1_\phi, h'\pi_\phi(b^*)1_\phi)\\ &= (P^*h'\pi_\phi(b^*)1_\phi, \pi_\phi(a^*)1_\phi).\end{aligned}$$

Hence

$$(h'P^*\pi_\phi(b^*)1_\phi, \pi_\phi(a^*)1_\phi) = (P^*h'\pi_\phi(b^*)1_\phi, \pi_\phi(a^*)1_\phi)$$

Since $P^*|\overline{\pi_\phi(A_2(\delta))1_\phi} = P^*, h'P^* \subset P^*h'$ and so $P^*h'P^* \subset P^{*2}h' \subset h'$. For $x' \in \pi_\phi(A)', P^*h'P^*x'1_\phi = P^*h'x'^*1_\phi = x'h'1_\phi = h'x'1_\phi$. Hence, $x'h' = h'x'$ and so $h' \in \pi_\phi(A)'' \cap \pi_\phi(A)'$, and so $h' = \lambda 1$ for some positive number. □

4.3.11 Proposition *$K_\beta(\alpha)$ is a Choquet simplex.*

Proof Let V be the real linear space spanned by $\{\phi \mid \phi \in K_\beta(\alpha)\}$; then from the considerations in the proof of 4.3.10, one can easily see that V is a vector lattice; hence by the well-known theorem ([200], [201]) of Choquet, $K_\beta(\alpha)$ is a Choquet simplex. □

4.3.12 Corollary *Let ϕ_1, ϕ_2 be two distinct extreme points in $K_\beta(\alpha)$; then ϕ_1, ϕ_2 are mutually centrally orthogonal in the second dual A^{**} of A.*

Proof Let $\phi = (\phi_1 + \phi_2)/2$; then there are two central positive elements h_1, h_2 in $\pi_\phi(A)''$ such that $\phi_1(a) = (\pi_\phi(a)h_1 1_\phi, 1_\phi)(a \in A)$ and $\phi_2(a) = (\pi_\phi(a)h_2 1_\phi, 1_\phi)$ $(a \in A)$. If $h_1 h_2 \neq 0$, then there is an element $\psi \in K_\beta(\alpha)$ such that $\psi \leqslant \lambda\phi_1$ and $\psi \leqslant \lambda\phi_2$ for some positive number λ, a contradiction. □

We shall discuss some alternative formulations of KMS states. Let $C_c^\infty(\mathbb{R})$

be the algebra of all infinitely differentiable functions on $\mathbb{R}$ with compact support and if the Fourier transform $\hat{f}$ of a function f belongs to $C_c^\infty(\mathbb{R})$ then the inverse Fourier transform $f(z)=(2\pi)^{-1/2}\int_{-\infty}^{\infty}\exp(\mathrm{i}xz)\hat{f}(x)\mathrm{d}x$ is an entire analytic function on the complex plane. Moreover the following proposition is known (cf. [202]).

4.3.13 Proposition (Paley–Wiener). *A function f is the inverse Fourier transform of a function $\hat{f}$ with support $[-M,M]$ if and only if f is entire analytic and for each integer n there exists a constant C_n such that*

$$|f(z)|\leqslant C_n(1+|z|)^{-n}\exp(M|\mathrm{Im}(z)|) \qquad \text{for } z\in\mathbb{C}.$$

4.3.14 Proposition *Let ϕ be a state on a C*-algebra A. Then, ϕ is a KMS state for a C*-dynamics $\{A,\alpha\}$ at β if and only if $\int_{-\infty}^{\infty}f(t)\phi(a\alpha_t(b))\mathrm{d}t=\int_{-\infty}^{\infty}f(t+\mathrm{i}\beta)\phi(\alpha_t(b)a)\mathrm{d}t$ for all $a,b\in A$ and all f with $\hat{f}\in C_c^\infty(\mathbb{R})$.*

Proof Suppose that ϕ is a KMS state for $\{A,\alpha\}$ at β. If $b\in A_1(\delta)$ with $\alpha_t=\exp(t\delta)$, $z\to\phi(a\alpha_z(b))$ is entire analytic and $\phi(a\alpha_t(b))=\phi(a\alpha_{\mathrm{i}\beta}\alpha_{-\mathrm{i}\beta}(\alpha_t(b)))=\phi(\alpha_{t-\mathrm{i}\beta}(b)a)$ for $t\in R$. The function $z\to f(z)\phi(\alpha_z(a)b)$ is entire analytic and decreases faster than $|\mathrm{Re}(z)|^{-2}$ as $|\mathrm{Re}(z)|\to\infty$, provided $\mathrm{Im}(z)\leqslant\beta$. Hence by Cauchy's theorem,

$$\int_{-\infty}^{\infty}f(t)\phi(a\alpha_t(b))\mathrm{d}t=\int_{-\infty}^{\infty}f(t+\mathrm{i}\beta)\phi(a\alpha_{t+\mathrm{i}\beta}(b))\mathrm{d}t=\int_{-\infty}^{\infty}f(t+\mathrm{i}\beta)\phi(\alpha_t(b)a)\mathrm{d}t.$$

By continuity, and the decreasing rate of f, one can easily conclude that $\int_{-\infty}^{\infty}f(t)\phi(a\alpha_t(b))\mathrm{d}t=\int_{-\infty}^{\infty}f(t+\mathrm{i}\beta)\phi(\alpha_t(b)a)\mathrm{d}t$ *for* $a,b\in A$.

Conversely suppose that

$$\int_{-\infty}^{\infty}f(t)\phi(a\alpha_t(b))\mathrm{d}t=\int_{-\infty}^{\infty}f(t+\mathrm{i}\beta)\phi(\alpha_t(b)a)\mathrm{d}t.$$

For $b\in A_1(\delta)$,

$$\begin{aligned}\int_{-\infty}^{\infty}f(t)\phi(a\alpha_t(b))\mathrm{d}t&=\int_{-\infty}^{\infty}f(t+\mathrm{i}\beta)\phi(\alpha_t(b)a)\mathrm{d}t\\&=\int_{-\infty}^{\infty}f(t)\phi(\alpha_{t-\mathrm{i}\beta}(b)a)\mathrm{d}t.\end{aligned}$$

Take (f_n) such that $\hat{f}_n\in C_c^\infty(\mathbb{R})$, $0\leqslant\hat{f}_n\leqslant 1$, $\hat{f}_n(x)=1$ if $|x|\leqslant n$ and $\hat{f}_n(x)=0$ if $|x|\geqslant n+1$. Then for any bounded continuous function g, $\lim_{n\to\infty}\int_{-\infty}^{\infty}f_n(x)g(x)\mathrm{d}x=g(0)$.

Hence $\phi(ab)=\phi(\alpha_{-\mathrm{i}\beta}(b)a)$. By replacing b by $\alpha_{\mathrm{i}\beta}(b)$, we have $\phi(a\alpha_{\mathrm{i}\beta}(b))=\phi(ba)$. □

Now let ϕ be an invariant state on A under α. Define the linear functionals μ_a, ν_a on $C_c^\infty(\mathbb{R})$ as follows.

$$\mu_a(\hat{f}) = \int_{-\infty}^{\infty} f(t)\phi(a^*\alpha_t(a))\mathrm{d}t \qquad \text{and} \qquad \nu_a(\hat{f}) = \int_{-\infty}^{\infty} f(t)\phi(\alpha_t(a)a^*)\mathrm{d}t.$$

Let $\{\pi_\phi, U_\phi, \mathscr{H}_\phi\}$ be the covariant representation of $\{A, \alpha\}$ constructed via ϕ. Let $U_\phi(t) = \int_{-\infty}^{\infty} \exp(\mathrm{i}tx)\mathrm{d}E(x)$ be the Stone representation of $t \to U_\phi(t)$. Then

$$\mu_a(\hat{f}) = \int_{-\infty}^{\infty} f(t)(U_\phi(t)a_\phi, a_\phi)\mathrm{d}t = \int_{-\infty}^{\infty}\int_{-\infty}^{\infty} f(t)\exp(\mathrm{i}tx)\mathrm{d}(E(x)a_\phi, a_\phi)\mathrm{d}t$$

$$= (2\pi)^{1/2}\int_{-\infty}^{\infty} \hat{f}(-x)\mathrm{d}(E(x)a_\phi, a_\phi),$$

and

$$\nu_a(\hat{f}) = \int_{-\infty}^{\infty} f(t)(a_\phi{}^*, U_\phi(t)a_\phi{}^*)\mathrm{d}t = \int_{-\infty}^{\infty}\int_{-\infty}^{\infty} f(t)\exp(-\mathrm{i}tx)\mathrm{d}(E(x)a_\phi{}^*, a_\phi{}^*)\mathrm{d}t$$

$$= (2\pi)^{1/2}\int_{-\infty}^{\infty} \hat{f}(x)\mathrm{d}(E(x)a_\phi{}^*, a_\phi{}^*).$$

Therefore μ_a, ν_a are bounded positive Radon measures on $\mathbb{R}$.

4.3.15 Proposition Let $\{A, \alpha\}$ be a C*-dynamics, ϕ an α-invariant state on A, and μ_a, ν_a positive Radon measures on $\mathbb{R}$ associated with ϕ. Then the following conditions are equivalent:

(1) ϕ is a KMS state for $\{A, \alpha\}$ at β.
(2) μ_a and ν_a are equivalent, with Radon–Nikodym derivative $\mathrm{d}\mu_a(x)/\mathrm{d}\nu_a(x) = \exp(-\beta x)$.

Proof (1)⇒(2): By Proposition 4.3.14,

$$\mu(\hat{f}) = \int_{-\infty}^{\infty} f(t)\phi(a^*\alpha_t(a))\mathrm{d}t = \int_{-\infty}^{\infty} f(t+\mathrm{i}\beta)\phi(\alpha_t(a)a^*)\mathrm{d}t.$$

$$f(t+\mathrm{i}\beta) = (2\pi)^{-1/2}\int_{-\infty}^{\infty} \hat{f}(x)\exp(\mathrm{i}x(t+\mathrm{i}\beta))\mathrm{d}x$$

$$= (2\pi)^{-1/2}\int_{-\infty}^{\infty} \hat{f}(x)\exp(\mathrm{i}xt)\exp(-\beta x)\mathrm{d}x$$

$$= (2\pi)^{-1/2}\int_{-\infty}^{\infty} \hat{f}(x)\exp(-\beta x)\exp(\mathrm{i}xt)\mathrm{d}x.$$

Hence $\mu_a(\hat{f}) = \int_{-\infty}^{\infty} f(t + i\beta)\phi(\alpha_t(a)a^*)dt = \nu_a(k \cdot \hat{f})$, where $k(x) = \exp(-\beta x)$ and so $d\mu_a(x)/d\nu_a(x) = \exp(-\beta x)$.

(2)⇒(1):

$$\int_{-\infty}^{\infty} f(t + i\beta)\phi(\alpha_t(a)a^*)dt = \nu_a(k \cdot \hat{f}) = \mu(\hat{f}) = \int_{-\infty}^{\infty} f(t)\phi(a^*\alpha_t(a))dt.$$

□

Now we shall show the following theorem.

4.3.16 Theorem *A state ϕ on a C*-algebra A is a KMS state for a C*-dynamics $\{A, \alpha\}$ at an inverse temperature β if and only if*

$$-i\beta\phi(a^*\delta(a)) \geqslant S(\phi(a^*a); \phi(aa^*)) \text{ for } a \in \mathscr{D}(\delta),$$

where $\exp(t\delta) = \alpha_t$ *and*

$$S(u; v) = \begin{cases} u \log(u/v) & (u > 0, v > 0) \\ 0 & (v = 0) \\ +\infty & (v = 0, u > 0). \end{cases}$$

Proof It is easily seen that the function $(u, v) \in \mathbb{R}^+ \times \mathbb{R}^+ \to S(u; v) = u \log(u/v)$ is continuous, jointly convex in (u, v) and homogeneous of degree 1 in (u, v), i.e. $S(\lambda u; \lambda v) = \lambda S(u, v)$ for $u, v, \lambda \in \mathbb{R}^+$. Hence $S(\sum \lambda_i u_i; \sum \lambda_i v_i) \leqslant \sum \lambda_i S(u_i; v_i)$ for all finite sequences $\{\lambda_i\}, \{\mu_i\}$ and $\{v_i\}$ in $\mathbb{R}^+$. If f_1 and f_2 are bounded continuous non-negative functions on $\mathbb{R}$, and μ is a positive finite Radon measure on R, then $S(\int f_1(t)d\mu(t); \int f_2(t)d\mu(t)) \leqslant \int S(f_1(t); f_2(t))d\mu(t)$. Now suppose that ϕ is a KMS state for $\{A, \alpha\}$ at β. Then $\phi(a^*a) = 0$ is equivalent to $\phi(aa^*)$, so that it is enough to assume that $\phi(a^*a) > 0$.

Let $U(i\beta) = \int_{-\infty}^{\infty} \exp(-\beta t)dE(t)$; then for $a \in A_2(\delta)$,

$$\begin{aligned} &S(\phi(a^*a); \phi(aa^*)) \\ &= S((a_\phi, a_\phi); (U(i\beta)a_\phi, a_\phi)) \\ &= S\left(\int d(E(t)a_\phi, a_\phi); \int \exp(-\beta t)d(E(t)a_\phi, a_\phi)\right) \\ &\leqslant \int S(1; \exp(-\beta t))d(E(t)a_\phi, a_\phi) = \int \log \frac{1}{\exp(-\beta t)} d(E(t)a_\phi, a_\phi) \\ &= \int \beta t\, d(E(t)a_\phi, a_\phi) = \beta(H_\phi a_\phi, a_\phi) = -i\beta\phi(a^*\delta(a)). \end{aligned}$$

For $a \in \mathscr{D}(\delta)$, take a sequence $\{a_n\}$ of elements in $A_2(\delta)$ such that $a_n \to a$ and $\delta(a_n) \to \delta(a)$; then $S(\phi(a_n{}^*a_n); \phi(a_n a_n{}^*)) \to S(\phi(a^*a); \phi(aa^*))$ and

$-\mathrm{i}\beta\phi(a_n{}^*\delta(a_n)) \to -\mathrm{i}\beta\phi(a^*\delta(a))$; hence $S(\phi(a^*a):\phi(aa^*)) \leqslant -\mathrm{i}\beta\phi(a^*\delta(a))$.

Next we shall show the converse. Suppose that $\beta \neq 0$. Since $-\mathrm{i}\beta\phi(a^*\delta(a))$ are real for $a \in \mathscr{D}(\delta)$, $\phi(\delta(a)) = 0$ for $a \in \mathscr{D}(\delta)$ (cf. the proof of 4.2.2), and so ϕ is invariant under α. If $\hat{f} \in C_c{}^\infty(\mathbb{R})$, it follows that $\alpha_f(a) = \int f(t)\alpha_t(a)\mathrm{d}t \in A_1(\delta)$. Let $\{\pi_\phi, U_\phi \mathscr{H}_\phi\}$ be the covariant representation of $\{A, \alpha\}$ constructed via ϕ. $U_\phi(f) = \int f(t)U_\phi(t)\mathrm{d}t = \int f(t)\mathrm{d}t \int \exp(\mathrm{i}xt)\mathrm{d}E(x) = (2\pi)^{1/2}\int \hat{f}(-x)\mathrm{d}E(x) = (2\pi)^{1/2}\hat{f}(-\hat{H}_\phi)$, where

$$H_\phi = \int_{-\infty}^{\infty} x\,\mathrm{d}E(x).$$

Define $h(x) = |\hat{f}(x)|^2$ and $k(x) = \exp(-\beta x)$; then

$$\begin{aligned}
-\mathrm{i}\beta\phi(\alpha_f(a)^*\delta(\alpha_f(a))) &= -\mathrm{i}\beta(\mathrm{i}H_\phi U_\phi(f)a_\phi, U_\phi(f)a_\phi)\\
&= 2\pi(\beta H_\phi \hat{f}(-H_\phi)\pi_\phi(a)1_\phi, \hat{f}(-H_\phi)\pi_\phi(a)1_\phi)\\
&= 2\pi(\beta H_\phi \overline{\hat{f}(-H_\phi)}\hat{f}(-H_\phi)\pi_\phi(a)1_\phi, \pi_\phi(a)1_\phi)\\
&= (2\pi)^{1/2}\mu_a((\log k)h).
\end{aligned}$$

By similar computations, one finds

$$\phi(\alpha_f(a)^*\alpha_f(a)) = (2\pi)^{1/2}\mu_a(h),\ \phi(\alpha_f(a)\alpha_f(a)^*) = (2\pi)^{1/2}\nu_a(h)$$

and

$$\begin{aligned}
-\mathrm{i}\beta\phi(\alpha_f(a)\delta(\alpha_f(a)^*)) &= -\mathrm{i}\beta(\mathrm{i}H_\phi U_\phi(\hat{f})\pi_\phi(a^*)1_\phi, U_\phi(\hat{f})\pi_\phi(a^*)1_\phi)\\
&= (\beta H_\phi |\hat{f}(H_\phi)|^2\pi_\phi(a^*)1_\phi, \pi_\phi(a^*)1_\phi)\\
&= -(2\pi)^{1/2}\nu((\log k)h).
\end{aligned}$$

Thus, $\mu_a((\log k)h) \geqslant S(\mu_a(h);\nu_a(h))$ and $-\nu_a((\log k)h) \geqslant S(\nu_a(h);\mu_a(h))$. Define $\bar{g}(h) = \sup(\operatorname{supp} h)$ and $\underline{g}(h) = \inf(\operatorname{supp} h)$. Assume that $\beta > 0$; then

$$-\bar{g}(h)\beta h(x) \leqslant h(x)\log k(x) = -h(x)\beta x \leqslant -\underline{g}(h)\beta h(x).$$

Hence,

$$\mu_a(-\underline{g}(h)\beta h(x)) = -\underline{g}(h)\beta\mu_a(h) \geqslant \mu_a(h\log k) \geqslant \mu_a(h)\log\frac{\mu_a(h)}{\nu_a(h)}$$

and so

$$-\underline{g}(h)\beta \geqslant \log\frac{\mu_a(h)}{\nu_a(h)} \qquad \text{and} \qquad \exp(-\underline{g}(h)\beta) \geqslant \frac{\mu_a(h)}{\nu_a(h)}.$$

and $\bar{g}(h)\beta h(x) \geqslant -h(x)\log k(x)$ and so

$$\nu_a(\bar{g}(h)\beta h) = \bar{g}(h)\beta\nu_a(h) \geqslant -\nu_a(h\log k) \geqslant \nu_a(h)\log\frac{\nu_a(h)}{\mu_a(h)}.$$

Hence

$$-\bar{g}(h)\beta \leqslant -\log\frac{\nu_a(h)}{\mu_a(h)} \quad \text{and} \quad \exp(-\bar{g}(h)\beta) \leqslant \frac{\mu_a(h)}{\nu_a(h)}.$$

Therefore

$$\exp(-\bar{g}(h)\beta) \leqslant \frac{\mu_a(h)}{\nu_a(h)} \leqslant \exp(-\underline{g}(h)\beta).$$

Since $\exp(-\bar{g}(h)\beta)h \leqslant kh \leqslant \exp(-\underline{g}(h)\beta h)$,

$$\exp(-\bar{g}(h)\beta)\nu_a(h) \leqslant \nu_a(kh) \leqslant \exp(-\underline{g}(h)\beta)\nu_a(h).$$

Let $\varepsilon > 0$ and let $\{h_n\}$ be a sequence of positive elements in $C_c^{\infty}(\mathbb{R})$ such that $\sum_n h_n = 1$ pointwise, and $|\exp(-\beta\underline{g}(h_n)) - \exp(-\beta\bar{g}(h_n))| \leqslant \varepsilon$. Then $|\mu_a(hh_n) - \nu_a(hh_nk)| \leqslant \varepsilon\nu_a(hh_n)$. By Lebesgue's theorem $|\mu_a(h) - \nu_a(hk)| \leqslant \varepsilon\nu_a(h)$. Hence $\mu_a(h) = \nu_a(hk)$ for all $h \in C_c^{\infty}(R)$. When $\beta < 0$, the reasoning is similar. Finally if $\beta = 0$, $\phi(a^*a)\log(\phi(a^*a)/\phi(aa^*)) \leqslant 0$ and so $\phi(a^*a) \leqslant \phi(aa^*)(a \in A)$. By interchanging a and a^*, $\phi(a^*a) = \phi(aa^*)$. □

4.3.17 Proposition *Suppose that $\{A, \alpha\}$ is weakly approximately inner and A has a tracial state; then the dynamics $\{A, \alpha\}$ has a KMS state ϕ_β at every inverse temperature $\beta(-\infty < \beta < +\infty)$.*

Proof Let τ be a tracial state on A and let $\{\pi_\tau, \mathscr{H}_\tau\}$ be the GNS representation of A via τ.

Since $\{A, \alpha\}$ is weakly approximately inner, there is a sequence of bounded *-derivations $\{\delta_n\}$ on A such that $(1 \pm \delta_n)^{-1} \to (1 \pm \delta)^{-1}$ (strongly), where $\alpha_t = \exp(t\delta)$. By 2.5.4, there exists a sequence of self-adjoint elements $\{h_n\}$ in the weak closure $\overline{\pi_\tau(A)}$ of $\pi_\tau(A)$ such that $\pi_\tau(\delta(a)) = \mathrm{i}[h_n, \pi_\tau(a)](a \in \mathscr{D}(\delta))$. Put

$$\phi_{\beta,n}(x) = \frac{(\pi_\tau(x)\exp(-\beta h_n)1_\tau, 1_\tau)}{\exp(-\beta h_n)1_\tau, 1_\tau)} \qquad (x \in A);$$

then $\{\phi_{\beta,n}\}$ is a sequence of states on A. Let ϕ_β be an accumulation point of $\{\phi_{\beta,n}\}$ in the state space of A. We shall show that ϕ_β is a KMS state for α at β. Let $\alpha_{n,t} = \exp(t\delta_n)$; then $\phi_{\beta,n}$ is a KMS state for α_n at β – in fact, let $F_{a,b,n}(z) = \phi_{\beta,n}(a\alpha_{n,z}(b)) = \phi_{\beta,n}(a(\exp(z\delta_n))(b))$ for $z \in \mathbb{C}$ and $a, b \in A$; then $F_{a,b,n}$ is entire analytic. Moreover,

$$\begin{aligned} &F_{a,b,n}(t+\mathrm{i}\beta) \\ &= \frac{(\pi_\tau(a)\exp(-\beta h_n)\exp(\mathrm{i}th_n)\pi_\tau(b)\exp(-\mathrm{i}th_n)\exp(\beta h_n)\exp(-\beta h_n)1_\tau, 1_\tau)}{(\exp(-\beta h_n)1_\tau, 1_\tau)} \\ &= \frac{(\pi_\tau(a)\exp(-\beta h_n)\pi_\tau(\alpha_{n,t}(b))1_\tau, 1_\tau)}{(\exp(-\beta h_n)1_\tau, 1_\tau)} = \frac{(\pi_\tau(\alpha_{n,t}(b))\pi_\tau(a)\exp(-\beta h_n)1_\tau, 1_\tau)}{(\exp(-\beta h_n)1_\tau, 1_\tau)} \end{aligned}$$

(because τ is a tracial state). Hence $F_{a,b,n}(t+i\beta)=\phi_{\beta,n}(\alpha_{n,t}(b)a)$ and $F_{a,b,n}(t)=\phi_{\beta,n}(a\alpha_{n,t}(b))$.

Moreover $|F_{a,b,n}(z)| \leqslant \|a\| \, \|b\| \exp|\beta| \, \|\delta_n\| (z \in S_\beta)$. Therefore $\phi_{\beta,n}$ is a KMS state for α_n at β. Moreover,

$$F_{a,b,n}(z)=\int_{-\infty}^{\infty} K_1(z,t)\phi_{\beta,n}(a\alpha_{n,t}(b))\mathrm{d}t$$

$$+\int_{-\infty}^{\infty} K_2(z,t)\phi_{\beta,n}(\alpha_{n,t}(b)a)\mathrm{d}t \qquad \text{for } z\in S_\beta{}^0.$$

Let A_0 be a C*-subalgebra of A generated by $\{\alpha_t(b)\}$; then A_0 is separable; hence there is a subsequence $\{n_j\}$ of $\{n\}$ such that $\phi_{\beta,n_j}(a\alpha_t(b)) \to \phi_\beta(a\alpha_t(b))$ $(t\in\mathbb{R})$.

$$|\phi_{\beta,n_j}(a\alpha_{n_j,t}(b))-\phi_\beta(a\alpha_t(b))|$$
$$\leqslant \|a\alpha_{n_j,t}(b)-a\alpha_t(b)\| + |\phi_{\beta,n_j}(a\alpha_t(b))-\phi_\beta(a\alpha_t(b))| \to 0 \qquad (n_j\to\infty),$$

and

$$|\phi_{\beta,n_j}(a\alpha_{n_j,t}(b))| \leqslant \|a\|\,\|b\| \qquad \text{and} \qquad |\phi_{\beta,n_j}(\alpha_{n_j,t}(b)a)| \leqslant \|a\|\,\|b\|.$$

Hence by the dominated convergence theorem, there is a bounded continuous function $F_{a,b}$ on the strip which is analytic in the interior of the strip such that

$$F_{a,b}(z)=\int_{-\infty}^{\infty} K_1(z,t)\phi_\beta(a\alpha_t(b))\mathrm{d}t+\int_{-\infty}^{\infty} K_2(z,t)\phi_\beta(\alpha_t(b)a)\mathrm{d}t$$

and

$$\lim_{y\to 0} F_{a,b}(t+iy)=\phi_\beta(a\alpha_t(b)), \lim_{y\to\beta} F_{a,b}(t+iy)=\phi_\beta(\alpha_t(b)a). \qquad \square$$

4.3.18 Proposition *Suppose that a C*-dynamics $\{A, \exp(t\delta)(t\in\mathbb{R})\}$ has a KMS state ϕ_β at β and there is a sequence $\{\delta_n\}$ of bounded *-derivations on A and a dense subset D of $\mathcal{D}(\delta)$ such that $\delta_n(a)\to\delta(a)$ $(a\in D)$; then A has a tracial state. In addition, if the dynamics $\{A, \exp(t\delta)(t\in\mathbb{R})\}$ is weakly approximately inner, then it has a KMS state ϕ_γ for every inverse temperature $\gamma(-\infty<\gamma<\infty)$.*

Proof We shall show later (§4.4) that if a C*-dynamics $\{A, \exp(t\delta)(t\in\mathbb{R})\}$ has a KMS state at β, then $\{A, \exp(t(\delta+\delta_0))(t\in\mathbb{R})\}$ with a bounded *-derivation δ_0 also has a KMS state at β. Consider a sequence $\{\exp(t(\delta-\delta_n))(t\in\mathbb{R})\}$ of one-parameter groups.

$$\|\{1-(\delta-\delta_n)\}^{-1}(a)-a\| = \|\{1-(\delta-\delta_n)\}^{-1}[a-\{1-(\delta-\delta_n)\}a]\|$$
$$\leqslant \|(\delta-\delta_n)a\| \to 0 \qquad (a\in D).$$

Since D is dense in A and $\|\{1-(\delta-\delta_n)\}^{-1}\| \leqslant 1$, $\{1-(\delta-\delta_n)\}^{-1} \to 1$ (strongly); hence $\exp(t(\delta-\delta_n))(x) \to x$ (uniformly on every compact subset of R) $(x \in A)$. Let $\psi_{\beta,n}$ be a KMS state for $\{A, \exp(t(\delta-\delta_n))(t \in R)\}$; then

$$F_{a,b,n}(z) = \int_{-\infty}^{\infty} K_1(z,t)\psi_{\beta,n}(a\exp(t(\delta-\delta_n))(b))\mathrm{d}t$$

$$+\int_{-\infty}^{\infty} K_2(z,t)\psi_{\beta,n}(\exp(t(\delta-\delta_n))(b)a)\mathrm{d}t$$

$$(0 < \mathrm{Im}(z) < \beta \text{ or } \beta < \mathrm{Im}(z) < 0).$$

Let ψ_β be an accumulation point of $\{\psi_{\beta,n}\}$ in the state space of A; then there is a bounded continuous function $F_{a,b}$ on the strip S_β which is analytic in the interior of S_β such that $F_{a,b}(t) = \psi_\beta(ab)$ and $F_{a,b}(t+\mathrm{i}\beta) = \psi_\beta(ba)$; hence $F_{a,b}(z) = \psi_\beta(ab) = \psi_\beta(ba)$ for $a, b \in A$ – i.e. ψ_β is a tracial state on A. In addition, if $\{A, \exp(t\delta(t \in \mathbb{R}))\}$ is weakly approximately inner, then by 4.3.14 it has a KMS state at each $\gamma(-\infty < \gamma < +\infty)$. □

4.3.19 Problem Can we conclude that any weakly approximately inner dynamics has a KMS state at every inverse temperature if it has a KMS state at some β?

4.3.20 Notes and remarks Proposition 4.3.10 and 4.3.11 were proved by Araki, Hugenholtz, Lanford, Ruelle and Takesaki during 1967 to 1970 [203], [204], [205], [184]. Theorem 4.3.16 is due to Roepstorff [206] and Araki and Sewell [192]. Proposition 4.3.17 is due to Powers and Sakai [145].

References [42], [72], [86], [111], [145], [170], [184], [192].

4.4 Bounded perturbations

One of the important problems in mathematical physics is the stability problem. Many important physical properties are stable under a slight change. In this section, we shall show these facts for C*-dynamical systems. As a slight change, we choose bounded perturbations. We shall show that the approximate innerness of a C*-dynamics, the existence of ground states, the existence of KMS states at an inverse temperature β are preserved under bounded perturbations.

Concerning KMS states, we establish much stronger conclusions. In particular, we will show that bounded perturbations keep the quasi-equivalence of KMS states at β which was proved by Araki (cf. Theorem 4.5.7) and the set of all KMS states at β, obtained by all bounded perturbations with norm less than a fixed positive number, is $\sigma(\mathcal{M}_*, \mathcal{M})$-relatively compact (cf. Theorem 4.5.7). This relative compactness is a powerful tool to prove the absence theorem for phase transition in C*-dynamics with bounded surface energy in a quite general setting (cf. §4.8).

Let δ_1, δ_2 be two *-derivations in a C*-algebra A. Then $\delta_1 + \delta_2$ is a linear operator in A with $\mathcal{D}(\delta_1 + \delta_2) = \mathcal{D}(\delta_1) \cap \mathcal{D}(\delta_2)$. If $\mathcal{D}(\delta_1) \cap \mathcal{D}(\delta_2)$ is dense in A, then $\delta_1 + \delta_2$ is again a *-derivation in A.

4.4.1 Proposition *Suppose that δ_1, δ_2 are two well-behaved *-derivations in A such that $\mathcal{D}(\delta_1) \cap \mathcal{D}(\delta_2)$ is dense in A. Then $\delta_1 + \delta_2$ is well-behaved.*

Proof For $x(=x^*) \in \mathcal{D}(\delta_1 \cap \delta_2)$, let ϕ be a state on A with $|\phi(x)| = \|x\|$. By 3.2.16, $\phi(\delta_1(x)) = \phi(\delta_2(x)) = 0$; hence $\phi((\delta_1 + \delta_2)(x)) = 0$. □

4.4.2 Corollary *Let δ_1 and δ_2 be two well-behaved *-derivations in A. If $\{1 \pm (\delta_1 + \delta_2)\}(\mathcal{D}(\delta_1) \cap \mathcal{D}(\delta_2))$ are dense in A, then $\delta_1 + \delta_2$ is a pre-generator.*

4.4.3 Proposition *If δ_1 is a pregenerator and δ_2 is a bounded *-derivation, then $\delta_1 + \delta_2$ is again a pre-generator.*

This is well-kno'vn from semi-group theory.

4.4.4 Proposition *Suppose that $\{A, \exp(t\delta)(t \in \mathbb{R})\}$ is a weakly approximately inner dynamics and δ_0 is a bounded *-derivation in A; then that dynamics $\{A, \exp(t(\delta + \delta_0))(t \in \mathbb{R})\}$ is again weakly approximately inner.*

Proof By Proposition 4.4.3, $\delta + \delta_0$ is a pre-generator. Since $\delta + \delta_0$ is closed, it is a generator. Take a positive number λ such that $\|\lambda\delta_0\| < \frac{1}{2}$. Let $\{\delta_n\}$ be a sequence of bounded *-derivations on A such that $(1 - \delta_n)^{-1} \to (1 - \delta)^{-1}$ (strongly). Then $(1 - \lambda\delta_n)^{-1} \to (1 - \lambda\delta)^{-1}$ (strongly) by the well-known theorem in semi-group theory.

$$(1 - \lambda\delta_n - \lambda\delta_0)^{-1} = (1 - \lambda\delta_n)^{-1}\{1 - \lambda\delta_0(1 - \lambda\delta_n)^{-1}\}^{-1}$$

$$= (1 - \lambda\delta_n)^{-1} \sum_{m=0}^{\infty} \{\lambda\delta_0(1 - \lambda\delta_n)^{-1}\}^m.$$

Similarly,

$$(1 - \lambda\delta - \lambda\delta_0)^{-1} = (1 - \lambda\delta)^{-1} \sum_{m=0}^{\infty} \{\lambda\delta_0(1 - \lambda\delta)^{-1}\}^m.$$

For $a \in A$,

$$\begin{aligned}&\|(1-\lambda\delta_n-\lambda\delta_0)^{-1}a-(1-\lambda\delta-\lambda\delta_0)^{-1}a\|\\&\leqslant\left\|(1-\lambda\delta_n)^{-1}\left\{\sum_{m=0}^{\infty}\{\lambda\delta_0(1-\lambda\delta_n)^{-1}\}^m-\sum_{m=0}^{\infty}\{\lambda\delta_0(1-\lambda\delta)^{-1}\}^m\right\}a\right\|\\&\quad+\left\|\{(1-\lambda\delta_n)^{-1}-(1-\lambda\delta)^{-1}\}\left\{\sum_{m=0}^{\infty}\{\lambda\delta_0(1-\lambda\delta)^{-1}\}^m\right\}a\right\|.\end{aligned}$$

For an arbitrary positive number ε, there is an m_0 such that

$$\sum_{m=m_0}^{\infty}\{\|\{\lambda\delta_0(1-\delta_n)^{-1}\}^m a\|+\|\{\lambda\delta_0(1-\delta)^{-1}\}^m a\|\}<\varepsilon,$$

for $\|\lambda\delta_0(1-\delta_n)^{-1}\|\leqslant\frac{1}{2}$ and $\|\lambda\delta_0(1-\delta)^{-1}\|<\frac{1}{2}$.

Since $\{\lambda\delta_0(1-\lambda\delta_n)^{-1}\}^m\to\{\lambda\delta_0(1-\lambda\delta)^{-1}\}^m$ (strongly) for each positive integer m we have

$$\overline{\lim_n}\,\|(1-\lambda\delta_n-\lambda\delta_0)^{-1}a-(1-\lambda\delta-\lambda\delta_0)^{-1}a\|\leqslant\varepsilon.$$

Since ε is arbitrary, $(1-\lambda\delta_n-\lambda\delta_0)^{-1}\to(1-\lambda\delta-\lambda\delta_0)^{-1}$ (strongly); hence $\{1-(\delta_n+\delta_0)\}^{-1}\to\{1-(\delta+\delta_0)\}^{-1}$ (strongly). □

4.4.5 Proposition *Let δ_1(resp. δ_2) be a *-derivation in A such that there is a sequence $\{\delta_{1,n}\}$ (resp. $\{\delta_{2,n}\}$) of bounded *-derivations on A such that $\delta_1(a)=\lim_n\delta_{1,n}(a)$ (resp. $\delta_2(a)=\lim_n\delta_{2,n}(a)$) for $a\in\mathscr{D}(\delta_1)$ (resp. $a\in\mathscr{D}(\delta_2)$). If $\{(1\pm(\delta_1+\delta_2)\}\mathscr{D}(\delta_1+\delta_2)$ are dense in A, then $\delta_1+\delta_2$ is a pre-generator and the dynamics $\{A,\exp t\overline{(\delta_1+\delta_2)}\ (t\in\mathbb{R})\}$ is weakly approximately inner.*

This proposition is easily proved by a slight modification of the proof of Proposition 4.4.4.

4.4.6 Proposition *Let $\{A,\exp(t\delta)(t\in\mathbb{R})\}$ be a C*-dynamics with a ground state and let δ_0 be an approximately bounded *-derivation in A – i.e. there is a sequence of bounded *-derivations on A such that $\delta_0(a)=\lim_n\delta_n(a)$ $(a\in\mathscr{D}(\delta_0))$. Then if $\{1\pm(\delta_0+\delta)\}\mathscr{D}(\delta_0+\delta)$ are dense in A, then the C*-dynamics $\{A,\exp(t\overline{(\delta_0+\delta)})\}\ (t\in\mathbb{R})\}$ has a ground state.*

Proof Let ϕ be a ground state for $\{A,\exp(t\delta)(t\in\mathbb{R})\}$ and let $\{\pi_\phi,U_\phi,\mathscr{H}_\phi\}$ be the covariant representation of $\{A,\exp(t\delta)(t\in\mathbb{R})\}$ constructed via ϕ. Let $U_\phi(t)=\exp t\,\mathrm{i}H_\phi$. Since δ_n is a bounded *-derivation on A, there is a self-adjoint element h_n in the weak closure of $\pi_\phi(A)$ such that $\pi_\phi(\delta_n(a))=\mathrm{i}[h_n,\pi_\phi(a)]$ $(a\in A)$. Let $H_n=H_\phi+h_n$; then H_n is lower bounded. Hence there is a real number λ_n such that $H_n-\lambda_n 1\geqslant 0$.

Let $V_n(t)=\exp\mathrm{i}t(H_n-\lambda_n 1)$; then $V_n(t)\in\pi_\phi(A)''$ since $U_\phi(t)\in\pi_\phi(A)''$ and so $(H_n-\lambda_n 1)\eta\pi_\phi(A)''$. For $a,b\in\mathscr{D}(\delta)$, a_ϕ, $(ba)_\phi\in\mathscr{D}(H_\phi)=\mathscr{D}(H_n)$. Hence

$$\lim_{t\to 0}\frac{V_n(t)\pi_\phi(b)V_n(-t)-\pi_\phi(b)}{t}a_\phi = \mathrm{i}H_n\pi_\phi(b)a_\phi - \mathrm{i}\pi_\phi(b)H_na_\phi = [\mathrm{i}H_n,\pi_\phi(b)]a_\phi$$

$$= [\mathrm{i}H_\phi + \mathrm{i}h_n, \pi_\phi(b)]a_\phi = \pi_\phi((\delta+\delta_n)(b))a_\phi.$$

Hence ϕ is a quasi-ground state for the dynamics $\{A, \exp(t(\delta+\delta_n))(t\in\mathbb{R})\}$ and so by 4.2.13, $\{a, \exp(t(\delta+\delta_n))(t\in\mathbb{R})\}$ has a ground state ϕ_n.

Since $\exp(t(\delta+\delta_0)) = \text{strong}\lim \exp(t(\delta+\delta_n))$, by a similar discussion with the proof of 4.2.5, $\{A, \exp(t(\delta+\delta_0))(t\in\mathbb{R})\}$ has a ground state. □

In previous discussions, we have shown that the approximate innerness of C*-dynamics holds under bounded perturbations, and if C*-dynamics $\{\mathscr{A},\alpha\}$ with $\alpha = \exp(t\delta)$ has a ground state ϕ and if δ_1 is a bounded *-derivation on $\mathscr{A}$, then C*-dynamics $\{\mathscr{A}, \exp(t(\delta+\delta_1))\ (t\in R)\}$ also has a ground state ψ and so the existence of a ground state is stable under bounded perturbations. However the states ϕ and ψ are not necessarily quasi-equivalent in general.

One can easily construct such an example even in the case when δ is a bounded *-derivation.

For example, let $B(\mathscr{H})$ be the C*-algebra of all bounded linear operators on a separable Hilbert space $\mathscr{H}$, and let h be a bounded self-adjoint operator on $\mathscr{H}$ such that h does not have any eigenvectors in $\mathscr{H}$. By Weyl's theorem, one can decompose h as follows: $h = h_1 + c$, where h_1 is a diagonal self-adjoint operator and c is a compact self-adjoint operator.

Let $h_1 = \sum_{j=1}^{\infty}\lambda_j e_j$, where $\{e_j\}$ are mutually orthogonal one-dimensional projections. By replacing c by $c + \lambda_1 e_1$, we may assume that $\lambda_1 = 0$. Let $h_2 = \sum_{\lambda_j\geqslant 0}\lambda_j e_j$ and $h_3 = \sum_{\lambda_j<0}\lambda_j e_j$, and put $\delta = \delta_{\mathrm{i}h_2}$ and $\delta_1 = \delta_{\mathrm{i}(h_3+c)}$.

Let $\phi(x) = (x\xi_1, \xi_1)$ $(x\in B(\mathscr{H}))$ with $\|\xi_1\| = 1$ and $e_1\xi_1 = \xi_1$. Then ϕ is a ground state for a C*-dynamics $\{B(\mathscr{H}), \exp(t\delta_{\mathrm{i}h_2})(t\in\mathbb{R})\}$ On the other hand, if a C*-dynamics $\{B(\mathscr{H}), \exp(\delta_{\mathrm{i}h_2+h_3+c})\ (t\in\mathbb{R})\}$ has a ground state ψ which is quasi-equivalent to ϕ, then one can easily see that there exists a non-zero vector η in $\mathscr{H}$ and a real number λ such that $(h+\lambda 1)\eta = 0$. This contradicts the fact that h has no eigenvectors in $\mathscr{H}$.

Next we shall study the relationship between KMS states and bounded perturbations. In this case, we have much stronger conclusions. To prove them, we need a long calculation. Therefore, we shall first state some of the main facts.

Let ϕ be a KMS state for C*-dynamics $\{\mathscr{A},\alpha\}$ with $\alpha_t = \exp(t\delta)$ $(t\in R)$ at an inverse temperature β, and let δ_1 be a bounded *-derivation on $\mathscr{A}$. Let $\{\pi_\phi, U_\phi, \mathscr{H}_\phi\}$ be the covariant representation of $\{\mathscr{A},\alpha\}$ constructed via ϕ. Then we shall show that C*-dynamics $\{\mathscr{A}, \exp(t(\delta+\delta_1))(t\in R)\}$ has a KMS state ψ at β such that $\psi(x) = (x\xi,\xi)$ $(x\in\mathscr{A})$ with $\xi\in\mathscr{H}_\phi$ and $\|\xi\| = 1$, and the states ϕ and ψ on $\mathscr{A}$ are equivalent. In fact, we shall show that ξ can be written in an exact form, by using δ and δ_1.

Moreover let $\mathcal{M}$ be the weak closure of $\pi_\phi(\mathcal{A})$ in $B(\mathcal{H}_\phi)$; then the set of all those KMS states obtained by all bounded perturbations with norm less than a fixed positive number is $\sigma(\mathcal{M}_*, \mathcal{M})$-relatively compact in the pre-dual space $\mathcal{M}_*$ of $\mathcal{M}$. The exact statement of these facts can be found in Theorem 4.4.7, the proof of which proceeds in the following manner.

First, we assume that $\delta_1 = \delta_{ik}$, where k is a geometric element of $\mathcal{A}^s$ with respect to δ. Then by using Theorem 1.17, we show that

$$\phi^k(x) = (\pi_\phi(x)\exp(-\beta((H_\phi + \pi_\phi(k))/2)1_\phi, \exp(-\beta((H_\phi + \pi_\phi(k))/2)1_\phi)/$$
$$(\exp(-\beta((H_\phi + \pi_\phi(k))/2)1_\phi, \exp(-\beta((H_\phi + \pi_\phi(k))/2)1_\phi) \qquad \text{for } x\in\mathcal{A}$$

is a KMS state for $\{\mathcal{A}, \exp(t(\delta + \delta_{ik}))(t\in\mathbb{R})\}$ at β.

Next we shall generalize Theorems 1.13 and 1.14 to analytic functions of several complex variables (Lemma 4.4.8). Then by using Lemma 4.4.8 and the KMS condition, we shall extend this form of a KMS state at β to every element $h\in\mathcal{A}^s$, and finally we shall show $\sigma(\mathcal{M}_*, \mathcal{M})$-relative compactness.

Let ϕ be a KMS state for $\{A, \alpha\}$ with $\alpha_t = \exp(t\delta)$ $(t\in R)$ at β and let $\{\pi_\phi, U_\phi, \mathcal{H}_\phi\}$ be the covariant representation of $\{A, \alpha\}$ constructed via ϕ and let $U_\phi(t) = \exp(itH_\phi)$. For $k(=k^*)\in A_2(\delta)$, by the theory of semi-groups (1.17), we have

$$\begin{aligned}\exp(-t(H_\phi + \pi_\phi(k))) \supseteqq \sum_{p=0}^{\infty} (-1)^p \int_{0\leqslant s_1\leqslant s_2\leqslant \cdots \leqslant s_p\leqslant t} & \exp(-s_1H_\phi)\pi_\phi(k) \\ & \times \exp((-s_2+s_1)H_\phi)\pi_\phi(k)\cdots\exp((-s_p+s_{p-1})H_\phi) \\ & \times \pi_\phi(k)\exp((-t+s_p)H_\phi)\,\mathrm{d}s_1\,\mathrm{d}s_2\cdots\mathrm{d}s_p\end{aligned}$$

Suppose that $\|\delta^n(k)\| \leqslant M^n\|k\|$ $(n=0,1,2,\ldots)$; then

$$\begin{aligned}\|\alpha_{is_j}(k)\| &= \|(\exp(is_j\delta)(k)\| = \left\|\sum_{n=0}^{\infty}\frac{\delta^n(k)(is_j)^n}{n!}\right\| \\ &= \sum_{n=0}^{\infty}\frac{M^n|s_j|^n}{n!}\|k\| = \exp(M|s_j|)\|k\| \qquad (j=1,2,\ldots,p).\end{aligned}$$

Hence

$$\begin{aligned}&\overline{\|\exp(-t(H_\phi+\pi_\phi(k))\exp(tH_\phi)\|} \\ &\leqslant \sum_{p=0}^{\infty}\int_{0\leqslant s_1\leqslant s_2\leqslant\cdots\leqslant s_p\leqslant t}\exp(s_1+s_2+\cdots+s_p)M\|k\|^p\,\mathrm{d}s_1\,\mathrm{d}s_2\cdots\mathrm{d}s_p \\ &\leqslant \sum_{p=0}^{\infty}\exp(ptM)\|k\|^p\int_{0\leqslant s_1\leqslant s_2\leqslant\cdots\leqslant s_p\leqslant t}\mathrm{d}s_1\,\mathrm{d}s_2\cdots\mathrm{d}s_p = \sum_{p=0}^{\infty}\exp(ptM)\|k\|^p\frac{t^p}{p!} \\ &= \exp((\exp Mt)\|k\|t).\end{aligned}$$

For $a \in \mathscr{D}(\delta + \delta_{\mathrm{i}k})$,

$$\frac{\mathrm{d}}{\mathrm{d}t}\exp(\mathrm{i}t(H_\phi + \pi_\phi(k)))\pi_\phi(a)\exp(-\mathrm{i}t(H_\phi + \pi_\phi(k)))|_{t=0} = \overline{\mathrm{i}[H_\phi + \pi_\phi(k), \pi_\phi(a)]}$$

$$= \pi_\phi((\delta + \delta_{\mathrm{i}k})(a))$$

in the strong operator topology of $B(H_\phi)$, and so

$$\pi_\phi(\exp(t(\delta + \delta_{\mathrm{i}k}))(a)) = \exp(\mathrm{i}t(H_\phi + \pi_\phi(k)))\pi_\phi(a)\exp(-\mathrm{i}t(H_\phi + \pi_\phi(k))).$$

Let

$$\phi^k(x) = (\pi_\phi(x)\exp(-\beta(H_\phi + \pi_\phi(k))/2)\exp(\beta H_\phi/2)1_\phi, \exp(-\beta(H_\phi + \pi_\phi(k))/2) \times \exp((\beta H_\phi)/2)1_\phi) \qquad (x \in A).$$

Then $\phi^k(x)/\phi^k(1)$ is a KMS state for $\{A, \exp t(\delta + \delta_{\mathrm{i}k})(t \in \mathbb{R})\}$ at β. In fact, for $a, b \in A_2(\delta + \delta_{\mathrm{i}k})$,

$$\begin{aligned}
\phi^k(a\exp(\mathrm{i}\beta(\delta + \delta_{\mathrm{i}k}))(b)) &= (\pi_\phi(a)\exp(-\beta(H_\phi + \pi_\phi(k)))\pi_\phi(b)\exp(\beta(H_\phi + \pi_\phi(k))) \\
&\quad \times (\exp(-\beta(H_\phi + \pi_\phi(k))/2)\exp(\beta H_\phi/2)1_\phi, \exp(-\beta(H_\phi + \pi_\phi(k))/2)\exp(\beta H_\phi/2)1_\phi) \\
&= (\overline{\exp(\beta H_\phi/2)\exp(-\beta(H_\phi + \pi_\phi(k))/2)}\pi_\phi(a)\overline{\exp(-\beta(H_\phi + \pi_\phi(k))/2)\exp(\beta H_\phi/2)} \\
&\quad \times \exp(-\beta H_\phi)\overline{\exp(\beta H_\phi/2)\exp(-\beta(H_\phi + \pi_\phi(k))/2)} \\
&\quad \times \pi_\phi(b)\overline{\exp(\beta(H_\phi + \pi_\phi(k))/2)\exp(-\beta H_\phi/2)}\exp(\beta H_\phi)1_\phi, 1_\phi) \\
&= (\overline{\exp(\beta H_\phi/2)\exp(-\beta(H_\phi + \pi_\phi(k))/2)}\pi_\phi(b)\overline{\exp(\beta(H_\phi + \pi_\phi(k))/2)\exp(-\beta H_\phi/2)} \\
&\quad \times \overline{\exp(\beta H_\phi/2)\exp(-\beta(H_\phi + \pi_\phi(k))/2)}\pi_\phi(a) \\
&\quad \times \overline{\exp(-\beta(H_\phi + \pi_\phi(k))/2)\exp(\beta H_\phi/2)}1_\phi, 1_\phi) \\
&= (\pi_\phi(b)\pi_\phi(a)\exp(-\beta(H_\phi + \pi_\phi(k))/2)\exp(\beta H_\phi/2)1_\phi, \\
&\quad \exp(-\beta(H_\phi + \pi_\phi(k))/2)\exp(\beta H_\phi/2)1_\phi)
\end{aligned}$$

(use that ϕ is a KMS state for $\{A, \exp(t\delta)\}$ at β).

Now we shall show that $\overline{\exp(\mathrm{i}z(H_\phi + \pi_\phi(k)))\exp(-\mathrm{i}zH_\phi)} \in \pi_\phi(A)$ for $z \in \mathbb{C}$. In fact,

$$\overline{\exp(\mathrm{i}z(H_\phi + \pi_\phi(k)))\exp(-\mathrm{i}zH_\phi)} = \pi_\phi\left(\sum_{p=0}^{\infty}(\mathrm{i}z)^p \int_{0 \leqslant s_1 \leqslant s_2 \leqslant \cdots \leqslant s_p \leqslant 1} \alpha_{s_1 z}(k)\alpha_{s_2 z}(k)\cdots \alpha_{s_p z}(k)\,\mathrm{d}s_1\,\mathrm{d}s_2\cdots\mathrm{d}s_p\right)$$

Hence

$$\overline{\exp(\mathrm{i}z(H_\phi + \pi_\phi(k)))\exp(-\mathrm{i}zH_\phi)} \in \pi_\phi(A)$$

and

$$\overline{\|\exp(\mathrm{i}z(H_\phi+\pi_\phi(k)))\exp(-\mathrm{i}zH_\phi)}\| \leqslant \exp(\exp(M|z|)\|k\|\,|z|).$$

For $b \in A_2(\delta+\delta_{\mathrm{i}k})$

$$\begin{aligned}
&(\pi_\phi(b)\exp(-\beta(H_\phi+\pi_\phi(k)))1_\phi, 1_\phi)\\
&\quad=(\exp(\beta(H_\phi+\pi_\phi(k))/2)\pi_\phi(b)\exp(-\beta(H_\phi+\pi_\phi(k))/2)\exp(-\beta(H_\phi+\pi_\phi(k))/2)1_\phi,\\
&\qquad \exp(-\beta(H_\phi+\pi_\phi(k))/2)1_\phi)\\
&\quad=(\pi_\phi(\exp(-\mathrm{i}\beta/2(\delta+\delta_{\mathrm{i}k}))(b))\exp(-\beta(H_\phi+\pi_\phi(k))/2))1_\phi,\\
&\qquad \exp(-\beta(H_\phi+\pi_\phi(k))/2)1_\phi)\\
&\quad=(\pi_\phi(b)\exp(-\beta(H_\phi+\pi_\phi(k))/2)1_\phi, \exp(-\beta(H_\phi+\pi_\phi(k))/2)1_\phi)
\end{aligned}$$

(the invariance of KMS states). Since $A_2(\delta+\delta_{\mathrm{i}k})$ is dense in A, $\phi^k(a)=(\pi_\phi(a)\exp(-\beta(H_\phi+\pi_\phi(k)))1_\phi, 1_\phi)$ $(a \in A)$.

$$\begin{aligned}
\phi^k(a)&=(\pi_\phi(a)\exp(-\beta(H_\phi+\pi_\phi(k)))1_\phi, 1_\phi)\\
&=(\pi_\phi(a)\exp(-\beta(H_\phi+\pi_\phi(k)))\exp(\beta H_\phi)1_\phi, 1_\phi)\\
&=(\pi_\phi(a)\overline{\exp(-\beta(H_\phi+\pi_\phi(k)))\exp(\beta H_\phi)}1_\phi, 1_\phi) \qquad (a \in A),
\end{aligned}$$

where $\overline{\exp(-\beta(H_\phi+\pi_\phi(k)))\exp(\beta H_\phi)} \in \pi_\phi(A)$.

Now we shall discuss the following theorem.

4.4.7 Theorem *Let ϕ be a KMS state for $\{A,\alpha\}$ at β, $S_\beta=\{z \mid 0 \leqslant \mathrm{Im}(z) \leqslant \beta$ for $\beta \geqslant 0$ (resp. $\beta \leqslant \mathrm{Im}(z) \leqslant 0$ for $\beta < 0$), $z \in \mathbb{C}\}$ and $\mathscr{M}^s$ the self-adjoint portion of the weak closure $\mathscr{M}$ of $\pi_\phi(A)$ in $\mathscr{H}_\phi$. Then there is a mapping $(z,h) \mapsto f(z,h)$ of $S_\beta \times \mathscr{M}^s$ into the pre-dual $\mathscr{M}_*$ of $\mathscr{M}$ satisfying the following conditions:*

(1) *$\|f(z,h)\| \leqslant \exp(|\beta|\,\|h\|)$ for $z \in S_\beta$.*

(2) *For $x \in \mathscr{M}$, $h \in \mathscr{M}^s$, $f(z,h)(x)$ is a bounded continuous function on S_β and is analytic in the interior $S_\beta{}^0$ of S_β.*

(3) *If a directed set $\{h_\alpha\}$ in $\mathscr{M}^s$ with $\|h_\alpha\| \leqslant M$ (M, a fixed number) converges to h in the strong operator topology of $B(\mathscr{H}_\phi)$, then $\{f(z,h_\alpha)\}$ converges to $f(z,h)$ in the norm of $\mathscr{M}_*$ uniformly on every compact subset of S_β.*

(4) *For $h \in \mathscr{M}^s$, $f(0,h)=\tilde{\phi}$, where $\tilde{\phi}(x)=(x1_\phi, 1_\phi)$ $(x \in \mathscr{M})$, and $f(\mathrm{i}\beta,h)$ is a faithful normal positive linear functional on $\mathscr{M}$, and*

$$\exp(\mathrm{i}t(H_\phi+h))\exp(-\mathrm{i}tH_\phi) \in \mathscr{M} \, (t \in \mathbb{R})$$

and moreover

$$f(t+\mathrm{i}\beta,h)(x)=f(\mathrm{i}\beta,h)(\exp(\mathrm{i}t(H_\phi+h))\exp(-\mathrm{i}tH_\phi)x) \qquad (t \in \mathbb{R})$$

and

$$f(t,h)(x)=f(0,h)(x\exp(\mathrm{i}t(H_\phi+h))\exp(-\mathrm{i}tH_\phi) \text{ for } t \in \mathbb{R}.$$

(5) *For* $h \in \mathcal{M}^s$, $1_\phi \in \mathcal{D}(\exp(-\beta(H_\phi + h)/2))$ *and*

$$f(\mathrm{i}\beta, h)(x) = (x \exp(-\beta(H_\phi + h)/2)1_\phi, \exp(-\beta(H_\phi + h)/2)1_\phi) \qquad (x \in \mathcal{M}).$$

Also

$$f(z,h)(1_{\mathcal{H}_\phi}) = (\exp(\mathrm{i}\mathrm{Re}(z))\exp(-\mathrm{Im}(z)(H_\phi + h)/2)1_\phi, \exp(-\mathrm{Im}(z)(H_\phi + h)/2)1_\phi)$$

for $z \in S_\beta$ *and* $x \in \mathcal{M}$.
And, if $\{h_\alpha\}$ *converges strongly to* h *with* $\|h_\alpha\| \leqslant M$, *then* $\{\exp(-\mathrm{Im}(z)(H_\phi + h_\alpha)/2)1_\phi\}$ $(0 \leqslant \mathrm{Im}(z) \leqslant \beta)$ *converges to* $\{\exp(-\mathrm{Im}(z)(H_\phi + h)/2)1_\phi\}$ *in the norm of* $\mathcal{H}_\phi$.

(6) *Let* δ_0 *be a bounded *-derivation on* A *and let* $\pi_\phi(\delta_\phi(a)) = \mathrm{i}[h, \pi_\phi(a)]$ $(a \in A)$ *with* $h \in \mathcal{M}^s$ (cf. 2.5.4); *then* $\psi(a) = f(\mathrm{i}\beta, h)(\pi_\phi(a))/f(\mathrm{i}\beta, h)(1_{\mathcal{H}_\phi})$ $(a \in A)$ *is a KMS state for* $\{A, \exp(t(\delta + \delta_0))(t \in \mathbb{R})\}$ *at* β, *where* $\alpha_t = \exp(t\delta)$ $(t \in R)$.

(7) *For* $h, k \in \mathcal{M}^s$, $\|f(z,h) - f(z,k)\| \leqslant |\beta| \exp(|\beta| \max\{\|h\|, \|k\|\})\|h - k\| (z \in S_\beta)$.

(8) *For* $\gamma > 0$, *let* $\Gamma_\gamma = \{f(\mathrm{i}\beta, h) \mid \|h\| \leqslant \gamma, h \in \mathcal{M}^s\}$; *then* Γ_γ *is relatively* $\sigma(\mathcal{M}_*, \mathcal{M})$*-compact in* $\mathcal{M}_*$. *Moreover let* $\bar{\Gamma}_\gamma$ *be the* $\sigma(\mathcal{M}*, \mathcal{M})$*-closure of* Γ_γ *in* $\mathcal{M}_*$; *then each* ξ *in* $\bar{\Gamma}_\gamma$ *is a faithful normal positive linear functional on* $\mathcal{M}$ *and for each* $\xi \in \bar{\Gamma}_\gamma$, *there is a bounded* $\mathcal{M}_*$*-valued continuous function,* $F_\xi(z)$ *on* S_β *satisfying the following properties:*
(i) *for each* $x \in \mathcal{M}$, $F_\xi(z)(x)$ *is bounded continuous on* S_β *and is analytic in the interior* $S_\beta{}^0$ *of* S_β, *and*

$$|F_\xi(z)(x)| \leqslant \exp(|\beta|\gamma)\|x\|, \qquad |F_\xi(t + \mathrm{i}\beta)(x)| \leqslant F_\xi(\mathrm{i}\beta)(x^*x)^{1/2}$$

and

$$|F_\xi(t)(x)| \leqslant F_\xi(0)(xx^*)^{1/2} (x \in \mathcal{M});$$

(ii) *for each* $x \in \mathcal{M}$, $F_\xi(t)(x)$ *and* $F_\xi(t + \mathrm{i}\beta)(x)$ *are differentiable for almost all* $t \in R$ *and*

$$\left|\frac{\mathrm{d}}{\mathrm{d}t} F_\xi(t)(x)\right| \leqslant \gamma\|x\| \text{a.e.} \qquad \text{and} \qquad \left|\frac{\mathrm{d}}{\mathrm{d}t} F_\xi(t + \mathrm{i}\beta)(x)\right| \leqslant \gamma \exp(|\beta|\gamma)\|x\| \text{a.e.};$$

(iii) $\xi(x) = F_\xi(\mathrm{i}\beta)(x)$ $(x \in \mathcal{M})$, *and* $\tilde{\phi}(x) = F_\xi(0)(x)$ $(x \in \mathcal{M})$.

(9) *More generally for* $\gamma > 0$, *let*

$$\Omega_\gamma = \{f(\mathrm{i}\beta, h) \mid f(\mathrm{i}\beta, h)(1_{\mathcal{H}_\phi}) \leqslant \exp(\gamma), f(\mathrm{i}\beta, h)(h^2)^{1/2} \leqslant \gamma \exp(\gamma)$$
$$\text{and } f(0,h)(h^2)^{1/2} \leqslant \gamma; h \in \mathcal{M}^s\};$$

then Ω_γ *is again a* $\sigma(\mathcal{M}_*, \mathcal{M})$*-relatively compact subset in* $\mathcal{M}_*$. *Furthermore, for each* $\xi \in \bar{\Omega}_\gamma$ (*the* $\sigma(\mathcal{M}_*, \mathcal{M})$*-closure of* Ω_γ *in* $\mathcal{M}_*$) *there is a bounded continuous function* F_ξ *on* S_β *satisfying the same properties with the* (8).

To prove the theorem, we shall provide some lemmas. Since discussions can be carried out symmetrically, we shall assume that $\beta \geqslant 0$.

4.4.8 Lemma *For any positive number* γ, *let*

$$S_{p,\gamma} = \{(z_1, z_2, \ldots, z_p) | 0 \leqslant \mathrm{Im}(z_1) \leqslant \mathrm{Im}(z_2) \leqslant \cdots \leqslant \mathrm{Im}(z_p) \leqslant \gamma, (z_1, z_2, \ldots, z_p) \in \mathbb{C}^p\}.$$

Then there are positive functions $K_j^{p,\gamma}(t_1, t_2, \ldots, t_p; z_1, z_2, \ldots, z_p)$ $(j = 1, 2, \ldots, p+1)$ *on* $\mathbb{R}^p \times S_{p,\gamma}{}^0$ ($S_{p,\gamma}{}^0$ *is the interior of* $S_{p,\gamma}$) *such that for any bounded continuous function* f *on* $S_{p,\gamma}$ *which is analytic with respect to p-variables on* $S_{p,\gamma}{}^0$ *we have*

$$\begin{aligned} f(z_1, z_2, \ldots, z_p) = &\int_{\mathbb{R}^p} K_1{}^{p,\gamma}(t_1, t_2, \ldots, t_p; z_1, z_2, \ldots, z_p) f(t_1, t_2, \ldots, t_p)\, \mathrm{d}t_1\, \mathrm{d}t_2 \cdots dt_p \\ &+ \int_{\mathbb{R}^p} K_2{}^{p,\gamma}(t_1, t_2, \ldots, t_p; z_1, z_2, \ldots, z_p) f(t_1, t_2, \ldots, t_{p-1}, t_p + \mathrm{i}\gamma)\, \mathrm{d}t_1\, \mathrm{d}t_2 \cdots \mathrm{d}t_p + \cdots \\ &+ \int_{\mathbb{R}^p} K_{p+1}{}^{p,\gamma}(t_1, t_2, \ldots, t_p; z_1, z_2, \ldots, z_p) f(t_1 + \mathrm{i}\gamma, t_2 + \mathrm{i}\gamma, \ldots, t_p + \mathrm{i}\gamma)\, \mathrm{d}t_1\, \mathrm{d}t_2 \cdots dt_p \end{aligned}$$

for $(z_1, z_2, \ldots, z_p) \in S_{p,\gamma}{}^0$, *where*

$$\sum_{j=1}^{p+1} \int_{\mathbb{R}^p} K_j{}^{p,\gamma}(t_1, t_2, \ldots, t_p; z_1, z_2, \ldots, z_p)\, \mathrm{d}t_1\, \mathrm{d}t_2 \cdots \mathrm{d}t_p = 1$$

for $(z_1, z_2, \ldots, z_p) \in S_{p,\gamma}{}^0$.

Proof If $p = 1$, it is known. Suppose that it is true for p and every positive number γ; then for

$$(z_1, z_2, \ldots, z_p, z_{p+1}) \in S_{p+1,\gamma},$$

$$\begin{aligned} &f(z_1, z_2, \ldots, z_p, z_{p+1}) \\ &= \int_{\mathbb{R}^p} K_1{}^{p,\mathrm{Im}(z_{p+1})}(t_1, t_2, \ldots, t_p; z_1, z_2, \ldots, z_p) f(t_1, t_2, \ldots, t_p, z_{p+1})\, \mathrm{d}t_1\, \mathrm{d}t_2 \cdots \mathrm{d}t_p \\ &\quad + \int_{\mathbb{R}^p} K_2{}^{p,\mathrm{Im}(z_{p+1})}(t_1, t_2, \ldots, t_p; z_1, z_2, \ldots, z_p) \\ &\qquad \times f(t_1, t_2, \ldots, t_{p-1}, t_p + \mathrm{i}\mathrm{Im}(z_{p+1}), z_{p+1})\, \mathrm{d}t_1\, \mathrm{d}t_2 \cdots \mathrm{d}t_p + \cdots \\ &\quad + \int_{\mathbb{R}^p} K_{p+1}{}^{p,\mathrm{Im}(z_{p+1})}(t_1, t_2, \ldots, t_p; z_1, z_2, \ldots, z_p) \\ &\qquad \times f(t_1 + \mathrm{i}\,\mathrm{Im}(z_{p+1}), \ldots, t_p + \mathrm{i}\mathrm{Im}(z_{p+1}), z_{p+1})\, \mathrm{d}t_1\, \mathrm{d}t_2 \cdots \mathrm{d}t_p. \end{aligned}$$

On the other hand,

$$\int_{\mathbb{R}^p} K_2{}^{p,\mathrm{Im}(z_{p+1})}(t_1, t_2, \ldots, t_p; z_1, z_2, \ldots, z_p)\, f(t_1, t_2, \ldots, t_{p-1}, t_p + \mathrm{i}\,\mathrm{Im}(z_{p+1}), z_{p+1})$$

$$\times\, \mathrm{d}t_1\, \mathrm{d}t_2 \cdots \mathrm{d}t_p$$

$$= \int_{\mathbb{R}^p} K_2{}^{p,\mathrm{Im}(z_{p+1})}(t_1, t_2, \ldots, t_p; z_1, z_2, \ldots, z_p)$$

$$\times f(t_1, t_2, \ldots, t_{p-1}, t_p - t_{p+1} + z_{p+1}, z_{p+1})\, \mathrm{d}t_1\, \mathrm{d}t_2 \cdots \mathrm{d}t_p$$

$$= \int_{\mathbb{R}^p} K_2{}^{p,\mathrm{Im}(z_{p+1})}(t_1, t_2, \ldots, t_{p-1}, t_p + t_{p+1}; z_1, z_2, \ldots, z_p)$$

$$\times f(t_1, t_2, \ldots, t_{p-1}, t_p + z_{p+1}, z_{p+1})\, \mathrm{d}t_1\, \mathrm{d}t_2 \cdots \mathrm{d}t_p$$

$$= \int_{\mathbb{R}^p} K_2{}^{p,\mathrm{Im}(z_{p+1})}(t_1, t_2, \ldots, t_{p-1}, t_p + t_{p+1}; z_1, z_2, \ldots, z_p)\, \mathrm{d}t_1 \cdots \mathrm{d}t_p$$

$$\times \Bigg(\int_{\mathbb{R}} K_1{}^{1,\gamma}(t_{p+1}, z_{p+1}) f(t_1, t_2, \ldots, t_{p-1}, t_p + t_{p+1}, t_{p+1})\, \mathrm{d}t_{p+1}$$

$$+ \int_{\mathbb{R}} K_2{}^{1,\gamma}(t_{p+1}, z_{z+1}) f(t_1, t_2, \ldots, t_{p-1}, t_p + t_{p+1} + \mathrm{i}\gamma, t_{p+1} + \mathrm{i}\gamma)\, \mathrm{d}t_{p+1} \Bigg)$$

$$= \int_{\mathbb{R}^{p+1}} K_2{}^{p,\mathrm{Im}(z_{p+1})}(t_1, t_2, \ldots, t_{p-1}, t_p; z_1, z_2, \ldots, z_p) K_1{}^{1,\gamma}(t_{p+1}, z_{p+1})$$

$$\times f(t_1, t_2, \ldots, t_p, t_{p+1})\, \mathrm{d}t_1\, \mathrm{d}t_2 \cdots \mathrm{d}t_p\, \mathrm{d}t_{p+1}$$

$$+ \int_{\mathbb{R}^{p+1}} K_2{}^{p,\mathrm{Im}(z_{p+1})}(t_1, t_2, \ldots, t_{p-1}, t_p; z_1, z_2, \ldots, z_p) K_2{}^{1,\gamma}(t_{p+1}, z_{p+1})$$

$$\times f(t_1, t_2, \ldots, t_{p-1}, t_p + \mathrm{i}\gamma, t_{p+1} + \mathrm{i}\gamma)\, \mathrm{d}t_1\, \mathrm{d}t_2 \cdots \mathrm{d}t_p\, \mathrm{d}t_{p+1}.$$

By using the similar discussion, we have

$$f(z_1, z_2, \ldots, z_p, z_{p+1}) = \int_{\mathbb{R}^{p+1}} \{K_1{}^{p,\mathrm{Im}(z_{p+1})}(t_1, t_2, \ldots, t_p; z_1, z_2, \ldots, z_p) K_1{}^{1,\gamma}$$

$$(t_{p+1}, z_{p+1}) + K_2{}^{p,\mathrm{Im}(z_{p+1})}(t_1, t_2, \ldots, t_p; z_1, z_2, \ldots, z_p) K_1{}^{1,\gamma}(t_{p+1}, z_{p+1})$$

$$+ \cdots + K_{p+1}{}^{p,\mathrm{Im}(z_{p+1})}(t_1, t_2, \ldots, t_p; z_1, z_2, \ldots, z_p) K_1{}^{1,\gamma}(t_{p+1}, z_{p+1})\}$$

$$\times f(t_1, t_2, \ldots, t_p, t_{p+1})\, \mathrm{d}t_1\, \mathrm{d}t_2 \cdots \mathrm{d}t_p\, \mathrm{d}t_{p+1} + \int_{\mathbb{R}^{p+1}} K_1{}^{p,\mathrm{Im}(z_{p+1})}$$

$$(t_1, t_2, \ldots, t_p; z_1, z_2, \ldots, z_p) K_2{}^{1,\gamma}(t_{p+1}, z_{p+1}) f(t_1, t_2, \ldots, t_p, t_{p+1} + \mathrm{i}\gamma)$$

$$\times\, \mathrm{d}t_1\, \mathrm{d}t_2 \cdots \mathrm{d}t_{p+1} + \cdots + \int_{\mathbb{R}^{p+1}} K_{p+1}{}^{p,\mathrm{Im}(z_{p+1})}(t_1, t_2, \ldots, t_p; z_1, z_2, \ldots, z_p)$$

$$\times K_2{}^{1,\gamma}(t_{p+1}, z_{p+1}) f(t_1 + \mathrm{i}\gamma, \ldots, t_{p+1} + \mathrm{i}\gamma)\, \mathrm{d}t_1\, \mathrm{d}t_2 \cdots \mathrm{d}t_{p+1}.$$

Put

$$K_1^{p+1,\gamma} = \left\{ \sum_{j=1}^{p+1} K_j^{p,\mathrm{Im}(z_{p+1})} \right\} K_1^{1,\gamma}(t_{p+1}, z_{p+1})$$

and

$$K_{j+1}^{p+1,\gamma} = K_j^{p,\mathrm{Im}(z_{p+1})} K_2^{1,\gamma} \qquad (j = 1, 2, \ldots, p+1);$$

then $K_j^{p+1,\gamma} \geqslant 0$ and we can easily see that

$$\sum_{j=1}^{p+2} \int_{R^{p+1}} K_j^{p+1,\gamma}(t_1, t_2, \ldots, t_{p+1}; z_1, z_2, \ldots, z_{p+1})\, \mathrm{d}t_1\, \mathrm{d}t_2 \cdots \mathrm{d}t_{p+1} = 1. \quad \square$$

4.4.9 Lemma *Suppose that* $\{\pi_\phi(k_\alpha)\}$ $(k_\alpha^* = k_\alpha \in A_2(\delta))$ *with* $\|\pi_\phi(k_\alpha)\| \leqslant M$ $(\alpha \in \Pi)$ *converges to an element* h *in* $\mathcal{M}$ *strongly and let* $f(z, \pi_\phi(k_\alpha))(x) = (x \exp(\mathrm{i}z(H_\phi + k_\alpha))1_\phi, 1_\phi)$ $(x \in \mathcal{M}, z \in \mathbb{C})$; *then* $\{f(z, \pi_\phi(k_\alpha))\}$ *converges to a* $\mathcal{M}_*$*-valued function (denoted by* $f(z, h)$*) in the norm of* $\mathcal{M}_*$*, uniformly on every compact subset of* S_β.

Proof By the KMS condition,

$$\begin{aligned}(\pi_\phi(a)\pi_\phi(\alpha_{t_1}(k_\alpha) \cdots \alpha_{t_j}(k_\alpha)\alpha_{t_{j+1}+i\beta}(k_\alpha) \cdots \alpha_{t_p+i\beta}(k_\alpha))1_\phi, 1_\phi) \\ = (\pi_\phi(a)\pi_\phi(\alpha_{t_1}(k_\alpha) \cdots \alpha_{t_j}(k_\alpha))1_\phi, \pi_\phi(\alpha_{t_p}(k_\alpha) \cdots \alpha_{t_{j+1}}(k_\alpha))1_\phi) \\ (j = 0, 1, 2, \ldots, p-1),\end{aligned}$$

where $\pi_\phi(\alpha_{t_0}(k_\alpha)) = 1$ and $a \in A_2(\delta)$. Since $\pi_\phi(\alpha_{t_j}(k_\alpha)) = U_\phi(t_j)\pi_\phi(k_\alpha)U_\phi(-t_j)$, $\{\pi_\phi(\alpha_{t_j}(k_\alpha))\}$ converges uniformly to $U_\phi(t_j)hU_\phi(-t_j)$ on every compact subset of R in the strong operator topology. Therefore, $\{\pi_\phi(\alpha_{t_1}(k_\alpha)) \cdots \pi_\phi(\alpha_{t_j}(k_\alpha))\}$ and $\{\pi_\phi(\alpha_{t_p}(k_\alpha)) \cdots \pi_\phi(\alpha_{t_{j+1}}(k_\alpha))\}$ converge uniformly to $U_\phi(t_1)hU_\phi(-t_1) \cdots U_\phi(t_j)hU_\phi(-t_j)$ and $U_\phi(t_p)hU_\phi(-t_p) \cdots U_\phi(t_{j+1})hU_\phi(-t_{j+1})$ on every compact subset of R^j and R^{p-j} in the strong operator topology respectively. Hence

$$\begin{aligned}\lim_\alpha \sup_{\|a\| \leqslant 1} |(\pi_\phi(a)\pi_\phi(\alpha_{t_1}(k_\alpha) \cdots \alpha_{t_j}(k_\alpha))1_\phi, \pi_\phi(\alpha_{t_p}(k_\alpha) \cdots \alpha_{t_{j+1}}(k_\alpha))1_\phi) \\ - (\pi_\phi(a)U_\phi(t_1)hU_\phi(-t_1) \cdots U_\phi(t_j)hU_\phi(-t_j)1_\phi, U_\phi(t_p)hU_\phi(-t_p) \cdots \\ U_\phi(t_{j+1})hU(-t_{j+1})1_\phi)| = 0\end{aligned}$$

uniformly on every compact subset of $\mathbb{R}^p$. Since $\|\pi_\phi(k_\alpha)\| \leqslant M$,

$$\begin{aligned}|(\pi_\phi(a)\pi_\phi(\alpha_{t_1}(k_\alpha)) \cdots \pi_\phi(\alpha_{t_j}(k_\alpha))1_\phi, \pi_\phi(\alpha_{t_p}(k_\alpha)) \cdots \pi_\phi(\alpha_{t_{j+1}}(k_\alpha))1_\phi)| \\ \leqslant \|a\| \, \|k_\alpha\|^p \leqslant M^p \|a\|.\end{aligned}$$

Therefore,

$$|(\pi_\phi(a)\pi_\phi(\alpha_{z_1}(k_\alpha)\alpha_{z_2}(k_\alpha) \cdots \alpha_{z_p}(k_\alpha))1_\phi, 1_\phi)| \leqslant M^p \|a\| \qquad \text{for } (z_1, z_2, \ldots, z_p) \in S_{p,\beta}.$$

Moreover by 4.4.8,

$$(x\exp(-\beta(H_\phi+\pi_\phi(k_\alpha)))1_\phi,1_\phi)=(x\exp(-\beta(H_\phi+\pi_\phi(k_\alpha)))\exp(\beta H_\phi)1_\phi,1_\phi)$$

$$=\left(x\left\{\sum_{p=0}^{\infty}(-1)^p\int_{0\leqslant s_1\leqslant s_2\leqslant\cdots\leqslant s_p\leqslant\beta}\alpha_{\mathrm{i}s_1}(k_\alpha)\alpha_{\mathrm{i}s_2}(k_\alpha)\cdots\alpha_{\mathrm{i}s_p}(k_\alpha)\,\mathrm{d}s_1\,\mathrm{d}s_2\cdots\mathrm{d}s_p\right\}1_\phi,1_\phi\right)$$

$$=\sum_{p=0}^{\infty}(-1)^p\int_{0\leqslant s_1\leqslant s_2\leqslant\cdots\leqslant s_p\leqslant\beta}\mathrm{d}s_1\,\mathrm{d}s_2\cdots\mathrm{d}s_p(x\alpha_{is_1}(k_\alpha)\alpha_{is_2}(k_\alpha)\cdots\alpha_{is_p}(k_\alpha)1_\phi,1_\phi)$$

$$=\sum_{p=0}^{\infty}(-1)^p\int_{0\leqslant s_1\leqslant s_2\leqslant\cdots\leqslant s_p\leqslant\beta}\mathrm{d}s_1\,\mathrm{d}s_2\cdots\mathrm{d}s_p$$

$$\times\left\{\int_{\mathbb{R}^p}K_1{}^{p,\beta}(t_1,t_2,\ldots,t_p;\mathrm{i}s_1,\mathrm{i}s_2,\ldots,\mathrm{i}s_p)(x\alpha_{t_1}(k_\alpha)\alpha_{t_2}(k_\alpha)\cdots\alpha_{t_p}(k_\alpha)1_\phi,1_\phi)\right.$$

$$\times\,\mathrm{d}t_1\,\mathrm{d}t_2\cdots dt_p+\int_{\mathbb{R}^p}K_2{}^{p,\beta}(t_1,t_2,\ldots,t_p;\mathrm{i}s_1,\mathrm{i}s_2,\ldots,\mathrm{i}s_p)$$

$$\times(x\alpha_{t_1}(k_\alpha)\alpha_{t_2}(k_\alpha)\cdots\alpha_{t_{p-1}}(k_\alpha)\alpha_{t_{p+\mathrm{i}\beta}}(k_\alpha)1_\phi,1_\phi)\,\mathrm{d}t_1\,\mathrm{d}t_2\cdots\mathrm{d}t_p+\cdots$$

$$+\int_{\mathbb{R}^p}K_{p+1}{}^{p,\beta}(t_1,t_2,\ldots,t_p;\mathrm{i}s_1,\mathrm{i}s_2,\ldots,\mathrm{i}s_p)(x\alpha_{t_1+\mathrm{i}\beta}(k_\alpha)\alpha_{t_2+\mathrm{i}\beta}(k_\alpha)$$

$$\left.\alpha_{t_{p+\mathrm{i}\beta}}(k_\alpha)1_\phi,1_\phi)\,\mathrm{d}t_1\,\mathrm{d}t_2\cdots\mathrm{d}t_p\right\}$$

$$=\sum_{p=0}^{\infty}(-1)^p\int_{0\leqslant s_1\leqslant s_2\leqslant\cdots\leqslant s_p\leqslant\beta}\mathrm{d}s_1\,\mathrm{d}s_2\cdots\mathrm{d}s_p$$

$$\times\left\{\int_{\mathbb{R}^p}K_1{}^{p,\beta}(t_1,t_2,\ldots,t_p;\mathrm{i}s_1,\mathrm{i}s_2,\ldots,\mathrm{i}s_p)(x\alpha_{t_1}(k_\alpha)\alpha_{t_2}(k_\alpha)\cdots\alpha_{t_p}(k_\alpha)1_\phi,1_\phi)\right.$$

$$\times\,\mathrm{d}t_1\,\mathrm{d}t_2\cdots\mathrm{d}t_p+\int_{\mathbb{R}^p}K_2{}^{p,\beta}(t_1,t_2,\ldots,t_p;\mathrm{i}s_1,\mathrm{i}s_2\cdots\mathrm{i}s_p)$$

$$\times(x\alpha_{t_1}(k_\alpha)\alpha_{t_2}(k_\alpha)\cdots\alpha_{t_{p+1}}(k_\alpha)1_\phi,\alpha_{t_p}(k_\alpha)1_\phi)\,\mathrm{d}t_1\,\mathrm{d}t_2\cdots\mathrm{d}t_p+\cdots$$

$$+\int_{\mathbb{R}^p}K_{p+1}{}^{p,\beta}(t_1,t_2,\ldots,t_p,\mathrm{i}s_1,\mathrm{i}s_2,\ldots,\mathrm{i}s_p)(x1_\phi,\alpha_{t_p}(k_\alpha)\alpha_{t_{p-1}}(k_\alpha)\cdots\alpha_{t_1}(k_\alpha)1_\phi)$$

$$\left.\times\,\mathrm{d}t_1\,\mathrm{d}t_2\cdots\mathrm{d}t_p\right\}.$$

Hence

$$|(\exp(-\beta(H_\phi+\pi_\phi(k_\alpha)))1_\phi,1_\phi)|\leqslant\sum_{p=0}^{\infty}\int_{0\leqslant s_1\leqslant s_2\leqslant\cdots\leqslant s_p\leqslant\beta}\mathrm{d}s_1\,\mathrm{d}s_2\cdots\mathrm{d}s_p$$

$$\times\left\{\int_{\mathbb{R}^p}K_1{}^{p,\beta}(t_1,t_2,\ldots,t_p;\mathrm{i}s_1,\mathrm{i}s_2,\ldots,\mathrm{i}s_p)\|x\|\,\|k_\alpha\|^p\,\mathrm{d}t_1\,\mathrm{d}t_2\cdots\mathrm{d}t_p\right.$$

$$+\int_{\mathbb{R}^p} K_2^{p,\beta}(t_1,t_2,\ldots,t_p;\mathrm{i}s_1,\mathrm{i}s_2,\ldots,\mathrm{i}s_p)\|x\|\,\|k_\alpha\|^p\,\mathrm{d}t_1\,\mathrm{d}t_2\cdots\mathrm{d}t_p$$

$$+\int_{\mathbb{R}^p} K_{p+1}^{p,\beta}(t_1,t_2,\ldots,t_p;\mathrm{i}s_1,\mathrm{i}s_2,\ldots,\mathrm{i}s_p)\|x\|\,\|k_\alpha\|^p\,\mathrm{d}t_1\,\mathrm{d}t_2\cdots\mathrm{d}t_p\Big\}$$

$$=\sum_{p=0}^{\infty}\|x\|\,\|k_\alpha\|^p\int_{0\leqslant s_1\leqslant s_2\leqslant\cdots\leqslant s_p\leqslant\beta}\mathrm{d}s_1\,\mathrm{d}s_2\cdots\mathrm{d}s_p$$

$$=\sum_{p=0}^{\infty}\frac{\beta^p}{p!}\|k_\alpha\|^p\|x\|=\exp(\beta\|k_\alpha\|)\|x\|\leqslant\exp(\beta M)\|x\|$$

Hence by 4.4.8 there is a bounded continuous function $F_{x,h,p}$ on $S_{p,\beta}$ ($x\in\mathcal{M}, h\in\mathcal{M}^s$) which is analytic with respect to p-variables in $S_{p,\beta}{}^0$ such that

$$\begin{aligned}&F_{x,h,p}(t_1,t_2,\ldots,t_j,t_{j+1}+\mathrm{i}\beta,\ldots,t_p+\mathrm{i}\beta)\\&\quad=(xU_\phi(t_1)hU_\phi(-t_1)\cdots U_\phi(t_j)hU_\phi(-t_j)1_\phi,\\&\qquad U_\phi(t_p)hU_\phi(-t_p)\cdots U_\phi(t_{j+1})hU_\phi(-t_{j+1})1_\phi)\qquad(j=0,1,2,\ldots,p-1),\end{aligned}$$

where $U_\phi(t_0)hU_\phi(-t_0)=1_{\mathscr{H}_\phi}$, and $\{(x\pi_\phi(\alpha_{z_1}(k_\alpha)\alpha_{z_2}(k_\alpha)\cdots\alpha_{z_p}(k_\alpha))1_\phi,1_\phi)\}$ converges to $F_{x,h,p}(z_1,z_2,\ldots,z_p)$ uniformly on every compact subset of $S_{p,\beta}$.

Moreover by Lebesgue's dominated convergence theorem,

$$\lim\sup_{\|x\|\leqslant 1}\Big|(x\exp(\mathrm{i}z(H_\phi+\pi_\phi(k_\alpha))1_\phi,1_\phi)-\sum_{p=0}^{\infty}(\mathrm{i}z)^p\int_{0\leqslant s_1\leqslant s_2\leqslant\cdots\leqslant s_p\leqslant\beta}F_{x,h,p}(s_1z,s_2z,\ldots,s_pz)\,\mathrm{d}s_1\,\mathrm{d}s_2\cdots\mathrm{d}s_p\Big|=0,$$

uniformly on every compact subset of $S_{p,\beta}$. Put

$$f(z,h)(x)=\sum_{p=0}^{\infty}(\mathrm{i}z)^p\int_{0\leqslant s_1\leqslant s_2\leqslant\cdots\leqslant s_p\leqslant\beta}F_{x,h,p}(s_1z,s_2z,\ldots,s_pz_p)\,\mathrm{d}s_1\,\mathrm{d}s_2\cdots\mathrm{d}s_p$$

for $x\in\mathcal{M}$; then $\|f(z,\pi_\phi(k_\alpha))-f(z,h)\|\to 0$, uniformly on every compact subset of S_β. □

Proof of Theorem 4.4.7. Condition (1): For $k(=k^*)\in A_2(\delta)$,

$$f(z,\pi_\phi(k))(x)=(x\exp(\mathrm{i}z(H_\phi+\pi_\phi(k)))\exp(-\mathrm{i}zH_\phi)1_\phi,1_\phi)\qquad(z\in\mathbb{C}),$$

so that it is a bounded analytic function on S_β. Moreover

$$\begin{aligned}|f(t+\mathrm{i}\beta,\pi_\phi(k))(x)|&=|(x\exp(\mathrm{i}t(H_\phi+\pi_\phi(k))\exp(-\beta(H_\phi+\pi_\phi(k)))1_\phi,1_\phi)|\\&=|(\exp(-\mathrm{i}tH_\phi)x\exp(\mathrm{i}tH_\phi)\exp(-\mathrm{i}tH_\phi)\exp(\mathrm{i}t(H_\phi+\pi_\phi(k)))\end{aligned}$$

$$\times \exp(-\beta(H_\phi + \pi_\phi(k)))1_\phi, 1_\phi)|$$
$$\leqslant \exp(\beta\|k\|)\|\exp(-\mathrm{i}tH_\phi)x\exp(\mathrm{i}tH_\phi)\exp(-\mathrm{i}tH_\phi)\exp(\mathrm{i}t(H_\phi+\pi_\phi(k)))\|$$

(cf. the proof of 4.4.9) for $x \in \mathcal{M}$.

Also,

$$|f(t, \pi_\phi(k))(x)| = |(x\exp(\mathrm{i}t(H_\phi + \pi_\phi(k)))1_\phi, 1_\phi)| \leqslant \|x\| \qquad (x \in \mathcal{M}).$$

Hence

$$|f(z, \pi_\phi(k))(x)| \leqslant \exp(|\beta|\,\|k\|)\|x\|, \qquad (x \in \mathcal{M}, z \in S_\beta).$$

By Kaplansky's density theorem and 4.4.9, we have

$$\|f(z, h)\| \leqslant \exp(|\beta|\,\|h\|) \qquad (z \in S_\beta).$$

Conditions (2) and (3) are clear from the proof of 4.4.9. and Kaplansky's density theorem.

Condition (4): For $k(=k^*) \in A_2(\delta)$,

$$f(0, \pi_\phi(k))(x) = (x1_\phi, 1_\phi) = \tilde{\phi}(x) \qquad (x \in \mathcal{M}),$$

and so $f(0, h) = \tilde{\phi}$ for $h \in \mathcal{M}^s$.

$$f(\mathrm{i}\beta, \pi_\phi(k))(x) = (x\exp(-\beta(H_\phi + \pi_\phi(k)))1_\phi, 1_\phi)(x \in \mathcal{M})$$

and so it is positive so that $f(\mathrm{i}\beta, h)$ is also positive $(h \in \mathcal{M}^s)$.

$$\begin{aligned} f(t + \mathrm{i}\beta, \pi_\phi(k))(x) &= (x\exp(\mathrm{i}(t + \mathrm{i}\beta)(H_\phi + \pi_\phi(k)))1_\phi, 1_\phi) \\ &= (\exp(-\mathrm{i}tH_\phi)\exp(\mathrm{i}t(H_\phi + \pi_\phi(k)))\exp(-\mathrm{i}t(H_\phi + \pi_\phi(k)))x \\ &\quad \times \exp(\mathrm{i}t(H_\phi + \pi_\phi(k)))\exp(-\beta(H_\phi + \pi_\phi(k)))1_\phi, 1_\phi) \\ &= (\exp(\mathrm{i}t(H_\phi + \pi_\phi(k)))\exp(-\mathrm{i}tH_\phi)\exp(\mathrm{i}t(H_\phi + \pi_\phi(k))) \\ &\quad \times \exp(-\mathrm{i}t(H_\phi + \pi_\phi(k)))x\exp(-\beta(H_\phi + \pi_\phi(k)))1_\phi, 1_\phi) \\ &= (\exp(\mathrm{i}t(H_\phi + \pi_\phi(k)))\exp(-\mathrm{i}tH_\phi)x\exp(-\beta(H_\phi + \pi_\phi(k)))1_\phi, 1_\phi) \\ &= f(\mathrm{i}\beta, \pi_\phi(k))(\exp(\mathrm{i}t(H_\phi + \pi_\phi(k))\exp(-\mathrm{i}tH_\phi)x) \quad (x \in \mathcal{M}) \end{aligned}$$

(use the invariance of KMS state and

$$\exp(-\mathrm{i}tH_\phi)\exp(\mathrm{i}t(H_\phi + \pi_\phi(k))) \in \mathcal{M}).$$

On the other hand,

$$\begin{aligned} f(t, \pi_\phi(k))(x) &= (x\exp(\mathrm{i}t(H_\phi + \pi_\phi(k)))\exp(-\mathrm{i}tH_\phi)1_\phi, 1_\phi) \\ &= f(0, \pi_\phi(k))(x\exp(\mathrm{i}t(H_\phi + \pi_\phi(k)))\exp(-\mathrm{i}tH_\phi)). \end{aligned}$$

From the strong continuity of $h \mapsto \exp(\mathrm{i}t(H_\phi + h))$ $(h \in \mathcal{M}^s)$ and (3), one can easily conclude

$$f(t + \mathrm{i}\beta, h)(x) = f(\mathrm{i}\beta, h)(\exp(\mathrm{i}t(H_\phi + h))\exp(-\mathrm{i}tH_\phi)x)$$

and

$$f(t,h)(x)=f(0,h)(x\exp(\mathrm{i}t(H_\phi+h))\exp(-\mathrm{i}tH_\phi)) \qquad (h\in\mathcal{M}^s, x\in\mathcal{M}).$$

Suppose that $f(\mathrm{i}\beta,h)(x^*x)=0$; then

$$\begin{aligned}|f(t+\mathrm{i}\beta,h)(x^*x)| &= |f(\mathrm{i}\beta,h)(\exp(\mathrm{i}t(H_\phi+h))\exp(-\mathrm{i}tH_\phi)x^*x)|\\ &\leqslant f(\mathrm{i}\beta,h)(\exp(\mathrm{i}t(H_\phi+h))\exp(-\mathrm{i}tH_\phi)x^*x\exp(\mathrm{i}tH_\phi)\\ &\quad\times\exp(-\mathrm{i}t(H_\phi+h)))^{1/2}f(\mathrm{i}\beta,h)(x^*x)^{1/2}=0.\end{aligned}$$

Since $f(z,h)(x^*x)$ is analytic on $S_\beta{}^0$, $f(0,h)(x^*x)=\tilde{\phi}(x^*x)=0$, and so $x=0$. Hence $f(\mathrm{i}\beta,h)$ is a faithful normal positive linear functional on $\mathcal{M}$.

Condition (5): For $h\in\mathcal{M}^s$, let $\{k_\alpha\}$ be a directed set of self-adjoint elements in $A_2(\delta)$ such that $\|k_\alpha\|\leqslant\|h\|$ and $\pi_\phi(k_\alpha)\to h$ (strongly). Let $H_\phi+\pi_\phi(k_\alpha)=\int_{-\infty}^{\infty}\lambda\mathrm{d}E_\alpha(\lambda)$ (resp. $H_\phi+h=\int_{-\infty}^{\infty}\lambda\mathrm{d}E(\lambda)$) be the spectral decomposition of $H_\phi+\pi_\phi(k_\alpha)$ (resp. $H_\phi+h$). Since

$$\begin{gathered}(1\pm\mathrm{i}(H_\phi+\pi_\phi(k_\alpha))^{-1}\to(1\pm\mathrm{i}(H_\phi+h))^{-1} \text{ (strongly)},\\ \exp(\mathrm{i}t(H_\phi+\pi_\phi(k_\phi((k_\alpha)))\to\exp(\mathrm{i}t(H_\phi+h))\end{gathered}$$

uniformly on every compact subset of $\mathbb{R}$ in the strong operator topology. Therefore,

$$\lim_\alpha\int_{-\infty}^{\infty}\exp(\mathrm{i}t\lambda)\,\mathrm{d}(E_\alpha(\lambda)1_\phi,1_\phi)=\int_{-\infty}^{\infty}\exp(\mathrm{i}t\lambda)\,\mathrm{d}(E(\lambda)1_\phi,1_\phi).$$

Let $f\in L^1(R)$; then

$$\begin{aligned}&\int_{-\infty}^{\infty}f(t)\,\mathrm{d}t\int_{-\infty}^{\infty}\exp(\mathrm{i}t\lambda)\,\mathrm{d}(E_\alpha(\lambda)1_\phi,1_\phi)\\ &\quad\to\int_{-\infty}^{\infty}f(t)\,\mathrm{d}t\int_{-\infty}^{\infty}\exp(\mathrm{i}t\lambda)\mathrm{d}(E(\lambda)1_\phi,1_\phi).\end{aligned}$$

Since

$$\begin{aligned}&\int_{-\infty}^{\infty}f(t)\,\mathrm{d}t\int_{-\infty}^{\infty}\exp(\mathrm{i}t\lambda)\,\mathrm{d}(E_\alpha(\lambda)1_\phi,1_\phi)\\ &\quad=\int_{-\infty}^{\infty}\mathrm{d}(E_\alpha(\lambda)1_\phi,1_\phi)\int_{-\infty}^{\infty}f(t)\exp(\mathrm{i}t\lambda)\,\mathrm{d}t\\ &\quad\to\int_{-\infty}^{\infty}f(t)\,\mathrm{d}t\int_{-\infty}^{\infty}\exp(\mathrm{i}t\lambda)\,\mathrm{d}(E(\lambda)1_\phi,1_\phi)\\ &\quad=\int_{-\infty}^{\infty}\mathrm{d}(E(\lambda)1_\phi,1_\phi)\int_{-\infty}^{\infty}f(t)\exp(\mathrm{i}t\lambda)\,\mathrm{d}t,\end{aligned}$$

we have

$$\lim_{\alpha} \int_{-\infty}^{\infty} \hat{f}(\lambda)\, \mathrm{d}(E_\alpha(\lambda)1_\phi, 1_\phi) = \int_{-\infty}^{\infty} \hat{f}(\lambda)\mathrm{d}(E(\lambda)1_\phi, 1_\phi) \qquad \text{for all } f \in L^1(R),$$

where $\hat{f}$ is the Fourier transform of f. Since $\widehat{L^1(\mathbb{R})}$ is norm-dense in $C_0(\mathbb{R})$, the directed set $\{\mu_\alpha(\lambda) = (E_\alpha(\lambda)1_\phi, 1_\phi)\}$ of probability Radon measures converges to the probability Radon measure $\{\mu(\lambda) = (E(\lambda)1_\phi, 1_\phi)\}$ in $\sigma(C_0(\mathbb{R})^*, C_0(\mathbb{R}))$. Hence there is a subsequence $\{\mu_n\}$ of $\{\mu_\alpha\}$ such that $\mu_n \to \mu$ in $\sigma(C_0(\mathbb{R})^*, C_0(\mathbb{R}))$.

Let $\{f_m\}$ be a monotone increasing sequence of positive continuous functions on $\mathbb{R}$ such that $0 \leqslant f_m \leqslant 1, f_m(\lambda) = 1$ for λ with $|\lambda| \leqslant m$ and $f_m(\lambda) = 0$ for λ with $|\lambda| \geqslant m+1$. Then

$$\begin{aligned}\int_{-\infty}^{\infty} f_m(\lambda)|\exp(\mathrm{i}z\lambda)|\,\mathrm{d}\mu_n(\lambda) &= \int_{-\infty}^{\infty} f_m(\lambda)\exp(-\operatorname{Im}(z)\lambda)\,\mathrm{d}\mu_n(\lambda)\\ &\to \int_{-\infty}^{\infty} f_m(\lambda)|\exp(\mathrm{i}z\lambda)|\,\mathrm{d}\mu(\lambda).\end{aligned}$$

On the other hand,

$$\begin{aligned}\int_{-\infty}^{\infty} f_m(\lambda)|\exp(\mathrm{i}z\lambda)|\,\mathrm{d}\mu_n(\lambda) &= \int_{-\infty}^{\infty} f_m(\lambda)\exp(-\operatorname{Im}(z)\lambda)\,\mathrm{d}(E_n(\lambda)1_\phi, 1_\phi)\\ &\leqslant \int_{-\infty}^{\infty} \exp(-\operatorname{Im}(z)\lambda)\,\mathrm{d}(E_n(\lambda)1_\phi, 1_\phi)\\ &= (\exp(-\operatorname{Im}(z)(H_\phi + \pi_\phi(k_n)))1_\phi, 1_\phi)\\ &\leqslant \exp(|\beta|\,\|k_n\|) \leqslant \exp(|\beta|\,\|h\|) \quad \text{for } z \in S_\beta.\end{aligned}$$

Therefore

$$\int_{-\infty}^{\infty} f_m(\lambda)\exp(-\operatorname{Im}(z)\lambda)\,\mathrm{d}\mu(\lambda) \leqslant \exp(|\beta|\,\|h\|) \qquad (z \in S_\beta).$$

By Lebesgue's monotone convergence theorem,

$$\begin{aligned}\lim_{m\to\infty} \int_{-\infty}^{\infty} f_m(\lambda)\exp(-\operatorname{Im}(z)\lambda)\,\mathrm{d}\mu(\lambda) &= \int_{-\infty}^{\infty} \exp(-|\operatorname{Im}(z)|\lambda)\,\mathrm{d}\mu(\lambda)\\ &\leqslant \exp(|\beta|\,\|h\|) \qquad (z \in S_\beta).\end{aligned}$$

Therefore $F(z) = \int_{-\infty}^{\infty} \exp(\mathrm{i}z\lambda)\,\mathrm{d}\mu(\lambda)$ is defined on S_β. Let

$$F_n(z) = \int_{-n}^{n} \exp(\mathrm{i}z\lambda)\,\mathrm{d}\mu(\lambda) \qquad \text{for } z \in S_\beta;$$

then

$$|\exp(iz\lambda)| \leqslant \exp(|\beta|n) \qquad \text{for } z \in S_\beta,$$

so that $F_n(z)$ is a bounded continuous function on S_β and is analytic in the interior of S_β. Since $F(z) = \lim_n F_n(z)$ for $z \in S_\beta$, $F(z)$ is analytic in the interior of S_β. Since $|F_n(z)| \leqslant \exp(|\beta| \, \|h\|)$, it is bounded on S_β.

$$F_n(z) = \int_{-\infty}^{\infty} K_1(t,z)F_n(t)\,dt + \int_{-\infty}^{\infty} K_2(t,z)F_n(t+i\beta)\,dt \qquad \text{for } z \in S_\beta{}^0.$$

Hence

$$F(z) = \int_{-\infty}^{\infty} K_1(t,z)F(t)\,dt + \int_{-\infty}^{\infty} K_2(t,z)F(t+i\beta)\,dt \qquad \text{for } z \in S_\beta{}^0.$$

Therefore $F(z)$ is continuous on S_β.

$$F(t) = \int_{-\infty}^{\infty} \exp(it\lambda)\,d(E(\lambda)1_\phi, 1_\phi) = (\exp(it(H_\phi + h))1_\phi, 1_\phi)$$

$$= (\exp(it(H_\phi + h))\exp(-itH_\phi)1_\phi, 1_\phi) = f(t,h)(1_{\mathscr{H}_\phi}) \qquad (t \in R).$$

Hence $F(z) = f(z,h)(1_{\mathscr{H}_\phi})$ $(z \in S_\beta)$ and so

$$f(is,h)(1_{\mathscr{H}_\phi}) = \lim_\alpha f(is, \pi_\phi(k_\alpha))(1_{\mathscr{H}_\phi}) = \lim_\alpha (\exp(-s(H_\phi + \pi_\phi(k_\alpha)))1_\phi, 1_\phi)$$

$$= \lim_\alpha \|\exp(-s(H_\phi + \pi_\phi(k_\alpha))/2)1_\phi\|^2 = F(is)$$

$$= \int_{-\infty}^{\infty} \exp(-s\lambda)\,d\mu(\lambda) = \int_{-\infty}^{\infty} \exp(-s\lambda)\,d(E(\lambda)1_\phi, 1_\phi)$$

$$= \|\exp(-s(H_\phi + h)/2)1_\phi\|^2 \qquad (0 \leqslant s \leqslant \beta).$$

Therefore $1_\phi \in \mathscr{D}(\exp - (s(H_\phi + h)/2))$ $(0 \leqslant s \leqslant \beta)$ and

$$\lim_\alpha \|\exp(-s(H_\phi + \pi_\phi(k_\alpha))/2)1_\phi\| = \|\exp(-s(H_\phi + h)/2)1_\phi\|.$$

Moreover

$$(\exp(-s(H_\phi + \pi_\phi(k_\alpha))/2)1_\phi, x1_\phi) = f(is/2, \pi_\phi(k_\alpha))(x^*) \to f(i(s/2), h)(x^*) \qquad (x \in \mathscr{M}).$$

$z \mapsto \exp(iz(H_\phi + h))1_\phi = \int_{-\infty}^{\infty} \exp(iz\lambda)dE(\lambda)1_\phi$ is bounded continuous on $S_{\beta/2}$ and analytic in the interior of $S_{\beta/2}$. Put $G_x(z) = (\exp(iz(H_\phi + h))1_\phi, x1_\phi)$ $(x \in \mathscr{M})$; then

$$G_x(t) = (\exp(it(H_\phi + h))1_\phi, x1_\phi) = (x^* \exp(it(H_\phi + h))\exp(-itH_\phi)1_\phi, 1_\phi)$$

$$= f(0,h)(x^* \exp(it(H_\phi + h))\exp(-itH_\phi)) = f(t,h)(x^*)$$

Hence by the analyticity,

$$G_x(z) = f(z, h)(x^*) \qquad (z \in S_{\beta/2})$$

and so

$$\lim_\alpha f\left(\mathrm{i}\frac{s}{2}, \pi_\phi(k_\alpha)\right)(x^*) = f\left(\mathrm{i}\frac{s}{2}, h\right)(x^*) = (\exp(-(s/2)(H_\phi + h))1_\phi, x1_\phi)$$
$$(x \in \mathscr{M};\ 0 \leqslant s \leqslant \beta)$$

Therefore

$$(\exp(-(s/2)(H_\phi + \pi_\phi(k_\alpha))1_\phi, x1_\phi) \to (\exp(-s(H_\phi + h)/2)1_\phi, x1_\phi) \qquad (x \in \mathscr{M})$$

so that $\exp(-s(H_\phi + \pi_\phi(k_\alpha))/2)1_\phi \to \exp(-s(H_\phi + h)/2)1_\phi$ in the norm of $\mathscr{H}_\phi$. We have seen that

$$(x\exp(-\beta(H_\phi + \pi_\phi(k_\alpha))1_\phi, 1_\phi)$$
$$= (x\exp(-\beta(H_\phi + \pi_\phi(k_\alpha))/2)1_\phi, (\exp(-\beta(H_\phi + \pi_\phi(k_\alpha))/2)1_\phi)$$
$$(x \in \mathscr{M})$$

and so

$$\lim_\alpha f(\mathrm{i}\beta, \pi_\phi(k_\alpha))(x)$$
$$= f(\mathrm{i}\beta, h)(x) = (x\exp(-\beta(H_\phi + h)/2)1_\phi, \exp(-\beta(H_\phi + h)/2)1_\phi)$$
$$(x \in \mathscr{M}).$$

Also,

$$f(z, h)(1_{\mathscr{H}_\phi}) = (\exp(\mathrm{i}\,\mathrm{Re}(z)(H_\phi + h))\exp(-\mathrm{Im}(z)(H_\phi + h)/2)1_\phi,$$
$$\exp(-\mathrm{Im}(z)(H_\phi + h)/2)1_\phi)$$

for $z \in S_\beta$. The rest is clear.

Condition 6: Let $k_\alpha \in A_2(\delta)$ such that $\| k_\alpha \| \leqslant \| h \|$ and $\pi_\phi(k_\alpha) \to h$ (strongly). Then by the previous considerations, $f(\mathrm{i}\beta, \pi_\phi(k_\alpha))(\pi_\phi(a))/f(\mathrm{i}\beta, \pi_\phi(k_\alpha))(1_{\mathscr{H}_\phi})$ $(a \in A)$ is a KMS state for $\{A, \exp(t(\delta + \delta_{\mathrm{i}k_\alpha}))\}$ $(t \in R)$ at β.

Since $\| f(\mathrm{i}\beta, \pi_\phi(k_\alpha)) - f(\mathrm{i}\beta, h) \| \to 0$ and

$$\exp(\mathrm{i}t(H_\phi + \pi_\phi(k_\alpha)))\pi_\phi(a)\exp(-\mathrm{i}t(H_\phi + \pi_\phi(k_\phi)))$$
$$\to \exp(-\mathrm{i}t(H_\phi + h))\pi_\phi(a)\exp(\mathrm{i}t(H_\phi + h)) \qquad \text{(strongly)},$$

$$f(\mathrm{i}\beta, h)(\pi_\phi(a)\exp(t(\delta + \delta_0))(b)))/f(\mathrm{i}\beta, h)(1_{\mathscr{H}_\phi})$$
$$= f(\mathrm{i}\beta, h)(\pi_\phi(a)\exp(\mathrm{i}t(H_\phi + h))\pi_\phi(b)\exp(-\mathrm{i}t(H_\phi + h))/f(\mathrm{i}\beta, h)(1_{\mathscr{H}_\phi})$$
$$= \lim_\alpha f(\mathrm{i}\beta, \pi_\phi(k_\alpha))(\pi_\phi(a)\exp(\mathrm{i}t(H_\phi + \pi_\phi(k_\alpha)))\pi_\phi(b)$$
$$\times \exp(-\mathrm{i}t(H_\phi + \pi_\phi(k_\alpha)))/f(\mathrm{i}\beta, \pi_\phi(k_\alpha))(1_{\mathscr{H}_\phi})$$

and analogously

$$\begin{aligned} & f(\mathrm{i}\beta, h)(\pi_\phi(\exp(t(\delta+\delta_0))(b)a))/f(\mathrm{i}\beta,h)(1_{\mathscr{H}_\phi}) \\ &\quad = \lim_\alpha f(\mathrm{i}\beta, \pi_\phi(k_\alpha))(\pi_\phi(\exp(t(\delta+\delta_{\mathrm{i}k_\alpha}))(b)a)/f(\mathrm{i}\beta,\pi_\phi(k_\alpha))(1_{\mathscr{H}_\phi}). \end{aligned}$$

Therefore one can easily see that $f(\mathrm{i}\beta, h)(\pi_\phi(a))/f(\mathrm{i}\beta,h)(1_{\mathscr{H}_\phi})$ is a KMS state for $\{A, \exp(t(\delta+\delta_0))\}$ at β.

Condition (7): For $h, k \in A_2(\delta)^s$,

$$\begin{aligned} \| f(z,\pi_\phi(h)) - f(z,\pi_\phi(k))\| &= \sup_{\|x\|\leqslant 1} | f(z,\pi_\phi(h))(x) - f(z,\pi_\phi(k))(x)| \\ &\leqslant \sup_{\|x\|\leqslant 1} |(x\exp(\mathrm{i}z(H_\phi+\pi_\phi(k)))1_\phi, 1_\phi) - (x\exp(\mathrm{i}z(H_\phi+\pi_\phi(h)))1_\phi,1_\phi)|, \end{aligned}$$

and

$$\begin{aligned} & |(\pi_\phi(a)\pi_\phi(\alpha_{t_1}(h)\cdots\alpha_{t_j}(h)\alpha_{t_{j+1}+\mathrm{i}\beta}(h)\cdots\alpha_{t_p+\mathrm{i}\beta}(h))1_\phi,1_\phi) \\ &\qquad - (\pi_\phi(a)\pi_\phi(\alpha_{t_1}(k)\cdots\alpha_{t_j}(k)\alpha_{t_{j+1}+\mathrm{i}\beta}(k)\cdots\alpha_{t_p+\mathrm{i}\beta}(k))1_\phi,1_\phi)| \\ &\quad = |(\pi_\phi(a)\pi_\phi(\alpha_{t_1}(h)\cdots\alpha_{t_j}(h))1_\phi, \pi_\phi(\alpha_{t_p}(h)\cdots\alpha_{t_{j+1}}(h))1_\phi) \\ &\qquad - (\pi_\phi(a)\pi_\phi(\alpha_{t_1}(k)\cdots\alpha_{t_j}(k))1_\phi, \pi_\phi(\alpha_{t_p}(k)\cdots\alpha_{t_{j+1}}(k))1_\phi)| \\ &\quad \leqslant \|\pi_\phi(a)\|\,\|h-k\|\,\|h\|^{p-1} + \|\pi_\phi(a)\|\,\|h-k\|\,\|h\|^{p-2}\|k\| + \cdots \\ &\qquad + \|\pi_\phi(a)\|\,\|h-k\|\,\|k\|^{p-1} \leqslant p\|\pi_\phi(a)\|\max\{\|h\|,\|k\|\}^{p-1}\|h-k\|. \end{aligned}$$

Hence,

$$\begin{aligned} & \left| \int_{0\leqslant s_1\leqslant s_2\leqslant\cdots\leqslant s_p\leqslant 1} \{(\pi_\phi(a)\pi_\phi(\alpha_{s_1z}(h)\alpha_{s_2z}(h)\cdots\alpha_{s_pz}(h))1_\phi,1_\phi) \right. \\ &\qquad \left. - (\pi_\phi(a)\pi_\phi(\alpha_{s_1z}(k)\alpha_{s_2z}(k)\cdots\alpha_{s_pz}(k))1_\phi,1_\phi)\}\,\mathrm{d}s_1\cdots\mathrm{d}s_p \right| \\ & = \left| \int_{0\leqslant s_1\leqslant s_2\leqslant\cdots\leqslant s_p\leqslant 1} \mathrm{d}s_1\cdots\mathrm{d}s_p \left[\int_{\mathbb{R}^p} K_1{}^{p,\beta}(t_1,t_2,\ldots t_p; s_1z, s_2z, \ldots s_pz) \right.\right. \\ &\qquad \times \left\{ \left(\pi_\phi(a)\pi_\phi(\alpha_{t_1}(h)\cdots\alpha_{t_p}(h))1_\phi,1_\phi \right) \right. \\ &\qquad \left.\left. - (\pi_\phi(a)\pi_\phi(\alpha_{t_1}(k)\cdots\alpha_{t_p}(k))1_\phi, 1_\phi \right)\right\}\mathrm{d}t_1\,\mathrm{d}t_2\cdots\mathrm{d}t_p \\ &\qquad + \int_{\mathbb{R}^p} K_2{}^{p,\beta}(t_1,t_2,\ldots,t_p; s_1z,s_2z,\ldots s_pz) \\ &\qquad \times \{(\pi_\phi(a)\pi_\phi(\alpha_{t_1}(h)\cdots\alpha_{t_{p-1}}(h)\alpha_{t_p+\mathrm{i}\beta}(h))1_\phi,1_\phi) \\ &\qquad - (\pi_\phi(a)\pi_\phi(\alpha_{t_1}(k)\cdots\alpha_{t_{p-1}}(k)\alpha_{t_p+1\beta}(k))1_\phi,1_\phi)\}\,\mathrm{d}t_1\,\mathrm{d}t_2\cdots\mathrm{d}t_p + \cdots \end{aligned}$$

$$+\int_{\mathbb{R}^p} K_{p+1}{}^{p,\beta}(t_1,t_2,\ldots,t_p;s_1z,s_2z,\ldots,s_pz)$$
$$\times\{(\pi_\phi(a)\pi_\phi(\alpha_{t_1+i\beta}(h)\cdots\alpha_{t_p+i\beta}(h)))1_\phi,1_\phi)$$
$$-(\pi_\phi(a)\pi_\phi(\alpha_{t_1+\beta}(k)\cdots\alpha_{t_p+i\beta}(k))1_\phi,1_\phi)\}\,dt_1\cdots dt_p\Bigg]\Bigg|$$

$$\leqslant\int_{0\leqslant s_1\leqslant s_2\leqslant\cdots\leqslant 1} ds_1\,ds_2\cdots ds_p\Bigg[\int_{\mathbb{R}^p} K_1{}^{p,\beta}(t_1,t_2,\ldots,t_p;$$
$$s_1z,s_2z,\ldots,s_pz)p\,\|\pi_\phi(a)\|\max\{\|h\|,\|k\|\}^{p-1}\|h-k\|\,d_{t_1}\cdots dt_p$$
$$+\int_{\mathbb{R}^p} K_2{}^{p,\beta}(t_1,t_2,\ldots,t_p;s_1z,s_2z,\ldots,s_pz)p\,\|\pi_\phi(a)\|\max\{\|h\|,\|k\|\}^{p-1}$$
$$\times\|h-k\|dt_1\,dt_2\cdots dt_p+\cdots\int_{\mathbb{R}^p} K_{p+1}{}^{p,\beta}(t_1,t_2,\ldots,t_p;s_1z,s_2z,\ldots s_pz)p$$
$$\times\|\pi_\phi(a)\|\max\{\|h\|,\|k\|\}^{p-1}\|h-k\|\}\,dt_1\,dt_2\cdots dt_p\Bigg]$$
$$=\int_{0\leqslant s_1\leqslant s_2\leqslant\cdots\leqslant 1} p\,\|\pi_\phi(a)\|\max\{\|h\|,\|k\|\}^{p-1}\|h-k\|\,ds_1\cdots ds_p$$
$$=\frac{p}{p!}\|\pi_\phi(a)\|\max\{\|h\|,\|k\|\}^{p-1}\|h-k\|.$$

Hence

$$|(\pi_\phi(a)\exp(iz(H_\phi+h))1_\phi,1_\phi)-(\pi_\phi(a)\exp(iz(H_\phi+k))1_\phi,1_\phi)|$$
$$\leqslant\sum_{p=1}^{\infty}|iz|^p\Bigg|\int_{0\leqslant s_1\leqslant s_2\leqslant\cdots\leqslant 1}\{(\pi_\phi(a)\pi_\phi(\alpha_{s_1}z(h)\cdots\alpha_{s_p}z(h))1_\phi,1_\phi)$$
$$-(\pi_\phi(a)\pi_\phi(\alpha_{s_1}z(k)\cdots\alpha_\phi(\alpha_{s_p}z(k))1_\phi,1_\phi)\}\,ds_1\,ds_2\cdots ds_p\Bigg|$$
$$=\sum_{p=1}^{\infty}|iz|^p\frac{\|\pi_\phi(a)\|}{(p-1)!}\max\{\|h\|,\|k\|\}^{p-1}\|h-k\|$$
$$=|iz|\,\|\pi_\phi(a)\|\,\|h-k\|\exp(|iz|\max\{\|h\|,\|k\|\})$$
$$\leqslant|\beta|\,\|\pi_\phi(a)\|\,\|h-k\|\exp(|\beta|\max\{\|h\|,\|k\|\}).$$

Hence

$$\|f(z,\pi_\phi(h))-f(z,\pi_\phi(k))\|\leqslant|\beta|\exp(|\beta|\max\{\|h\|,\|k\|\})\|h-k\|$$
$$\text{for } z\in S_\beta.$$

For $h,k\in\mathscr{M}^s$, from the considerations in the proof of 4.4.9 and the similar considerations with the above, one can conclude that

$$\| f(z,h) - f(z,k) \| \leqslant |\beta| \exp(|\beta| \max\{\|h\|, \|k\|\}) \|h - k\| \qquad \text{for } z \in S_\beta.$$

Condition (8): Let $h_n \in \mathcal{M}^s$ with $\|h_n\| \leqslant r(n = 1, 2, \ldots,)$. For $x \in \mathcal{M}$, $|f(z, h_n)(x)| \leqslant \exp(|\beta|r)\|x\|$ for $z \in S_\beta$.

$$\left| \frac{d}{dt} f(t, h_n)(x) \right| = \left| \frac{d}{dt} f(0, h_n)(x \exp(it(H_\phi + h)) \exp(-itH_\phi)) \right|$$

$$= \left| \frac{d}{dt} (x \exp(it(H_\phi + h_n)) \exp(-itH_\phi) 1_\phi, 1_\phi) \right|$$

$$= |(x \exp(it(H_\phi + h_n)) h_n 1_\phi, 1_\phi)| \leqslant \|x\| \|h_n\|$$

and

$$\left| \frac{d}{dt} f(t + i\beta, h_n)(x) \right| = \left| \frac{d}{dt} f(i\beta, h_n)(\exp(it(H_\phi + h_n)) \exp(-itH_\phi)x) \right|$$

$$= |f(i\beta, h_n)(\exp(it(H_\phi + h_n))(H_\phi + h_n - H_\phi)\exp(-itH_\phi)x)|$$

$$= |f(i\beta, h_n)(\exp(it(H_\phi + h_n))h_n \exp(-itH_\phi)x)|$$

$$= |f(i\beta, h_n)(\exp(it(H_\phi + h_n))\exp(-itH_\phi)\exp(itH_\phi)h_n \exp(-itH_\phi)x)|$$

$$\leqslant \|f(i\beta, h_n)\| \|h_n\| \|x\|.$$

Hence by the well-known theorem of analytic functions, there is a subsequence $\{f(z, h_{n_j})(x)\}$ of $\{f(z, h_n)(x)\}$ such that $\{f(z, h_{n_j})(x)\}$ converges to a bounded continuous function $F(z)(x)$ on S_β, which is analytic on the interior S_β^0 of S_β, uniformly on every compact subset of S_β, and $(d/dt)F(t)(x)$ and $(d/dt)F(t + i\beta)(x)$ are bounded measurables on $\mathbb{R}$, and moreover $|(d/dt)F(t)(x)| \leqslant \|x\|r$ a.e. and $|(d/dt)F(t + i\beta)(x)| \leqslant \exp(|\beta|r)r\|x\|$ a.e.

Now let $\Omega_{r,x}$ be the set of all bounded continuous functions g on S_β, which are analytic in the interior S_β^0 of S_β, and $g(t)$ and $g(t + i\beta)$ are almost everywhere differentiable with respect to t, and $|g(z)| \leqslant \exp(\gamma|\beta|)\|x\|$ on S_β, $|(d/dt)g(t)| \leqslant r\|x\|$ a.e. and $|(d/dt)g(t + i\beta)| \leqslant \exp(r|\beta|)r\|x\|$ a.e. Consider a mapping Φ of $\Omega_{r,x}$ into $\mathbb{C} \oplus \mathbb{C} \oplus \oplus L^\infty(\mathbb{R}) \oplus L^\infty(\mathbb{R})$ as follows: $\Phi(g) = (g(0), g(i\beta), l, m)$, where

$$l(t) = \frac{d}{dt} g(t) \qquad \text{and} \qquad m(t) = \frac{d}{dt} g(t + i\beta)(t \in R).$$

Then Φ is one-to-one. In fact, if $\Phi(g_1) = \Phi(g_2)$, then

$$\frac{d}{dt} g_1(t) = \frac{d}{dt} g_2(t) \text{ a.e.} \qquad \text{and} \qquad \frac{d}{dt} g_1(t + i\beta) = \frac{d}{dt} g_2(t + i\beta) \text{ a.e.}$$

Hence

$$(g_1 - g_2)(t) = c_1 \qquad \text{and} \qquad (g_1 - g_2)(t + i\beta) = c_2,$$

where c_1 and c_2 are constants. By the analycity of g_1 and g_2, $c_1 = c_2$ and so $g_1 - g_2 = c_1 = c_2$ on S_β; hence $g_1(0) = g_2(0)$ implies $g_1 = g_2$.

Next we shall show that $\Phi(\Omega_{r,x})$ is closed in $\mathbb{C} \oplus \mathbb{C} \oplus L^\infty(\mathbb{R}) \oplus L^\infty(\mathbb{R})$ with respect to the $\sigma(\mathbb{C} \oplus \mathbb{C} \oplus L^\infty(\mathbb{R}) \oplus L^\infty(\mathbb{R}), \mathbb{C} \oplus \mathbb{C} \oplus L^1(\mathbb{R}) \oplus L^1(\mathbb{R}))$-topology. Since $\Phi(\Omega_{r,x})$ is bounded in $\mathbb{C} \oplus \mathbb{C} \oplus L^\infty(\mathbb{R}) \oplus L^\infty(\mathbb{R})$, the closure $\overline{\Phi(\Omega_{r,x})}$ of $\Phi(\Omega_\gamma)$ is separable with respect to $\sigma(\mathbb{C} \oplus \mathbb{C} \oplus L^\infty(\mathbb{R}) \oplus L^\infty(\mathbb{R}),\ \mathbb{C} \oplus \mathbb{C} \oplus L^1(\mathbb{R}) \oplus L^1(\mathbb{R}))$.

Suppose that $\Phi(g_n) \to (c_1, c_2, p, q)$; then $g_n(0) \to c_1$, $g_n(i\beta) \to c_2$, $d/dt\, g_n(t) \to p(t)$ (with respect to $\sigma(L^\infty(\mathbb{R}), L^1(\mathbb{R}))$ and $d/dt\, g_n(t + i\beta)(t) \to q(t)$ (with respect to $\sigma(L^\infty(\mathbb{R}), L^1(\mathbb{R})))$. Therefore,

$$\int_0^x \frac{d}{dt} g_n(t)\, dt = g_n(x) - g_n(0) \to \int_0^x p(t)\, dt$$

and

$$\int_0^x \frac{d}{dt} g_n(t + i\beta) = g_n(x + i\beta) - g_n(i\beta) \to \int_0^x q(x)\, dt$$

for $x \in R$. Hence

$$g_n(x) \to \int_0^x p(t)\, dt + c_1 \qquad \text{and} \qquad g_n(x + i\beta) \to \int_0^x q(x)\, dt + c_2.$$

Define

$$g(z) = \int_{-\infty}^{\infty} \left(\int_0^x p(t)\, dt + c_1 \right) K_1(z, x)\, dx + \int_{-\infty}^{\infty} \left(\int_0^x q(t)\, dt + c_2 \right) K_2(z, x)\, dx$$

for $z \in S_\beta{}^0$.

Since

$$g_n(z) = \int_{-\infty}^{\infty} g_n(x) K_1(z, x)\, dx + \int_{-\infty}^{\infty} g_n(x + i\beta) K_2(z, x)\, dx$$

for $z \in S_\beta{}^0$, $g_n(z) \to g(z)$ for $z \in S_\beta{}^0$.

Therefore g is bounded analytic on $S_\beta{}^0$ and $|g(z)| \leqslant \exp r|\beta|\, \|x\|$ on S_β. Since $\int_0^x p(t)\, dt + c_1$ and $\int_0^x q(t)\, dt + c_2$ are continuous on $\mathbb{R}$, g is bounded continuous on S_β. Hence $\Phi(\Omega_{r,x})$ is closed, and so $\Phi(\Omega_{r,x})$ is compact.

We shall define a compact Hausdorff topology on $\Omega_{r,x}$, by using the topology $\sigma(\Phi(\Omega_\gamma), \mathbb{C} \oplus \mathbb{C} \oplus L^1(\mathbb{R}) \oplus L^1(\mathbb{R}))$. Let $\Omega = \prod_{x \in \mathcal{M}} \Omega_{r,x}$ be the infinite product space of $\{\Omega_{r,x} | x \in \mathcal{M}\}$ with the weak topology; then Ω is a compact.

Let $\{h_\alpha\}$ be a directed set of self-adjoint elements in $\mathcal{M}$ with $\|h_\alpha\| \leqslant r$. Then $\{f(\cdot, h_\alpha)(x) | x \in \mathcal{M}\}$ is a family of points in Ω. Let $F(\cdot)(x)$ be an accumulation point of $\{f(\cdot, h_\alpha)(x) | x \in \mathcal{M}\}$ in Ω. Then clearly $|F(z)(x)| \leqslant \exp(\beta r)\|x\|$ $(z \in S_\beta)$, $|F(t)(x)| \leqslant r\|x\|$ and $|F(t + i\beta)(x)| \leqslant \exp(\beta r)\|x\|$ $(t \in R)$.

For each $x \in \mathcal{M}$, $F(z)(x)$ is a bounded continuous function on S_β, which is analytic in the interior $S_\beta{}^0$ of S_β, and $F(t)(x)$ and $F(t+\mathrm{i}\beta)(x)$ are differentiable for almost all $t \in \mathbb{R}$, and $|(\mathrm{d}/\mathrm{d}t)F(t)(x)| \leqslant r\|x\|$ a.e. and $|(\mathrm{d}/\mathrm{d}t)F(t+\mathrm{i}\beta)(x)| \leqslant r\exp(r|\beta|)\|x\|$ a.e.

Moreover $x \mapsto F(z)(x)$ is a bounded linear functional on $\mathcal{M}$ for each $z \in S_\beta$. Now we shall show that $F(z)$ is a normal linear functional on $\mathcal{M}$ for each $z \in S_\beta$. Suppose that $F(z_0)$ is not normal for some $z_0 \in S_\beta$. Then there is a directed set $\{x_\gamma\}$ of elements in $\mathcal{M}$ with $\|x_\gamma\| \leqslant 1$ such that $x_\gamma \to 0$ (*-strongly) and $F(z_0)(x_\gamma) \nrightarrow 0$. Then there is a cofinal directed subset $\{\gamma_\xi\}$ of $\{\gamma\}$ and a positive number ε such that $|F(z_0)(x_{\gamma_\xi})| \geqslant \varepsilon$ for all γ_ξ. On the other hand $\phi(x_{\gamma\xi}{}^* x_{\gamma\xi}) \to 0$. Hence there is a subsequence $\{x_{\gamma n}\}$ of $\{x_\gamma\}$ such that $\phi(x_{\gamma_n}{}^* x_{\gamma_n}) \to 0$ and $|F(z_0)(x_{\gamma_n})| \geqslant \varepsilon$ for all n.

$$\begin{aligned}|f(t,h_\alpha)(x)| &= |f(0,h_\alpha)(x\exp(\mathrm{i}t(H_\phi+h_\alpha))\exp(-\mathrm{i}tH_\phi))| \\ &= \tilde{\phi}(x\exp(\mathrm{i}t(H_\phi+h_\alpha))\exp(-\mathrm{i}tH_\phi))| \\ &\leqslant \tilde{\phi}(xx^*)^{1/2} = f(0,h_\alpha)(xx^*)^{1/2} = F(0)(xx^*)^{1/2}.\end{aligned}$$

Hence

$$|F(t)(x)| \leqslant F(0)(xx^*)^{1/2} = \tilde{\phi}(xx^*)^{1/2} \qquad \text{for } x \in \mathcal{M}.$$

Therefore $|F(t)(x_{\gamma_n})| \leqslant \phi(x_{\gamma_n} x_{\gamma_n}{}^*)^{1/2} \to 0$.

Since

$$\left|\frac{\mathrm{d}}{\mathrm{d}t}F(t)(x_{\gamma_n})\right| \leqslant r\|x_{\gamma_n}\| \leqslant r \text{ a.e.}$$

and

$$\left|\frac{\mathrm{d}}{\mathrm{d}t}F(t+\mathrm{i}\beta)(x_{\gamma_n})\right| \leqslant r\exp(r|\beta|) \text{ a.e.,}$$

there is a subsequence $\{F(z)(x_{\gamma_{n_j}})\}$ such that $\{F(z)(x_{\gamma_{n_j}})\}$ converges to a bounded continuous function $G(z)$ on S_β which is analytic on the interior $S_\beta{}^0$ of S_β, uniformly on every compact subset of S_β. Since $|F(t)(x_{\gamma_n})| \to 0$ $(t \in \mathbb{R})$, $G(t) = 0$ $(t \in \mathbb{R})$ and so $G(z) \equiv 0$ on S_β. This contradicts that $|F(z_0)(x_{\gamma_n})| \geqslant \varepsilon$ for all n. Therefore the $\sigma(\mathcal{M}^*, \mathcal{M})$-closure $\bar{\Gamma}_\gamma$ of Γ_γ in $\mathcal{M}_*$ is contained in $\mathcal{M}_*$, and so Γ_γ is relatively $\sigma(\mathcal{M}_*, \mathcal{M})$-compact in $\mathcal{M}_*$.

Condition (9): From the proof of (8), one can easily see that $\|h\|$ can be replaced by $f(0,h)(h^2)^{1/2}$ and $f(\mathrm{i}\beta,h)(h^2)^{1/2}$ – in fact,

$$\begin{aligned}\left|\frac{\mathrm{d}}{\mathrm{d}t}f(t+\mathrm{i}\beta,h)(x)\right| &= |f(\mathrm{i}\beta,h)(\exp(\mathrm{i}t(H_\phi+h))h\exp(-\mathrm{i}tH_\phi)x)| \\ &= |f(\mathrm{i}\beta,h)(\exp(\mathrm{i}t(H_\phi+h))h \\ &\qquad \times \exp(-\mathrm{i}tH_\phi)x\exp(\mathrm{i}t(H_\phi+h))\exp(-\mathrm{i}t(H_\phi+h)))|\end{aligned}$$

$$= |f(\mathrm{i}\beta, h)(h \exp(-\mathrm{i}tH_\phi)x \exp(\mathrm{i}t(H_\phi + h)))|$$
$$\leqslant f(\mathrm{i}\beta, h)(h^2)^{1/2} \|x\| \, \|f(\mathrm{i}\beta, h)\|.$$

□

Remark 1 Theorem 4.4.7 implies that bounded perturbations will not change the existence or absence of a phase transition in a C*-dynamics (cf. §4.7).

Remark 2 Concerning ground states, we do not have definite relations between the original dynamics and a perturbed one in general settings. By Proposition 4.4.6, bounded perturbations will retain the existence of ground states. It would be interesting to know under what conditions, pure ground states can be retained.

4.4.10 Proposition *Let $\{A, \exp(t\delta)(t \in \mathbb{R})\}$ be a C*-dynamics with a KMS state ϕ_β at β and let δ_0 be an approximately bounded *-derivation in A. If $\overline{\{1 \pm (\delta + \delta_0)\}\mathscr{D}(\delta_0 + \delta)}$ are dense in A, then the C*-dynamics $\{A, \exp(t(\delta + \delta_0))$ $(t \in R)\}$ has a KMS state ψ_β at β.*

This is easily proved by using Theorem 4.4.7 and the proof of 4.3.17.

In general, the ψ_β is not realized as a vector state on $\pi_\phi(A)$ in $\mathscr{H}_\phi$ – therefore there is no canonical one-to-one correspondence between ϕ_β and ψ_β.

4.4.11 Proposition *Let $\alpha_t = \exp(t\delta_1)(t \in \mathbb{R})$ (resp. $\gamma_t = \exp(t\delta_2)$) be two one-parameter groups of *-automorphisms on a C*-algebra A, and let $K_{1,\beta}$ (resp. $K_{2,\beta}$) be the set of all KMS states for $\{A, \exp(t\delta_1)(t \in \mathbb{R})\}$ (resp. $\{A, \exp(t\delta_2)(t \in \mathbb{R})\}$) at β, and let $\varepsilon(K_{1,\beta})$ (resp. $\varepsilon(K_{2,\beta})$) be the set of all extreme points in $K_{1,\beta}$ (resp. $K_{2,\beta}$). Suppose that there is a sequence $\{\delta_n\}$ of bounded *-derivations with $\|\delta_n\| \leqslant M(n = 1, 2, \ldots,)$ such that $\{1 - (\delta_1 + \delta_n)\}^{-1} \to (1 - \delta_2)^{-1}$ (strongly); then there is a one-to-one mapping ρ of $\varepsilon(K_{1,\beta})$ onto $\varepsilon(K_{2,\beta})$ such that factorial states ϕ and $\rho(\phi)$ on A are equivalent. In particular if $\{A, \exp(t\delta_1)(t \in \mathbb{R})\}$ has a unique KMS state at β, then $\{A, \exp(t\delta_2)(t \in \mathbb{R})\}$ also has a unique KMS state at β.*

Proof Let $\phi \in \varepsilon(K_{1,\beta})$ and let $\{\pi_\phi, U_\phi, \mathscr{H}_\phi\}$ be the covariant representation of $\{A, \exp(t\delta_1)\}$. Let $\mathscr{M}$ be the weak closure of $\pi_\phi(A)$ in $\mathscr{H}_\phi$. Let $\pi_\phi(\delta_n(a)) = \mathrm{i}[h_n, \pi_\phi(a)]$ $(a \in A)$ with $h_n (= h_n^*) \in \mathscr{M}$ and $\|h_n\| \leqslant M(n = 1, 2, \ldots)$. Consider $\psi_n(a) = f(\mathrm{i}\beta, h_n)(\pi_\phi(a))/f(\mathrm{i}\beta, h_n)(1_{\mathscr{H}_\phi})$ $(a \in A)$; then ψ_n is a KMS state for $\{A, \exp(t(\delta_1 + \delta_n))$ $(t \in \mathbb{R})\}$ at β. Let ψ be an accumulation point of $\{\psi_n\}$ in $\mathscr{M}_*$; then by 4.4.7, ψ is equivalent to ϕ and by the proof of 4.3.17, ψ is a KMS state for $\{A, \exp(t\delta_2)(t \in \mathbb{R})\}$ at β.

On the other hand, let ψ_1 be a KMS state for $\{A, \exp(t\delta_2(t \in \mathbb{R})\}$ at β

which is equivalent to ϕ; then ψ_1 is factorial; hence if $\psi_1 \neq \psi$, ψ_1 and ψ are disjoint, a contradiction. Now put $\rho(\phi) = \psi$; then one can easily see that ρ is a one-to-one mapping of $\varepsilon(K_{1,\beta})$ onto $\varepsilon(K_{2,\beta})$. □

4.4.12 Proposition *Let A be a C^*-algebra such that there is an increasing sequence $\{A_n\}$ of C^*-subalgebras as follows: $1 \in A_1 \subset A_2 \subset \cdots \subset A_n \subset \cdots$ and $\bigcup_{n=1}^{\infty} A_n$ is dense in A. Suppose that A_n has a unique tracial state τ_n $(n = 1, 2, \ldots,)$, so that there is a unique tracial state τ on A. Let δ be a $*$-derivation in A such that $\mathscr{D}(\delta) = \bigcup_{n=1}^{\infty} A_n$, and assume that there is a sequence $\{h_n\}$ of self-adjoint elements in A and a sequence $\{k_n\}$ of self-adjoint elements in A such that $\delta(a) = \mathrm{i}[h_n, a]$ $(a \in A_n, n = 1, 2, \ldots,)$ and $k_n \in A_n$, $\| h_n - k_n \| = \mathrm{O}(1)$ $(n = 1, 2, \ldots,)$; then the C^*-dynamics $\{A, \exp(t\bar{\delta})(t \in \mathbb{R})\}$ has a unique KMS state ϕ_β at each $\beta(-\infty < \beta < +\infty)$.*

Proof By 4.1.9, $\exp(t\bar{\delta}) = \text{strong} \lim \exp(t\delta_{ih_n})$. Put $\delta_n = \delta + \delta_i(k_n - h_n)$; then $\delta_n(a) = \mathrm{i}[k_n, a]$ $(a \in A_n)$. Since A_n has a unique tracial state, $\{A_n, \exp(t\delta_{ik_n})(t \in R)\}$ has a unique KMS state ϕ_n such that

$$\phi_n(a) = \tau(a \exp(-\beta k_n))/\tau(\exp(-\beta k_n)) \qquad (a \in A_n).$$

Let ϕ be a factorial KMS state for $\{A, \exp(t\bar{\delta})(t \in \mathbb{R})\}$ at β (4.3.17) and let $\{\pi_\phi, U_\phi, \mathscr{H}_\phi\}$ be the covariant representation of $\{A, \exp(t\delta)(t \in \mathbb{R})\}$ constructed via ϕ. Let $\mathscr{M}$ be the weak closure of $\pi_\phi(A)$ in $\mathscr{H}_\phi$. Then

$$f(\mathrm{i}\beta, \pi_\phi(k_n - h_n))(\pi_\phi(a))/f(\mathrm{i}\beta, \pi_\phi(k_n - h_n))(1_{\mathscr{H}_\phi}) = \phi_n(a) \qquad (a \in A_n).$$

Since $\| k_n - h_n \| = \mathrm{O}(1)$, by 4.4.7 $\{f(\mathrm{i}\beta, \pi_\phi(k_n - h_n))\}$ is relatively $\sigma(\mathscr{M}_*, \mathscr{M})$-compact in $\mathscr{M}_*$. By Eberlein's theorem, the $\sigma(\mathscr{M}_*, \mathscr{M})$-compactness is equivalent to the $\sigma(\mathscr{M}_*, \mathscr{M})$-sequential compactness, so that there is a subsequence $\{f(\mathrm{i}\beta, \pi_\phi(k_{n_j} - h_{n_j}))\}$ of $\{f(\mathrm{i}\beta, \pi_\phi(k_n - h_n))\}$ which converges to some normal positive linear functional ψ in the $\sigma(\mathscr{M}_*, \mathscr{M})$-topology. In particular,

$$\omega(a) = \psi(\pi_\phi(a))/\psi(1_{\mathscr{H}_\phi}) = \lim_{n_j} \tau(a \exp(-\beta k_{n_j}))/\tau(\exp(-\beta k_{n_j})) \qquad (a \in A).$$

Clearly, the state ω on A is equivalent to ϕ. Next, let ξ be another factorial KMS state for $\{A, \exp(t\bar{\delta})(t \in R)\}$ at β. Then by a similar discussion, there is a state ω_1 on A such that

$$\omega_1(a) = \lim_{m_j} \tau(a \exp(-\beta h_{m_j})/\tau(\exp(-\beta h_{m_j})) \qquad (a \in A)$$

and ω_1 is equivalent to ξ, where (m_j) is a subsequence of (n_j). Therefore $\omega = \omega_1$, and so ϕ and ξ are equivalent. Hence by 4.3.12, $\phi = \xi$. Since extreme points of $K_\beta(\exp(t\bar{\delta}))$ are factorial (4.3.10), K_β consists of one point. □

4.4.13 Notes and remarks Propositions 4.4.6 and 4.4.10 are due to Powers and Sakai [145]. Theorem 4.4.7.(3), (5) and (6) are due to Araki [3], and the rest of Theorem 4.4.7 is due to the author. Proposition 4.4.12 was first proved by Araki [5] under the assumption that A_n $(n = 1, 2, \ldots,)$ are finite-dimensional. For commutative normal *-derivations, it was previously proved by the author [168]. Kishimoto [104] gave a simpler proof of Araki's theorem. Proposition 4.4.12 in the present general form is new.

References [3], [5], [104], [168], [169], [170].

4.5 UHF algebras and normal *-derivations

At the beginning of §4.1, we showed that the study of quantum lattice systems is included in the study of *-derivations in the following setting. A C*-algebra $\mathscr{A}$ is said to be a uniformly hyperfinite C*-algebra (UHF algebra) if there is an increasing sequence $\{\mathscr{A}_n\}$ of finite type-I subfactors (i.e. finite-dimensional full matrix algebras) such that $1 \in \mathscr{A}_1 \subset \mathscr{A}_2 \subset \cdots \subset \mathscr{A}_n \subset \cdots$ and the uniform closure of $\bigcup_{n=1}^{\infty} \mathscr{A}_n$ is $\mathscr{A}$. Let $\mathscr{D}(\delta) = \bigcup_{n=1}^{\infty} \mathscr{A}_n$ and let δ be a *-derivation in $\mathscr{A}$ with the domain $\mathscr{D}(\delta)$. The study of such *-derivations includes the study of general quantum lattice systems. Consequently, a C*dynamics $\{\mathscr{A}, \alpha\}$ with a UHF algebra $\mathscr{A}$ includes the dynamics of a quantum lattice system. It also includes a quasi-free dynamics in Fermion field theory.

In this section, we shall discuss a C*dynamics $\{\mathscr{A}, \alpha\}$ with a UHF algebra $\mathscr{A}$ and *-derivations in $\mathscr{A}$. We shall also discuss two important problems (the Powers–Sakai conjecture and the core problem) in the theory of unbounded derivations in detail.

4.5.1 Theorem *Let $\{A, \alpha\}$ be a C*-dynamics with a UHF algebra A. Then there is an increasing sequence of finite type I subfactors $\{A_n\}$ such that $1 \in A_n$, $A_n \subset A_{n+1}$, $\bigcup_{n=1}^{\infty} A_n$ is dense in A and every element of $\bigcup_{n=1}^{\infty} A_n$ is analytic with respect to α.*

To prove the theorem, we shall provide a lemma.

4.5.2 Lemma *Suppose that there is a type I_n-subfactor B $(n < +\infty)$ of A whose unit is also the unit of A and let $\{e_{ij} | i, j = 1, 2, \ldots, n\}$ be a matrix unit of B; then for an arbitrary $\varepsilon > 0$, there is a type I_n-subfactor N of $A(\delta)$ and its matrix unit $\{f_{ij} | i, j = 1, 2, \ldots, n\}$ such that $\| e_{ij} - f_{ij} \| < \varepsilon$ $(i, j = 1, 2, \ldots, n)$, where $\alpha_t = \exp(t\delta)$.*

Proof For $\varepsilon' > 0$ with $\varepsilon' = \min(\varepsilon/n!(21)^{n+2}), (1/n!(21)^{n+2})$, take a projection

f_1 in $A(\delta)$ such that $\|e_{11} - f_1\| < \varepsilon'$ (cf. 3.4.8). Then $\|(1-e_{11})-(1-f_1)\| < \varepsilon'$. Take an element $b \in A(\delta)$ such that $0 < b < 1$ and $\|e_{22} - b\| < \varepsilon'$. Then,

$$\begin{aligned}\|e_{22}-(1-f_1)b(1-f_1)\| &\leqslant \|e_{22}-(1-e_{11})b(1-e_{11})\| \\ &\quad + \|(1-e_{11})b(1-e_{11})-(1-f_1)b(1-f_1)\| < \varepsilon' \\ &\quad + \|(1-e_{11})b(1-e_{11})-(1-f_1)b(1-e_{11})\| \\ &\quad + \|(1-f_1)b(1-e_{11})-(1-f_1)b(1-f_1)\| < \varepsilon' \\ &\quad + \varepsilon' + \varepsilon' = 3\varepsilon'.\end{aligned}$$

Therefore by the similar discussion with the proof of 3.4.8, there is a projection f_2 in $A(\delta)$ such that $f_2 \leqslant 1 - f_1$ and $\|e_{22} - f_2\| < 7.3\varepsilon'$.

Continuing this process one can construct a family of mutually orthogonal projections $\{f_1, f_2, \ldots, f_n\}$ in $A(\delta)$ such that $\|e_{ii} - f_i\| < (j-1)(21)^j\varepsilon'$ $(j = 2, 3, \ldots, n)$, and

$$\left\|\left(1 - \sum_{j=1}^{n} e_{jj}\right) - \left(1 - \sum_{j=1}^{n} f_j\right)\right\| < \sum_{j=1}^{n} \|e_{jj} - f_j\| < n!(21)^n\varepsilon'.$$

Since $1 - \sum_{j=1}^{n} e_{jj} = 0$, $\sum_{j=1}^{n} f_j = 1$. Next for e_{j1}, take an element $b_j \in A(\delta)$ with $\|b_j\| \leqslant 1$ such that $\|e_{j1} - b_j\| < \varepsilon'$; then

$$\begin{aligned}\|e_{j1} - f_j b_j f_1\| &\leqslant \|e_{j1} - e_{jj} b_j e_{11}\| + \|e_{jj} b_j e_{11} - f_j b_j f_1\| \\ &< \varepsilon' + \|e_{jj} b_j e_{11} - f_j b_j e_{11}\| + \|f_j b_j e_{11} - f_j b_j f_1\| \\ &< \varepsilon' + n!(21)^n\varepsilon' + n!(21)^n\varepsilon' < 3(n!(21)^n\varepsilon').\end{aligned}$$

Let $f_j b_j f_1 = v_j h_j$ be the polar decomposition of $f_j b_j f_1$, where $h_j = \{(f_j b_j f_1)^*(f_j b_j f_1)\}^{1/2}$. Then

$$\|(e_{j1})^*(e_{j1}) - (f_j b_j f_1)^*(f_1 b_j f_j)\| = \|e_{11} - h_j^2\| < 2{\cdot}3{\cdot}n!(21)^n\varepsilon'.$$

Since

$$\|(f_j - h_j^2)\| = \|f_j - e_{11}\| + \|e_{11} - h_j^2\| < n!(21)^{n+1}\varepsilon' < \tfrac{1}{4}$$

and since the support of $h_j^2 \leqslant f_1$, the support projection of $h_j^2 = f_1$.

$$1 - f_1 + h_j^2 \in A(\delta) \qquad \text{and} \qquad 1 - f_1 + h_j^2 = 1 - (f_1 - h_j^2).$$

For $z \in \mathbb{C}$ with $|z| < \min(r(f_1), r(h^2))$, consider

$$\begin{aligned}\alpha_z(1 - f_1 + h_j^2) &= 1 - \alpha_z(f_1 - h_j^2) = 1 - \sum_{n=0}^{\infty} \frac{\delta^n(f_1 - h_j^2)}{n!} z^n \\ &= 1 - (f_1 - h_j^2) - \sum_{n=1}^{\infty} \frac{\delta^n(f_1 - h_j^2)}{n!} z^n.\end{aligned}$$

Then there is a positive number β such that

$$\left\|\sum_{n=1}^{\infty} \frac{\delta^n(f_1 - h_j^2)}{n!} z^n\right\| < \tfrac{1}{3}.$$

for z with $|z| < \beta$ and $\beta < \min(r(f_1), r(h_j^2))$. Hence

$$\|\alpha_z(f_1 - h_j^2)\| < \tfrac{1}{4} + \tfrac{1}{3} = \tfrac{7}{12} \text{ for } z \text{ with } |z| < \beta.$$

Therefore,

$$\{1-\alpha_z(f_1-h_j^2)\}^{-1/2} = 1 + \sum_{n=1}^{\infty} (-1)^n \frac{(-1)^n(\frac{1}{2}+1)\cdots(\frac{1}{2}+n-1)}{n!}\{\alpha_z(f_1-h_j^2)\}^n.$$

The right side of this equality is uniformly convergent on $\{z\in\mathbb{C}\,|\,|z|<\beta\}$ and $\{\alpha_z(f_1-h_j^2)\}^n$ is analytic on $\{z\in\mathbb{C}\,|\,|z|<\beta\}$; hence $\{1-\alpha_z(f_1-h_j^2)\}^{-1/2}$ is analytic on $\{z\in\mathbb{C}\,|\,|z|<\beta\}$ and so $\{1-f_1+h_j^2\}^{-1/2}\in A(\delta)$. Therefore

$$f_j b_j f_1(1-f_1+h_j^2)^{-1/2} = v_j f_1 \in A(\delta),$$

and

$$(v_j f_1)^*(v_j f_1) = f_1 v_j^* v_j f_1$$

and

$$\begin{aligned}
\|v_j f_1 v_j^* - f_j\| &\leqslant \|f_j b_j f_1(1-f_1+h_j^2)^{-1/2}(1-f_1+h_j)^{-1/2} f_1 b_j^* f_j - f_j\| \\
&\leqslant \|f_j b_j f_1\{(1-f_1+h_j^2)^{-1}-1\} f_1 b_j^* f_j\| \\
&\quad + \|f_j b_j f_1 f_1 b_j^* f_j - e_{j1}e_{1j}\| + \|e_{j1}e_{1j} - f_j\| \\
&\leqslant \|f_j b_j f_1\{(1-f_1+h_j^2)^{-1}-1\} f_1 b_j^* f_j\| \\
&\quad + \|f_j b_j f_1 b_j^* f_j - f_j b_j f_1 e_{1j}\| + \|f_j b_j f_1 e_{1j} - e_{j1}e_{1j}\| \\
&\quad + \|e_{j1}e_{1j} - f_j\| < \|(1-f_1+h_j^2)^{-1} - 1\| + 9n!(21)^n\varepsilon' \\
&< \|(1-f_1+h_j^2)^{-1}-1\| + n!(21)^{n+1}\varepsilon' < \tfrac{1}{3} + \tfrac{1}{2} < 1.
\end{aligned}$$

Since $v_j f_1 v_j^* \leqslant f_j$ and $v_j f_1 v_j^*$ is a projection, $v_j f_1 v_j^* = f_j$. Put $w_j = v_j f_1$ $(j = 1, 2, \ldots, n)$; then $w_j \in A(\delta)$, and $w_j^* w_j = f_1$ and $w_j w_j^* = f_j$. Put $f_{ij} = w_i w_j^*$; then $\{f_{ij}\,|\,i, j = 1, 2, \ldots, n\}$ is a matrix unit such that $f_{ij}\in A(\delta)$. Moreover

$$\begin{aligned}
\|e_{ij} - f_{ij}\| &= \|e_{i1}e_{j1}^* - w_i w_j^*\| \leqslant \|e_{i1}e_{j1}^* - w_i e_{j1}^*\| + \|w_i(e_{j1}^* - w_j^*)\| \\
&\leqslant \|e_{i1} - w_i\| + \|e_{j1} - w_j\| \leqslant \|e_{i1} - f_i b_i f_1\| + \|f_i b_i f_1 - v_i f_1\| \\
&\quad + \|e_{j1} - f_j b_i f_1\| + \|f_j b_j f_1 - v_j f_1\| \leqslant 3n!(21)^n\varepsilon' + \|h_i - f_1\| \\
&\quad + 3n!(21)^n\varepsilon' + \|h_j - f_1\| \leqslant 6n!(21)^n\varepsilon' + \|h_i^2 - f_1\| \\
&\quad + \|h_j^2 - f_1\| < 18n!(21)^{n+1}\varepsilon'
\end{aligned}$$

for $i, j = 1, 2, \ldots, n$. □

Proof of Theorem 4.5.1 Let $\{B_n\}$ be an increasing sequence of finite type I subfactors in A such that $1\in B_n$, $B_n \subset B_{n+1}$ and $\bigcup_{n=1}^{\infty} B_n$ is dense in A. Let $\{e_{ij}\,|\,i, j = 1, 2, \ldots, n_1\}$ be a matrix unit of B_1; then for $\varepsilon > 0$ there is a matrix unit $\{f_{ij}\,|\,i, j = 1, 2, \ldots, n_1\}$ in $A(\delta)$ such that $\|e_{ij} - f_{ij}\| < \varepsilon/n_1^2$ $(i, j = 1, 2, \ldots, n_1)$. Let A_1 be the type I_{n_1} subfactor of A generated by $\{f_{ij}\,|\,i, j = 1, 2, \ldots, n_1\}$.

Then clearly

$$\sup_{\substack{x \in B_1 \\ \|x\| \leq 1}} \inf_{\substack{y \in A_1 \\ \|y\| \leq 1}} \|x - y\| < \varepsilon.$$

Since $A_1 \otimes A'_1 = A$, where A'_1 is the commutant of A_1 in A and A'_1 is *-isomorphic to B'_1, there is a finite type $\mathrm{I_m}$-subfactor N in A'_1 such that

$$\sup_{\substack{x \in B_2 \\ \|x\| \leq 1}} \inf_{\substack{y \in A_1 \otimes N \\ \|y\| \leq 1}} \|x - y\| < \varepsilon/4$$

Since $A(\delta) = A_1 \otimes (A(\delta) \cap A'_1)$, $A'_1 \cap A(\delta)$ is dense in A'_1; hence by the above lemma, there is a finite type $\mathrm{I_m}$-subfactor N_1 in $A'_1 \cap A(\delta)$ such that

$$\sup_{\substack{x \in N \\ \|x\| \leq 1}} \inf_{\substack{y \in N_1 \\ \|y\| \leq 1}} \|x - y\| < \varepsilon/4(n_1)^2$$

Then

$$\sup_{\substack{x \in B_2 \\ \|x\| \leq 1}} \inf_{\substack{y \in A_1 \otimes N_1 \\ \|y\| \leq 1}} \|x - y\| < \varepsilon/2.$$

Put $A_1 \otimes N_1 = A_2$. Continuing this process one can construct an increasing sequence $\{A_n\}$ of finite type-I subfactors in A such that $1 \in A_n$, $A_n \subset A_{n+1}$ $(n = 1, 2, \ldots)$. Moreover,

$$\sup_{\substack{x \in B_n \\ \|x\| \leq 1}} \inf_{\substack{y \in A_n \\ \|y\| \leq 1}} \|x - y\| < \varepsilon/n \text{ and } A_n \subset A(\delta).$$

Hence $\bigcup_{n=1}^{\infty} A_n$ is dense in A. □

By a minor change of the above discussions, one can easily prove the following corollary.

4.5.3 Corollary *Let δ be a closed *-derivation in a UHF algebra A; then there is an increasing sequence $\{A_n\}$ of finite type-I subfactors in $\mathscr{D}(\delta)$ such that $1 \in A_n$, $A_n \subset A_{n+1}$ and $\bigcup_{n=1}^{\infty} A_n$ is dense in A.*

Now we shall define a normal *-derivation in UHF algebras more restrictively than in general cases.

4.5.4 Definition *Let δ be a *-derivation in a UHF algebra A. δ is said to be normal if there is an increasing sequence $\{A_n\}$ of finite type I subfactors in A such that $1 \in A_n$, $A_n \subset A_{n+1}$ and $\mathscr{D}(\delta) = \bigcup_{n=1}^{\infty} A_n$.*

From 4.5.1 and 4.5.3, one can see that a normal *-derivation in a UHF algebra will play a key role in the study of unbounded derivations in UHF algebras. Most unbounded derivations in quantum lattice systems and Fermion field theory have normal *-derivations as their cores.

Let δ be a normal *-derivation in a UHF algebra A and let $\mathscr{D}(\delta) = \bigcup_{n=1}^{\infty} A_n$. Let $\{e_{ij} | i, j = 1, 2, \ldots, n_1\}$ be a matrix unit of A_1. Set

$$ih_1 = \sum_{i=1}^{n_1} \delta(e_{i1})e_{1i};$$

then one can easily see that $\delta(a) = [ih_1, a]$ for $a \in A_1$. Since

$$\left(\sum_{i=1}^{n_1} \delta(e_{i1})e_{1i}\right)^* = \sum_{i=1}^{n_1} e_{1i}{}^*\delta(e_{i1}{}^*) = \sum_{i=1}^{n_1} e_{i1}\delta(e_{1i}) \qquad \text{and} \qquad e_{i1}e_{1i} = e_{ii},$$

$$0 = \delta\left(\sum_{i=1}^{n_1} e_{i1}e_{1i}\right) = \sum_{i=1}^{n_1} \delta(e_{i1})e_{1i} + \sum_{i=1}^{n_1} e_{i1}\delta(e_{1i});$$

hence

$$\sum_{i=1}^{n_1} \delta(e_{i1})e_{1i} = -\sum_{i=1}^{n_1} e_{i1}\delta(e_{1i}).$$

Therefore h_1 is a self-adjoint element in A. Similarly we have a sequence (h_n) of self-adjoint elements in A such that $\delta(a) = \mathrm{i}[h_n, a]$ $(a \in A_n)$ $(n = 1, 2, \ldots,)$. Moreover if δ is a generator and $A_n \subset A(\delta)$, then $h_n \in A(\delta)$.

4.5.5 Proposition *Let δ be a normal *-derivation in a UHF algebra A such that $\mathscr{D}(\delta) = \bigcup_{n=1}^{\infty} A_n$. Then for $\varepsilon > 0$, there is a normal *-derivation δ_ε in A such that $\mathscr{D}(\delta_\varepsilon) = \bigcup_{n=1}^{\infty} A_n$, $\delta_\varepsilon(\mathscr{D}(\delta_\varepsilon)) \subset \mathscr{D}(\delta_\varepsilon)$ $(n = 1, 2, \ldots,)$ and $\delta - \delta_\varepsilon$ is a bounded *-derivation in A with $\|\delta - \delta_\varepsilon\| < \varepsilon$.*

Proof Let τ be the unique tracial state on A and let P_n be the canonical conditional expectation of A onto A_n defined by $\tau(xy) = \tau(P_n(x)y)$ for $x \in A$ and $y \in A_n$. Let $\delta(a) = \mathrm{i}[h_n, a]$ $(a \in A_n)$. Take P_{n_1} with $n_1 \geqslant 1$ such that $\|h_1 - P_{n_1}(h_1)\| < \varepsilon/2^2$ and P_{n_2} with $n_2 \geqslant 2$ such that $\|(h_2 - h_1) - P_{n_2}(h_2 - h_1)\| < \varepsilon/2^3$. Continuing this process, take P_{n_j} with $n_j > j$ such that

$$\|(h_j - h_{j-1}) - P_{n_j}(h_j - h_{j-1})\| < \frac{\varepsilon}{2^{j+1}}$$

Put $l_j = (h_j - h_{j-1}) - P_{n_j}(h_j - h_{j-1})$; then $\sum_{j=1}^{\infty} \|l_j\| < \varepsilon/2$, where $h_0 = 0$, and so $\sum_{j=1}^{\infty} l_j = d$ is an element of A. Moreover for $a \in A_{j_0}$,

$$\left[\mathrm{i}\sum_{j=1}^{\infty} l_j, a\right] = \mathrm{i}\sum_{j=1}^{\infty} [(h_j - h_{j-1}) - P_{n_j}(h_j - h_{j-1}), a]$$

$$= \mathrm{i}\sum_{j=1}^{j_0} [(h_j - h_{j-1}), a] - \mathrm{i}\sum_{j=1}^{\infty} [P_{n_j}(h_j - h_{j-1}), a].$$

For $j > j_0$,

$$[P_{n_j}(h_j - h_{j-1}), a] = (P_{n_j}(h_j - h_{j-1}))a - a(P_{n_j}(h_j - h_{j-1}))$$
$$= P_n([h_j - h_{j-1}, a]) = 0.$$

Hence

$$\mathrm{i}\left[\sum_{j=1}^{\infty} l_j, a\right] = \mathrm{i}[h_{j_0}, a] - \mathrm{i}\left[\sum_{j=1}^{j_0} P_{n_j}(h_j - h_{j-1}), a\right]$$

$$= \mathrm{i}\left[h_{j_0} - \sum_{j=1}^{j_0} P_{n_j}(h_j - h_{j-1}), a\right]$$

Now put $\delta_\varepsilon = \delta - \delta_{id}$; then

$$\mathscr{D}(\delta_\varepsilon) = \bigcup_{n=1}^{\infty} A_n \qquad \text{and} \qquad \delta_\varepsilon(a) = \mathrm{i}\left[\sum_{j=1}^{j_0} P_{n_j}(h_j - h_{j-1}), a\right] \qquad (a \in A_{j_0})$$

and so $\delta_\varepsilon(\mathscr{D}(\delta_\varepsilon)) \subset \mathscr{D}(\delta_\varepsilon)$ and $\|\delta - \delta_\varepsilon\| = \|\delta_{id}\| < \varepsilon$. □

Remark This proposition implies that a normal *-derivation can be perturbed to a normal *-derivation *with a finite range interaction* by a bounded *-derivation. Since bounded perturbations do not change many physical phenomena (for example, phase transition [(cf. 4.4, 4.7)]), one may reduce the study of normal *-derivations to the one of normal *-derivations with finite range interaction in many cases.

4.5.6 Proposition *Let δ be a normal *-derivation in a UHF algebra A with $\delta(\mathscr{D}(\delta)) \subset \mathscr{D}(\delta)$. Then there is an increasing sequence A_n of finite type I subfactors in $\mathscr{D}(\delta)$ and two normal *-derivations δ_1 and δ_2 such that $\mathscr{D}(\delta_1) = \mathscr{D}(\delta_2)$, $\delta_1(A_{2n}) \subset A_{2n}$ $(n = 1, 2, \ldots,)$, $\delta_2(A_{2n+1}) \subset A_{2n+1}$ $(n = 1, 2, \ldots,)$ and $\delta = \delta_1 + \delta_2$.*

Proof Let $\mathscr{D}(\delta) = \bigcup_{n=1}^{\infty} B_n$; then $\delta(B_1) \subset B_{n_1}$, $\delta(B_{n_1}) \subset B_{n_2}, \ldots,$ and $B_{n_1} \subset B_{n_2} \subset \cdots$.

Set $B_{n_j} = A_j$; then $\delta(A_j) \subset A_{j+1}$ $(j = 1, 2, \ldots,)$. Let $\delta(a) = \mathrm{i}[h_n, a]$ $(a \in A_n)$, and define

$$\delta_1(a) = \mathrm{i}\left[h_1 + \sum_{n=1}^{\infty} (h_{2n+1} - P_{2n+1}(h_{2n+1})), a\right]$$

and

$$\delta_2(a) = \mathrm{i}\left[\sum_{n=1}^{\infty} \{P_{2n+1}(h_{2n+1}) - h_{2n-1}\}, a\right] \qquad \left(a \in \bigcup_{n=1}^{\infty} A_n\right).$$

For $a \in A_{2n_0}$,

$$\delta_1(a) = \mathrm{i}\left[h_1 + \sum_{n=1}^{n_0-1} (h_{2n+1} - P_{2n+1}(h_{2n+1})), a\right] \in A_{2n_0},$$

since $h_n \in A_{n+1}$. For $a \in A_{2n_0+1}$,

$$\delta_2(a) = \mathrm{i}\left[\sum_{n=1}^{n_0} \{P_{2n+1}(h_{2n+1}) - h_{2n-1}\}, a\right] \in A_{2n_0+1}.$$

Moreover if $a \in A_{2n_0}$, then

$$\delta_1(a)+\delta_2(a) = \mathrm{i}\left[h_1 + \sum_{n=1}^{n_0-1}\{h_{2n+1} - P_{2n+1}(h_{2n+1})\}, a\right] + \mathrm{i}\left[\sum_{n=1}^{n_0}\{P_{2n+1}(h_{2n+1}) - h_{2n-1}\}, a\right]$$
$$= \mathrm{i}[h_1 + P_{2n_0+1}(h_{2n_0+1}) - h_1, a] = \mathrm{i}[P_{2n_0+1}(h_{2n_0+1}), a]$$
$$= \mathrm{i}P_{2n_0+1}([h_{2n_0+1}, a]) = \mathrm{i}P_{2n_0+1}([h_{2n_0}, a]) = \delta(a).$$

If $\in A_{2n_0+1}$, then $a \in A_{2n_0+2}$; hence; $(\delta_1+\delta_2)(a) = \delta(a)$. □

4.5.7 Definition *Let δ be a normal *-derivation in a UHF algebra A. δ is said to be commutative if one can choose a mutually commuting family $\{h_n\}$ of self-adjoint elements in A and an increasing sequence $\{A_n\}$ of finite type I subfactors in A such that $\bigcup_{n=1}^{\infty} A_n = \mathscr{D}(\delta)$ and $\delta(a) = \mathrm{i}[h_n, a]$ $(a \in A_n)$ $(n = 1, 2, \ldots,)$.*

Remark The notion of commutative normal *-derivations is a generalization of classical lattice systems.

In 4.5.6, $\delta_1(A_{2n}) \subset A_{2n}$ (resp. $\delta_2(A_{2n+1}) \subset A_{2n+1}$); therefore, one can choose a sequence $\{h_{2n}\}$ (resp. $\{h_{2n+1}\}$) of self-adjoint elements in A such that $h_{2n} \in A_{2n}$ (resp. $h_{2n+1} \in A_{2n+1}$) and $\delta_1(a) = \mathrm{i}[h_{2n}, a]$ $(a \in A_{2n})$ (resp. $\delta_2(a) = \mathrm{i}[h_{2n+1}, a]$ $(a \in A_{2n+1})$). Since $h_{2(n+1)} - h_{2n} \in A_{2(n+1)} \cap A'_{2n}$ (resp. $h_{2(n+1)+1} - h_{2n+1} \in A_{2(n+1)+1} \cap A'_{2n+1}$), $\{h_{2n}\}$ (resp. $\{h_{2n+1}\}$) is a mutually commuting family. Hence δ_1 (resp. δ_2) is commutative. Therefore 4.5.5 and 4.5.6 imply that any normal *-derivation can be written as a sum of two commutative normal *-derivations after bounded perturbation. On the other hand, bounded perturbations will not change many important physical phenomena. For example, a phase transition will not be changed by bounded perturbation (cf. 4.4, 4.7). Moreover one can analyse the commutative normal derivations than more easily the general normal derivations (cf. 4.6). Therefore, 4.5.6 may be useful for the study of normal derivations. A more powerful tool would be supplied if a normal *-derivation becomes a commutative normal *-derivation after bounded perturbation. Therefore an answer to the following problems would be interesting.

4.5.8 Problem *Let δ be a normal *-derivation in a UHF algebra A. Then can one choose normal *-derivations δ_1 and δ_2 in A such that $\mathscr{D}(\delta) = \mathscr{D}(\delta_1) = \mathscr{D}(\delta_2)$, and $\delta = \delta_1 + \delta_2$, with δ_1 commutative and δ_2 bounded?*

In the following, we shall show that all C*-dynamics appearing in quantum lattice systems and Fermion field theory are approximately inner. Let δ be a normal *-derivation in a UHF algebra A with $\mathscr{D}(\delta)=\bigcup_{n=1}^{\infty}A_n$. Let $\{h_n\}$ be a sequence of self-adjoint elements in A such that $\delta(a)=\mathrm{i}[h_n, a]$ $(a\in A_n)$. Then by 3.2.22, δ is well-behaved and so $\|(1\pm\delta)(a)\|\geqslant\|a\|$ $(a\in\mathscr{D}(\delta))$ (3.2.19). Therefore, δ is a pre-generator if and only if $(1\pm\delta)\mathscr{D}(\delta)$ are dense in A. In quantum lattice systems, it is not so easy to check the density of $(1\pm\delta)\mathscr{D}(\delta)$ directly; instead, the analytic method is often used. Suppose that $\delta(\mathscr{D}(\delta))\subset\mathscr{D}(\delta)$ (finite range interaction); then one can consider iterations δ^n $(n=1,2,3,\ldots,)$ on $\mathscr{D}(\delta)$. An element a in $\mathscr{D}(\delta)$ is said to be analytic with respect to δ if there exists a positive number r such that $\sum_{n=0}^{\infty}(\|\delta^n(a)\|/n!)r^n<+\infty$ $(\delta^0(a)=a)$. Let $A(\delta)$ be the set of all analytic elements in $\mathscr{D}(\delta)$. If $A(\delta)$ is dense in A, then by 3.4.5, δ is a pre-generator. If a quantum lattice system has a translation-invariant, finite range interaction, then $A(\delta)=\mathscr{D}(\delta)$ and so δ is a pre-generator (cf. [159]). In particular, the Examples (4) and (5) in §3.1 are pre-generators. For more general normal *-derivations, we shall use bounded perturbations (4.5.5).

Since the property of 'pre-generator' is invariant under bounded perturbations, it is sufficient to study the generation problem for normal *-derivations under the assumption of $\delta(\mathscr{D}(\delta))\subset\mathscr{D}(\delta)$.

Next suppose that a normal *-derivation δ in A is a pre-generator; then

$$\begin{aligned}\|(1\pm\delta_{\mathrm{i}h_n})^{-1}(1\pm\delta)a-(1\pm\delta)^{-1}(1\pm\delta)a\| &\leqslant \|(1\pm\delta_{\mathrm{i}h_n})^{-1}\{(1\pm\delta)a-(1\pm\delta_{\mathrm{i}h_n})\}a\|\\ &\leqslant\|(\delta-\delta_{\mathrm{i}h_n})(a)\|=0 \qquad (a\in A_n).\end{aligned}$$

Hence $(1\pm\delta_{\mathrm{i}h_n})^{-1}\to(1\pm\bar{\delta})^{-1}$ (strongly) so that $\exp(t\bar{\delta})=$ strong lim $\exp(t\delta_{\mathrm{i}h_n})$ – namely $\{\exp(t\bar{\delta})|t\in\mathbb{R}\}$ is approximately inner. Therefore it is approximately inner whenever a quantum lattice system defines a C*-dynamics.

Next, we shall consider quasi-free derivations in the Example (6) in §3.1. Let H be a self-adjoint operator in a Hilbert space $\mathscr{H}$ and let $H=H_1+T$ be the Weyl decomposition, where H_1 is a diagonalizable self-adjoint operator and T is of Hilbert–Schmidt class; then

$$H_1=\sum_{i=1}^{\infty}\lambda_iE_i, \quad (\lambda_i\in R) \quad E_iE_j=0 \quad (i\neq j), \quad E_i^2=E_i, \quad E_i^*=E_i$$

and $\dim(E_i)=1$ $(i,j=1,2,\ldots,)$. Let $P_n=\sum_{i=1}^{n}E_i$ and $\mathscr{H}_0=\bigcup_{n=1}^{\infty}P_n\mathscr{H}$. Let δ be the restriction of δ_{iH} to $\mathscr{A}_0(\mathscr{H}_0)$; then δ is a normal *-derivation. Since $\overline{H|\mathscr{H}_0}=H$, $\bar{\delta}=\bar{\delta}_{\mathrm{i}H}$, and moreover $\overline{(1\pm\delta)\mathscr{D}(\delta)}\supset a((1\pm\mathrm{i}H)(\mathscr{D}(H)))=a(\mathscr{H})$; hence $\overline{(1\pm\delta)\mathscr{D}(\delta)}=\mathscr{A}(\mathscr{H})$, and so $\delta_{\mathrm{i}H}$ is a pre-generator and $\{\exp(t\bar{\delta}_{\mathrm{i}H})|t\in R\}$ is approximately inner. More generally, if S is a symmetric operator, we can easily see that $\delta_{\mathrm{i}S}$ has a normal *-derivation δ as a core (i.e. $\bar{\delta}=\bar{\delta}_{\mathrm{i}S}$), by using the polar decomposition of S. If S has no self-adjoint extension, then $\delta_{\mathrm{i}S}$ does not necessarily have a generator extension.

We have seen that all C*-dynamical generators appearing in quantum lattice systems and Fermion field theory are approximately inner. On the other hand, if a C*-dynamics $\{A, \alpha\}$ (A, UHF algebra) is approximately inner, then it has many physical properties, so that a physical theory can be developed for approximately inner dynamics. Therefore the following conjecture is very important.

4.5.9 The Powers–Sakai conjecture *Any C*-dynamics $\{A, \alpha\}$ with a UHF algebra A is approximately inner.*

Let $\alpha_t = \exp(t\delta)$; then the following problem is important.

4.5.10 The core problem *Does every generator δ have a normal *-derivation δ_1 as a core (i.e. $\bar{\delta}_1 = \delta$)?*

An affirmative solution to the core problem would imply an affirmative solution to the conjecture. Therefore the affirmative solution to the core problem would be more desirable than the affirmative solution to the conjecture from the point of view of quantum lattice systems. In the following, we shall discuss both the conjecture and the core problem. Let $\alpha_t = \exp(t\delta)$. From 4.5.1, there is an increasing sequence $\{A_n\}$ of finite type I subfactors in A such that $A_n \subset A_{n+1}$, $\bigcup_{n=1}^{\infty} A_n \subset A(\delta)$ and $\bigcup_{n=1}^{\infty} A_n$ is dense in A. Let $\{h_n\}$ be a sequence of self-adjoint elements in A such that $\delta(a) = \mathrm{i}[h_n, a]$ $(a \in A_n)$ $(n = 1, 2, 3, \ldots,)$. If $(1-\delta)\bigcup_{n=1}^{\infty} A_n$ is dense in A, then δ is the closure of restriction of δ to $\bigcup_{n=1}^{\infty} A_n$ and so has a normal *-derivation as a core; hence $\exp(t\delta) = \text{strong}\lim \exp(t\delta_{\mathrm{i}h_n})$ – namely α is approximately inner. However, from the considerations on quasi-free derivations, one can easily see that the density of $(1-\delta)\bigcup_{n=1}^{\infty} A_n$ cannot be expected under a general selection of $\bigcup_{n=1}^{\infty} A_n$. An important, difficult problem is whether or not one can choose $\{A_n\}$ such that $(1-\delta)\bigcup_{n=1}^{\infty} A_n$ is dense in A. On the other hand, if the conjecture is true, then a general $\{A_n\}$ may have all the necessary information to prove it in a sense. In fact, if $(1-\delta_{\mathrm{i}l_n})^{-1} \to (1-\delta)^{-1}$ (strongly) for some (l_n), then by the density of $\bigcup_{n=1}^{\infty} A_n$, there is a sequence (s_n) of self-adjoint elements in $\bigcup_{n=1}^{\infty} A_n$ such that $\|l_n - s_n\| < 1/n$. Then,

$$
\begin{aligned}
&\|\{(1-\delta_{\mathrm{i}s_n})^{-1} - (1-\delta)^{-1}\}(a)\| \\
&\leqslant \|\{(1-\delta_{\mathrm{i}s_n})^{-1} - (1-\delta_{\mathrm{i}l_n})^{-1}\}(a)\| + \|\{(1-\delta_{\mathrm{i}l_n})^{-1} - (1-\delta)^{-1}\}(a)\| \\
&\leqslant \|(1-\delta_{\mathrm{i}s_n})^{-1}\{(1-\delta_{\mathrm{i}l_n}) - (1-\delta_{\mathrm{i}s_n})\}(1-\delta_{\mathrm{i}l_n})^{-1}(a)\| \\
&\quad + \|\{(1-\delta_{\mathrm{i}l_n})^{-1} - (1-\delta)^{-1}\}(a)\| \leqslant \|(\delta_{\mathrm{i}l_n} - \delta_{\mathrm{i}s_n})(1-\delta_{\mathrm{i}l_n})^{-1}(a)\| \\
&\quad + \|\{(1-\delta_{\mathrm{i}l_n})^{-1} - (1-\delta)^{-1}\}(a)\| \\
&\leqslant \frac{2\|a\|}{n} + \|\{(1-\delta_{\mathrm{i}l_n})^{-1} - (1-\delta)^{-1}\}(a)\| \to 0.
\end{aligned}
$$

Hence there is a sequence $\{s_n\}$ in $\bigcup_{n=1}^{\infty} A_n$ such that

$$(1-\delta_{is_n})^{-1}\to(1-\delta)^{-1} \qquad \text{(strongly)}.$$

4.5.11 Proposition $(1-\delta_{il_n})^{-1}\to(1-\delta)^{-1}$ *(strongly) if and only if for each $a\in\mathcal{D}(\delta)$, there is a sequence $\{a_n\}$ in A such that $a_n\to a$ and $\delta_{il_n}(a_n)\to\delta(a)$.*

Proof Suppose that $(1-\delta_{il_n})^{-1}\to(1-\delta)^{-1}$; then $\delta_{il_n}(1-\delta_{il_n})^{-1}=(1-\delta_{il_n})^{-1}-1\to\delta(1-\delta)^{-1}$ (strongly). For $a\in\mathcal{D}(\delta)$, take $b\in A$ such that $a=(1-\delta)^{-1}b$ (note $(1-\delta)^{-1}A=\mathcal{D}(\delta)$), and put $a_n=(1-\delta_{il_n})^{-1}b$; then $a_n\to a$ and $\delta_{il_n}(a_n)\to\delta(a)$. Conversely, for $a\in\mathcal{D}(\delta)$,

$$\begin{aligned}\|(1-\delta_{il_n})^{-1}(1-\delta)a-(1-\delta)^{-1}(1-\delta)a\| &= \|(1-\delta_{il_n})^{-1}(1-\delta)a-a\|\\ &\leqslant \|(1-\delta_{il_n})^{-1}(1-\delta_{il_n})a_n-a\|+\|(1-\delta_{il_n})^{-1}\{\delta_{il_n}(a_n)-\delta(a)+a-a_n\}\|\\ &\leqslant 2\|a-a_n\|+\|\delta_{il_n}(a_n)-\delta(a)\|\to 0.\end{aligned}$$

Since $(1-\delta)\mathcal{D}(\delta)=A$, $(1-\delta_{il_n})^{-1}\to(1-\delta)^{-1}$ (strongly). □

Now take $x_n\in\bigcup_{r=1}^{\infty}A_r$ such that $\|x_n-a_n\|<1/n\{\|l_n\|+1\}$; then $\|a-x_n\|\to 0$ and $\|\delta(a)-\delta_{il_n}(x_n)\|\leqslant\|\delta(a)-\delta_{il_n}(a_n)\|+\|\delta_{il_n}(x_n)-\delta_{il_n}(a_n)\|\to 0$. Therefore we have:

4.5.12 Proposition $\{\exp(t\delta)|t\in\mathbb{R}\}$ *is approximately inner if and only if for a general increasing sequence $\{A_n\}$ of finite type I subfactors in A with $\overline{\bigcup_{n=1}^{\infty}A_n}=A$, there is a sequence $\{s_n\}$ of self-adjoint elements in $\bigcup_{n=1}^{\infty}A_n$ such that for each $a\in\mathcal{D}(\delta)$, there exists a sequence (a_n) in $\bigcup_{n=1}^{\infty}A_n$ with $a_n\to a$ and $\delta_{is_n}(a_n)\to\delta(a)$.*

However this proposition is not so useful, because it does not suggest how to construct $\{s_n\}$ and $\{a_n\}$. In the following, we shall formulate a new problem to attack the conjecture.

Let $\mathcal{D}=\bigcup_{n=1}^{\infty}A_n$, where $\{A_n\}$ is the one given in 4.5.1. Since $\mathcal{D}\subset A(\delta)$, $\delta^n(\mathcal{D})\subset A(\delta)$ $(n=1,2,3,\ldots,)$. Let B be the linear subspace of A spanned by $\bigcup_{n=1}^{\infty}\delta^n(\mathcal{D})$ $(\delta^0(\mathcal{D})=\mathcal{D})$.

4.5.13 Proposition $(1-\delta)B$ *is dense in A.*

Proof Suppose that $(1-\delta)B$ is not dense in A; then there is an element f in the dual A^* of A such that $f((1-\delta)B)=0$. Therefore $f(x)=f(\delta(x))(x\in B)$ and so $f(a)=f(\delta^m(a))(a\in\bigcup_{n=1}^{\infty}A_n;\ m=1,2,3,\ldots,)$. Since $a\in A(\delta)$, there is an $r>0$ such that

$$\sum_{m=0}^{\infty} \frac{\|\delta^m(a)\|}{m!} r^m < +\infty.$$

$$f(\alpha_t(a)) = \sum_{m=0}^{\infty} \frac{f(\delta^m(a))}{m!} t^m = \exp(t) f(a) \qquad (|t| \leqslant r).$$

$t \to f(\alpha_t(a))$ and $t \to \exp(t) f(a)$ are real analytic on $\mathbb{R}$; hence

$$f(\alpha_t(a)) = \exp(t) f(a) \qquad \left(t \in R; a \in \bigcup_{n=1}^{\infty} A_n \right).$$

$|f(\alpha_t(a))| \leqslant \|f\| \, \|a\|$; hence $f(a) = 0$ and so $f = 0$. □

4.5.14 Proposition *Let $\{k_n\}$ be a sequence of self-adjoint elements in A and suppose that for each $a \in B$, there is a sequence $\{a_n\}$ of elements in A such that $a_n \to a$ and $\delta_{ik_n}(a_n) \to \delta(a)$; then* $\exp(t\delta) =$ *strong* $\lim \exp(t\delta_{ik_n})$.

Proof

$$\begin{aligned}
&\|(1-\delta_{ik_n})^{-1}(1-\delta)(a) - (1-\delta)^{-1}(1-\delta)(a)\| = \|(1-\delta_{ik_n})^{-1}(1-\delta)(a) - a\| \\
&\quad = \|(1-\delta_{ik_n})^{-1}\{(1-\delta_{ik_n})(a_n) + (1-\delta)(a) - (1-\delta_{ik_n})(a_n)\} - a\| \\
&\quad = \|a_n - a + (1-\delta_{ik_n})^{-1}\{a - a_n + \delta_{ik_n}(a_n) - \delta(a)\}\| \\
&\quad \leqslant 2\|a_n - a\| + \|\delta_{ik_n}(a) - \delta(a_n)\| \to 0.
\end{aligned}$$

Hence by 4.5.13 and 4.1.2, $\exp(t\delta) =$ strong lim $\exp(t\delta_{ik_n})$. □

4.5.15 Corollary *If there is a sequence $\{k_n\}$ of self-adjoint elements in A such that for $a \in \bigcup_{n=1}^{\infty} A_n$, $\delta^m(a) = \lim_{n\to\infty} \delta_{ik_n}{}^m(a)(m = 1, 2, 3, \ldots,)$, then* $\exp(t\delta) =$ strong lim $\exp(t\delta_{ik_n})$.

Proof $\delta_{ik_n}{}^m(a) \to \delta^m(a)$ and $\delta_{ik_n}(\delta_{ik_n}{}^m(a)) \to \delta(\delta^m(a))(m = 0, 1, 2, \ldots,)$. Hence for $a \in B$ there is a sequence of elements $\{a_n\}$ in A such that $a_n \to a$ and $\delta_{ik_n}(a_n) \to \delta(a)$. By 4.5.14, strong $\lim \exp(t\delta_{ik_n}) = \exp(t\delta)$. □

An important fact is that all generators considered in Examples (4)–(6) in §3.1 satisfy the assumption of this corollary. For completeness, we shall show that fact below for quasi-free derivations in which case the assertion may not be trivial. Let $\mathscr{H}$ be a Hilbert space and H be a self-adjoint operator in $\mathscr{H}$. Let $A(H)$ be the set of all analytic vectors in $\mathscr{H}$ with respect to H, and let $\{V_n\}$ be an increasing sequence of finite-dimensional subspaces of $A(H)$ such that $\bigcup_{n=1}^{\infty} V_n$ is dense in $\mathscr{H}$. Let $E_n =$ the linear span of $\bigcup_{m=0}^{n} H^m V_n$ where $H^0 V_n = V_n$; then E_n is finite-dimensional and so $\mathscr{A}(E_n)$ is a full matrix algebra; hence there is an element k_n in $\mathscr{A}(\mathscr{H})$ such that $\delta_{iH}(a) = \mathrm{i}[k_n, a](a \in \mathscr{A}(E_n))$;

then

$$\delta_{iH}{}^{m}(a) = \delta_{ik_n}{}^{m}(a)(a \in \mathscr{A}(V_n);\ m = 1, 2, \ldots, n;\ n = 1, 2, 3, \ldots,).$$

Therefore $\delta_{ik_n}{}^{m}(a) \to \delta^m(a)(n \to \infty; m = 1, 2, \ldots,)$ for $a \in \bigcup_{n=1}^{\infty} \mathscr{A}(V_n)$.

If $\delta(\bigcup_{n=1}^{\infty} A_n) \subset \bigcup_{n=1}^{\infty} A_n$; then one can easily see that the assumptions of 4.5.15 are satisfied. Therefore if there is an increasing sequence $\{A_n\}$ such that $\bigcup_{n=1}^{\infty} A_n \subset A(\delta)$, $\delta(\bigcup_{n=1}^{\infty} A_n) \subset \bigcup_{n=1}^{\infty} A_n$ and $\bigcup_{n=1}^{\infty} A_n$ is dense in A, then all the assumptions of 4.5.15 are satisfied. By 4.5.5, one can change δ to δ_1 by a bounded perturbation such that $\delta_1(\bigcup_{n=1}^{\infty} A_n) \subset \bigcup_{n=1}^{\infty} A_n$. However bounded perturbations do not generally preserve the analyticity – namely, generally $\bigcup_{n=1}^{\infty} A_n \not\subset A(\delta_1)$. Therefore the following open problem is interesting.

4.5.16 Problem *Let δ be a generator in a UHF algebra A and let $\{A_n\}$ be an increasing sequence of finite type I subfactors in A such that $\bigcup_{n=1}^{\infty} A_n \subset A(\delta)$ and $\bigcup_{n=1}^{\infty} A_n$ is dense in A; then can we choose a bounded perturbation δ_0 such that*

$$(\delta + \delta_0)\left(\bigcup_{n=1}^{\infty} A_n\right) \subset \bigcup_{n=1}^{\infty} A_n \quad \text{and} \quad \bigcup_{n=1}^{\infty} A_n \subset A(\delta + \delta_0)?$$

Changing the subject, let δ be a generator in a UHF algebra A. For each $a \in A(\delta)$, there is a positive number $r(a)$ such that

$$\sum_{n=0}^{\infty} \frac{\|\delta^n(a)\|}{n!} |z|^n < +\infty \qquad (|z| < r(a)).$$

Define

$$\alpha_z(a) = \sum_{n=0}^{\infty} \frac{\delta^n(a)}{n!} z^n \qquad (|z| < r(a)).$$

Then $\alpha_z(ab) = \alpha_z(a)\alpha_z(b)(|z| < \min(r(a), r(b)))$. For any projection $p \in A(\delta)$, $Sp(\alpha_z(p)) = Sp_A(p)$ in A, so that $Sp_A(\alpha_z(a)) = Sp_A(a)$ for $a(= a^*) \in \bigcup_{n=1}^{\infty} A_n$.

4.5.17 Problem *Can we conclude that $Sp(\alpha_z(a)) = Sp(a)$ for $a(= a^*) \in A(\delta)$ $(|z| < r(a))$?*

Remark The problem has a negative answer in the general case as mentioned before (cf. §3.4).

4.5.18 Problem *If Problem 4.5.17 has a negative answer, under what conditions can one conclude $\{\alpha_z\}$ preserves the spectrum of $A(\delta)$?*

4.5.19 Problem *Can we choose an increasing sequence $\{A_n\}$ of finite type I subfactors in $A(\delta)$ such that there exists a positive number r_0 for which $\bigcup_{n=1}^{\infty} A_n$ is dense in A and $\sum_{n=0}^{\infty}(\|\delta^n(a)\|/n!)t^n < +\infty$ $(0 \leqslant t < r_0)$ for $a \in \bigcup_{n=1}^{\infty} A_n$?*

The proof of 4.5.1 does not guarantee the existence of an increasing sequence $\{A_n\}$ as the above. On the other hand, in quantum lattice models, we can often choose such a fixed number [159].

In Theorem 4.5.1, we have shown that for a C*-dynamics $\{\mathscr{A}, \exp(t\delta)(t\in\mathbb{R})\}$ with a UHF algebra $\mathscr{A}$, there is an increasing sequence of finite type I subfactors $\{\mathscr{A}_n\}$ such that $1 \in \mathscr{A}_1 \subset \mathscr{A}_2 \subset \cdots \subset \mathscr{A}_n \subset \cdots$, $\bigcup_{n=1}^{\infty} \mathscr{A}_n$ is dense in $\mathscr{A}$ and every element of $\bigcup_{n=1}^{\infty} \mathscr{A}_n$ is analytic with respect to δ. Furthermore, by Proposition 4.5.5, there is a bounded *-derivation δ_{ih} such that $(\delta + \delta_{ih})\mathscr{D} \subset \mathscr{D}$, where $\mathscr{D} = \bigcup_{n=1}^{\infty} \mathscr{A}_n$. It is clear that every element of $\mathscr{D}$ is a C^∞-vector with respect to $\delta + \delta_{ih}$, but it is not necessarily analytic with respect to $\delta + \delta_{ih}$, because bounded perturbations cannot keep their analyticity in general. If one can choose a δ_{ih} such that every element of $\mathscr{D}$ is analytic with respect to $\delta + \delta_{ih}$, then the C*-dynamics $\{\mathscr{A}, \exp(t(\delta + \delta_{ih}))(t\in\mathbb{R})\}$ is approximately inner, because $\mathscr{D}$ is dense in $\mathscr{A}$ and every element of $\mathscr{D}$ is analytic with respect to the restriction of $\delta + \delta_{ih}$ to $\mathscr{D}$, for $(\delta + \delta_{ih})\mathscr{D} \subset \mathscr{D}$ (3.4.5). Therefore we can conclude that the C*-dynamics $\{\mathscr{A}, \exp(t\delta)(t\in R)\}$ is approximately inner. Finally we shall prove the following proposition.

4.5.20 Proposition *If one can choose an analytic element with respect to δ as h in the above consideration, then every element of $\mathscr{D}$ is analytic with respect to $\delta + \delta_{ih}$.*

Proof Let τ be the tracial state on $\mathscr{A}$, and let $\{\pi_\tau, U_\tau, \mathscr{H}_\tau\}$ be the covariant representation of $\{\mathscr{A}, \exp(t\delta)(t\in\mathbb{R})\}$ constructed via τ. Since $\mathscr{A}$ is simple, the representation π_τ of $\mathscr{A}$ is faithful. Put $\alpha_t = \exp(t\delta)$ and let $U_\tau(t) = \exp(\mathrm{i}tH_\tau)$. Then by the theory of semi-groups (cf. 1.17, and §4.4),

$$\begin{aligned}&\overline{\exp \mathrm{i}z(H_\tau + \pi_\tau(h))\exp(-\mathrm{i}zH_z)}\\ &\quad = 1 + \sum_{p=1}^{\infty} (\mathrm{i}z)^p \int_{0\leqslant s_1 \leqslant s_2 \leqslant \cdots \leqslant 1} \pi_\tau(\alpha_i s_1 z(h))\pi_\tau(\alpha_i s_2 z(h))\cdots\\ &\qquad \times \pi_\tau(\alpha_i s_p z(h))\mathrm{d}s_1\,\mathrm{d}s_2 \cdots \mathrm{d}s_p.\end{aligned}$$

Since $h \in A(\delta)$, there is a positive number r_0 such that $\pi_\tau(\alpha_z(h))$ is analytic on $D_{r_0}{}^0$ and continuous on D_{r_0}, where D_{r_0} is the closed disk with a radius r_0 at the center 0 in the complex plane and $D_{r_0}{}^0$ is the interior of D_{r_0}.

$$\|\overline{\exp \mathrm{i}z(H_\tau + \pi_\tau(h))\exp(-\mathrm{i}zH_\tau)}\| \leqslant \sum_{p=0}^{\infty} r_0{}^p M^p \frac{1}{p!} = \exp(r_0 M),$$

where

$$M = \sup_{z \in D_{r_0}} \| \pi_\tau(\alpha_z(h)) \|.$$

Hence $z \to \overline{\exp iz(H_\tau + \pi_\tau(h))\exp(-izH_\tau)}$ is analytic on $D_{r_0}{}^0$ and continuous on D_{r_0}. For $a \in \mathscr{D}$,

$$\begin{aligned}
\pi_\tau(\exp z(\delta + \delta_{ih})(a)) &= \overline{\exp iz(H_\tau + \pi_\tau(h))\pi_\tau(a)\exp - iz(H_\tau + \pi_\tau(h))} \\
&= \overline{\exp iz(H_\tau + \pi_\tau(h))\exp(-izH_\tau)} \cdot \overline{\exp izH_\tau \pi_\tau(a)\exp(-izH_\tau)} \\
&\quad \times \overline{\exp izH_\tau \exp - iz(H_\tau + \pi_\tau(h))} \\
&= \overline{\exp iz(H_\tau + \pi_\tau(h))\exp - izH_\tau}\,\pi_\tau(\alpha_z(a))\,\overline{\exp izH_\tau \exp - iz(H_\tau + \pi_\tau(h))}
\end{aligned}$$

Since $a \in \mathscr{D}$, there is a positive number r_1 such that $z \to \pi_\tau(\alpha_z(a))$ is analytic on $D_{r_1}{}^0$. Put $r = \min(r_0, r_1)$; then, $z \to \pi_\tau(\exp z(\delta + \delta_{ih})(a))$ is analytic on $D_r{}^0$. Since π_τ is an isometry, $\exp z(\delta + \delta_{ih})(a)$ is analytic on $D_r{}^0$, and so a is analytic with respect to $\delta + \delta_{ih}$. This completes the proof. □

4.5.21 Notes and remarks Theorem 4.5.1 was first proved by the author [166] in 1974. Except for the proof of analyticity the technique used in the proof was first invented by Glimm [207] for the study of UHF algebras. The Powers–Sakai conjecture has been studied by many researchers since the beginning of the theory of unbounded *-derivations in 1974. But, as yet, we do not have any definitive result on it. Proposition 4.5.11 was essentially due to Jaffe and Glimm [215]. However the present form was proved by Herman [76].

References [146], [166], [170], [175].

4.6 Commutative normal *-derivations in UHF algebras

A classical lattice system is usually treated in a commutative algebra of functions. However it is more convenient to deal with it as a special case of quantum lattice systems, because we can then define the time automorphism group, and the Gibbs states can be computed as KMS states with respect to this time automorphism group. A commutative normal *-derivation in a UHF algebra is a generalization of a classical lattice system and they have better properties than a non-commutative one.

In this section, we shall establish the basic properties of commutative derivations.

Let δ be a commutative normal *-derivation in a UHF-algebra A – i.e. one can choose a sequence $\{h_n\}$ of self-adjoint elements in A such that $\delta(a) = i[h_n, a]$ ($a \in A_n$; $n = 1, 2, \ldots,$), $h_m h_n = h_n h_m$ ($m, n = 1, 2, \ldots,$). Let B_n be the

C*-subalgebra of A generated by A_n and $h_1, h_2, \ldots, h_n$; then by 4.1.11 there is a unique generator δ_1 such that $\delta \subset \delta_1$ and $\exp(t\delta_1)(b) = \exp(t\delta i h_n)(b)$ $(b \in B_n; n = 1, 2, \ldots,)$. Let C_n be the C*-subalgebra of A generated by A_n and h_n; then $\delta_1(C_n) \subset C_n$ and $C_n \subset B_n$, so that $\exp(t\delta_1)(c) = \exp(t\delta i h_n)(c)$ $(c \in C_n; n = 1, 2, \ldots,)$. Now let ψ_β be a KMS state for $\{A, \exp(t\delta_1)(t \in R)\}$ at β. Then,

$$\begin{aligned} \psi_\beta(ab\exp(\beta h_n)) &= \psi_\beta(a\exp(\beta h_n)\exp(-\beta h_n)b\exp(\beta h_n)) \\ &= \psi_\beta(a\exp(\beta h_n)\exp(i\beta\delta_1)(b)) \\ &= \psi_\beta(ba\exp(\beta h_n)) \qquad (a \in A, b \in B_n). \\ \psi_\beta(\exp(\beta h_n)a) &= \psi_\beta((\exp(\beta h_n)a\exp(-\beta h_n)\exp(\beta h_n)) \\ &= \psi_\beta(a\exp(-\beta h_n)\exp(\beta h_n)\exp(\beta h_n)) \\ &= \psi_\beta(a\exp(\beta h_n)) \end{aligned}$$

(note $\exp(\beta h_n) \in B_n$); therefore $\psi_\beta(ka) = \psi_\beta(ak)$ $(a \in A)$, where k is any element of the C*-subalgebra of A generated by $\exp(\beta h_n)$. Define

$$\phi(b) = \psi_\beta(b\exp(\beta h_n))/\psi_\beta(\exp(\beta h_n)) \qquad (b \in B_n);$$

then ϕ is a tracial state on B_n, and

$$\psi_\beta(b) = \phi(b\exp(-\beta h_n))/\phi(\exp(-\beta h_n)) \qquad (b \in B_n).$$

Now let Ω_n be the set of all tracial states on B_n and let $\partial(\Omega_n)$ be the set of all extreme points in Ω_n (i.e. all factorial, tracial states on B_n); there is a unique probability Radon measure μ_n on Ω_n such that $\mu_n(\partial(\Omega_n)) = 1$ and $\phi = \int_{\partial(\Omega_n)} \sigma \, d\mu_n(\sigma)$. Hence we have

$$\begin{aligned} \psi_\beta(b) &= \int_{\partial(\Omega_n)} \sigma(b\exp(-\beta h_n))\,d\mu_n(\sigma) \bigg/ \int_{\partial(\Omega_n)} \sigma(\exp(-\beta h_n))\,d\mu_n(\sigma) \\ &= \int_{\partial(\Omega_n)} \frac{\sigma(b\exp(-\beta h_n))}{\sigma(\exp(-\beta h_n))} \left\{ \frac{\sigma(\exp(-\beta h_n))}{\displaystyle\int_{\partial(\Omega_n)} \sigma(\exp(-\beta h_n))\,d\mu_n(\sigma)} \right\} d\mu_n(\sigma) \qquad (b \in B_n). \end{aligned}$$

Define

$$d\nu_n(\sigma) = \left\{ \frac{\sigma(\exp(-\beta h_n))}{\displaystyle\int_{\partial(\Omega_n)} \sigma(\exp(-\beta h_n))\,d\mu_n(\sigma)} \right\} d\mu_n(\sigma);$$

then

$$\psi_\beta(b) = \int_{\partial(\Omega_n)} \frac{\sigma(b\exp(-\beta h_n))}{\sigma(\exp(-\beta h_n))} d\nu_n(\sigma) \qquad (b \in B_n).$$

Therefore we have the following theorem.

4.6.1 Theorem *Let ψ_β be a KMS state for $\{A, \exp(t\delta_1)(t \in \mathbb{R})\}$ at β; then there is a unique sequence $\{\nu_n\}$ of probability Radon measures on the compact spaces Ω_n consisting of all tracial states on B_n with $\nu_n(\partial(\Omega_n)) = 1$ such that*

$$\psi_\beta(b) = \int_{\partial(\Omega_n)} \frac{\sigma(b\exp(-\beta h_n))}{\sigma(\exp(-\beta h_n))} \mathrm{d}\nu_n(\sigma) \qquad (b \in B_n; n = 1, 2, \ldots,).$$

Moreover if $\{\tilde{h}_n\}$ is another sequence of self-adjoint elements in B_n such that $\delta_{\mathrm{i}h_n}(b) = \delta_{\mathrm{i}\tilde{h}_n}(b)$ $(b \in B_n)$, then

$$\frac{\sigma(b\exp(-\beta h_n))}{\sigma(\exp(-\beta h_n))} = \frac{\sigma(b\exp(-\beta \tilde{h}_n))}{\sigma(\exp(-\beta \tilde{h}_n))} \qquad (b \in B_n, \sigma \in \partial(\Omega_n))$$

and so ν_n does not depend on a special choice of (h_n).

Quite similarly, there is a unique sequence $\{\xi_n\}$ of probability Radon measures on the compact space Γ_n consisting of all tracial states on C_n with $\xi_n(\partial(\Gamma_n)) = 1$ such that

$$\psi_\beta(c) = \int_{\partial(\Gamma_n)} \frac{\sigma(c\exp(-\beta h_n))}{\sigma(\exp(-\beta h_n))} \mathrm{d}\xi_n(\sigma) \qquad (c \in C_n; n = 1, 2, \ldots,)$$

Moreover if $\{\tilde{h}_n\}$ is another sequence of self-adjoint elements in C_n such that $\delta_{\mathrm{i}h_n}(c) = \delta_{\mathrm{i}\tilde{h}_n}(c)$ $(c \in C_n)$, then

$$\frac{\sigma(c\exp(-\beta h_n))}{\sigma(\exp(-\beta h_n))} = \frac{\sigma(c(\exp(-\beta \tilde{h}_n))}{\sigma(\exp(-\beta h_n))} \qquad (c \in C_n, \sigma \in \partial(\Gamma_n))$$

and so ξ_n does not depend on a special choice of $(\tilde{h}_n)$.

Proof It is enough to show that

$$\frac{\sigma(b\exp(-\beta h_n))}{\sigma(\exp(-\beta h_n))} = \frac{\sigma(b\exp(-\beta \tilde{h}_n))}{\sigma(\exp(-\beta \tilde{h}_n))} \qquad (b \in B_n, \sigma \in \partial(\Omega_n)).$$

Since σ is factorial, σ is multiplicative on the center of B_n and since $\exp(-\beta(\tilde{h}_n - h_n))$ belongs to the center of B_n,

$$\begin{aligned}\sigma(b\exp(-\beta\tilde{h}_n)) &= \sigma(b\exp(-\beta h_n)\exp(-\beta(\tilde{h}_n - h_n))) \\ &= \sigma(b\exp(-\beta h_n))\sigma(\exp(-\beta(\tilde{h}_n - h_n)));\end{aligned}$$

hence

$$\frac{\sigma(b\exp(-\beta h_n))}{\sigma(\exp(-\beta h_n))} = \frac{\sigma(b\exp(-\beta \tilde{h}_n))}{\sigma(\exp(-\beta \tilde{h}_n))} \qquad (b \in B_n, \sigma \in \partial(\Omega_n)).$$ □

4.6.2 Proposition *Let δ be a commutative normal *-derivation in a UHF algebra A such that $\mathscr{D}(\delta) = \bigcup_{n=1}^\infty A_n$, $\delta(a) = \mathrm{i}[h_n, a]$ $(a \in A_n; n = 1, 2, \ldots,)$ $h_m h_n = h_n h_m$ $(m, n = 1, 2, \ldots,)$* and $h_n \in \bigcup_{m=1}^\infty A_m$ $(n = 1, 2, \ldots,)$. *Then for each*

KMS state ϕ_β *for* $\{A, \exp(t\delta_1)\}$ *at* β *there is a unique sequence* $\{\tilde{h}_n\}$ *of self-adjoint elements in* B_n *such that* $\delta_{i\tilde{h}_n}(a) = \delta_1(a)$ $(a \in B_n; n = 1, 2, \ldots,)$ *and* $\phi_\beta(a) = \tau(a \exp(-\beta\tilde{h}_n))$ $(a \in B_n; n = 1, 2, \ldots,)$.

Quite similarly there is a unique $(\tilde{\tilde{h}}_n)$ *of self-adjoint elements in* C_n *such that* $\delta_{i\tilde{\tilde{h}}_n}(a) = \delta_1(a)$ $(a \in C_n; n = 1, 2, \ldots,)$ *and* $\phi_\beta(a) = \tau(a \exp(-\beta\tilde{\tilde{h}}_n))$ $(a \in C_n; n = 1, 2, \ldots,)$.

Proof Since $h_n \in \bigcup_{m=1}^{\infty} A_m$ $(n = 1, 2, \ldots,)$, B_n is finite-dimensional; hence $B_n = \sum_{j=1}^{m_n} B_n p_{n,j}$, where the $p_{n,j}$ are the minimal central projection of B_n. Since $B_n p_{n,j}$ is a full matrix algebra and is invariant under $\exp(t\delta_1)$, by the unicity of KMS state at β,

$$\frac{\phi_\beta(ap_{n,j})}{\phi_\beta(p_{n,j})} = \frac{\tau(ap_{n,j}\exp(-\beta h_n)p_{n,j})}{\tau(\exp(-\beta h_n)p_{n,j})} = \frac{\tau(a\exp(-\beta h_n)p_{n,j})}{\tau(\exp(-\beta h_n)p_{n,j})} \qquad (a \in B_n).$$

Put

$$\tilde{h}_n = h_n + \sum_{j=1}^{m_n} \log\left(\frac{\phi_\beta(p_{n,j})}{\tau(\exp(-\beta h_n)p_{n,j})}\right)p_{n,j};$$

then $\delta_{i\tilde{h}_n}(a) = \delta_1(a) (a \in B_n)$. Moreover

$$\begin{aligned}
\tau(a\exp(-\beta\tilde{h}_n)) &= \tau\left(a\exp(-\beta h_n)\sum_{j=1}^{m_n}\frac{\phi_\beta(p_{n,j})}{\tau(\exp(-\beta h_n)p_{n,j})}p_{n,j}\right) \\
&= \sum_{j=1}^{m_n}\tau(a\exp(-\beta h_n)p_{n,j})\frac{\phi_\beta(p_{n,j})}{\tau(\exp(-\beta h_n)p_{n,j})} \\
&= \sum_{j=1}^{m_n}\phi_\beta(p_{n,j})\frac{\phi_\beta(ap_{n,j})}{\phi_\beta(p_{n,j})} \\
&= \phi_\beta(a) \qquad (a \in B_n).
\end{aligned}$$

□

Let δ be a commutative normal *-derivation in A such that $\delta(a) = i[h_n, a]$ $(a \in A_n; n = 1, 2, \ldots,)$, $h_m h_n = h_n h_m$ $(m, n, = 1, 2, \ldots,)$.

Now suppose that there exists a commutative C*-subalgebra C of A such that $h_n \in C$ $(n = 1, 2, \ldots,)$ and $C = (A_n \cap C) \otimes (A'_n \cap C)$ $(n = 1, 2, \ldots,)$, where A'_n is the commutant of A_n in A $(n = 1, 2, \ldots,)$. (Note: all classical lattice systems and Ising models satisfy this property.) Then $B_n = A_n \otimes (B_n \cap A'_n) = A_n \otimes Z_n$, where Z_n is the center of B_n, for B_n is the C*-subalgebra of A generated by A_n and $C = A_n \otimes (A'_n \cap C)$.

Let $Z_n = C(K_n)$; then $A_n \otimes Z_n = C(A_n, K_n) =$ the C*-algebra of all A_n-valued continuous functions on the compact space K_n. For $a \in C(A_n, K_n)$, define $\Phi_n(a)(t) = \tau(a(t))$ $(t \in K_n)$; then $\Phi_n(a) \in Z_n$. The Φ_n are norm-one projections of B_n onto Z_n satisfying the following conditions;

(1) $\Phi(a^*a) \geqslant 0$ $(a \in B_n)$;
(2) $\Phi(za) = z\Phi(a)$ $(z \in Z_n, a \in B_n)$;
(3) $\Phi(ab) = \Phi(ba)$ $(a, b \in B_n)$.

For $t \in K_n$, $a \mapsto \Phi(a)(t)$ is a factorial, tracial state on B_n; moreover if $t_1, t_2 \in K_n$ and $t_1 \neq t_2$, then they define different states; hence we can consider $K_n \subset \partial(\Omega_n)$. On the other hand, for $\sigma \in \partial(\Omega_n)$, σ is multiplicative on Z_n and so there is a point t in K_n such that $\sigma(a) = \Phi(a)(t)$ for $a \in Z_n$. Since A_n is a full matrix algebra, it has a unique tracial state and so $\sigma(a) = \Phi(a)(t)$ $(a \in B_n)$. Therefore $K_n = \partial(\Omega_n)$. Hence we have the following theorem.

4.6.3 Theorem *Suppose that δ is a normal commutative *-derivation in A such that $\delta(a) = \mathrm{i}[h_n, a]$ $(a \in A_n; n = 1, 2, \ldots,)$, $h_m h_n = h_n h_m$ $(m, n = 1, 2, \ldots,)$. Moreover suppose that there is a commutative C*-algebra C of A such that $h_n \in C$ $(n = 1, 2, \ldots,)$ and $C = (A_n \cap C) \otimes (A'_n \cap C)$ $(n = 1, 2, \ldots,)$. Let Z_n be the center of B_n and let $Z_n = C(K_n)$. Then for any KMS state ψ_β for $\{A, \exp(t\delta_1)\}$ there is a unique sequence $\{\mu_n\}$ of probability Radon measures on K_n such that*

$$\psi_\beta(a) = \int_{K_n} \frac{\Phi(a \exp(-\beta h_n))(t)}{\Phi(\exp(-\beta h_n))(t)} \mathrm{d}\mu_n(t) \qquad (a \in B_n)$$

Quite similarly, one can formulate the corresponding result for $\{C_n\}$.

4.6.4 Proposition *Under the assumptions of* 4.6.3, *for any KMS state ϕ_β for $\{A, \exp(t\delta_1)\}$ at β there is a sequence $\{\tilde{h}_n\}$ of self-adjoint elements in A such that $\delta(a) = \mathrm{i}[\tilde{h}_n, a]$ $(a \in A_n; n = 1, 2, \ldots,)$, $\tilde{h}_n \in C_n$ $(n = 1, 2, \ldots,)$ and*

$$\phi_\beta(x) = \lim_{n \to \infty} \frac{\tau(x \exp(-\beta \tilde{h}_n))}{\tau(\exp(-\beta \tilde{h}_n))} \qquad (x \in A).$$

Conversely for any sequence $\{\tilde{h}_n\}$ satisfying the conditions

(1) $\mathrm{i}[\tilde{h}_n, a] = \delta(a)$ $\qquad (a \in A_n; n = 1, 2, \ldots)$;
(2) $\tilde{h}_n \in C_n$ $(n = 1, 2, \ldots)$,

define

$$\phi_{\beta,n}(x) = \tau(x \exp(-\beta \tilde{h}_n))/\tau(\exp(-\beta \tilde{h}_n)) \qquad (x \in A);$$

then any accumulation point of $\{\phi_{\beta,n}\}$ in the state space of A is a KMS state for $\{A, \exp(t\delta_1\}$ at β.

Proof By a slight modification of the proof of 4.5.5, one can easily show that for any positive number $\varepsilon > 0$ there is a self-adjoint element d in C such that $(\delta - \delta_{\mathrm{id}})(\mathscr{D}(\delta)) \subset \mathscr{D}(\delta)$ and $\|\delta_{\mathrm{id}}\| < \varepsilon$. Therefore by 4.6.2, for any KMS state ψ_β for $\{A, \exp(t(\delta - \delta_{\mathrm{id}}))\}$ at β there is a sequence $\{k_n\}$ of self-adjoint elements in C such that $(\delta - \delta_{id})(a) = \mathrm{i}[k_n, a]$ $(a \in A_n; n = 1, 2, \ldots,)$ and

$$\psi_\beta(a) = \frac{\tau(a \exp(-\beta k_n))}{\tau(\exp(-\beta k_n))} \qquad (a \in A_n; n = 1, 2, \ldots).$$

In particular,

$$\psi_\beta(x) = \lim_{n \to \infty} \frac{\tau(x \exp(-\beta k_n))}{\tau(\exp(-\beta k_n))} \qquad (x \in A).$$

Since $\{\exp(t(\delta_1 - \delta_{id}))\}(c) = c$ for $c \in C$, the corresponding KMS state for $\{A, \exp(t\delta_1)\}$ at β is $(\psi_\beta(x \exp(-\beta d)))/(\psi_\beta(\exp(-\beta d)))$ $(x \in A)$ (4.4.7); hence

$$\frac{\phi_\beta(x \exp(-\beta d))}{\phi_\beta(\exp(-\beta d))} = \lim_n \frac{\tau(x \exp(-\beta(k_n + d)))}{\tau(\exp(-\beta(k_n + d)))} \qquad (x \in A),$$

where $\mathrm{i}[k_n + d, a] = \delta(a)$ $(a \in A_n; n = 1, 2, \ldots)$. The rest of the theorem is clear.

□

Let ϕ be a KMS state for $\{A, \exp(t\delta_1)(t \in \mathbb{R})\}$ at β, where $\delta_1(a) = \mathrm{i}[h_n, a]$ $(a \in A_n; n = 1, 2, \ldots)$ and $\{h_n\}$ is a mutually commuting family. Consider the covariant representation $\{\pi_\phi, U_\phi, \mathscr{H}_\phi\}$ of $\{A, \exp(t\delta_1)(t \in \mathbb{R})\}$ constructed via ϕ. Since $(\exp(t\delta_1))(h_n) = h_n$, we obtain $U_\phi(t)\pi_\phi(h_n)U_\phi(t)^* = \pi_\phi(h_n)$ $(n = 1, 2, \ldots)$.

Let D be the W*-subalgebra of $\mathscr{M}(= \pi_\phi(A)'')$ such that

$$D = \{a \in \mathscr{M} \mid U_\phi(t) a U_\phi(t)^* = a \quad (t \in R)\}.$$

Then for $h \in D$,

$$f(\mathrm{i}\beta, h)(x) = \left(x \exp\left(\frac{-\beta(H_\phi + h)}{2} \right) 1_\phi, \exp\left(\frac{-\beta(H_\phi + h)}{2} \right) 1_\phi \right)$$
$$= \left(x \exp\left(\frac{-\beta h}{2} \right) 1_\phi, \exp\left(\frac{-\beta h}{2} \right) 1_\phi \right) \qquad (x \in \mathscr{M}).$$

For $a \in A_2(\delta_1)$,

$$f(\mathrm{i}\beta, h)\left(\exp\left(\frac{\beta(h + H_\phi)}{2} \right) \pi_\phi(a) \exp\left(\frac{-\beta(h + H_\phi)}{2} \right) \right) = f(\mathrm{i}\beta, h)(\pi_\phi(a)).$$

$$\left(\exp\left(\frac{\beta H_\phi}{2} \right) \exp\left(\frac{\beta h}{2} \right) \pi_\phi(a) \exp\left(\frac{-\beta h}{2} \right) \exp\left(\frac{-\beta H_\phi}{2} \right) \exp\left(\frac{-\beta h}{2} \right) 1_\phi, \right.$$
$$\left. \exp\left(\frac{-\beta h}{2} \right) 1_\phi \right) = \left(\exp\left(\frac{\beta h}{2} \right) \pi_\phi(a) \exp\left(\frac{-\beta h}{2} \right) \exp\left(\frac{-\beta h}{2} \right) 1_\phi, \exp\left(\frac{-\beta h}{2} \right) 1_\phi \right)$$
$$= (\pi_\phi(a) \exp(-\beta h) 1_\phi, 1_\phi);$$

hence

$$(x\exp(-\beta h/2)1_\phi, \exp(-\beta h/2)1_\phi) = (x\exp(-\beta h)1_\phi, 1_\phi) \qquad (x \in \mathcal{M}).$$

4.6.5 Proposition *Let ϕ be a KMS state for $\{A, \exp(t\delta_1)(t \in \mathbb{R})\}$ at β, with $\delta_1(a) = \mathrm{i}[h_n, a]$ $(a \in A_n; n = 1, 2, \ldots)$ and a mutually commuting family $\{h_n\}$ $(h_n \in A^s)$. Let δ_2 be another commutative normal *-derivation in A such that $\mathcal{D}(\delta_2) = \bigcup_{n=1}^\infty A_n$, $\delta_2(a) = \mathrm{i}[k_n, a]$ $(a \in A_n; n = 1, 2, \ldots)$ and $\{k_n\}$ is a mutually commuting family of self-adjoint elements in A. Suppose that $h_n k_m = k_m h_n$ $(m, n = 1, 2, \ldots)$. If $(1_{\mathcal{H}} - \mathrm{i}\pi_\phi(k_n))^{-1} \to (1_{\mathcal{H}} - \mathrm{i}H)^{-1}$ strongly in $\mathcal{H}_\phi$ and $1_\phi \in \mathcal{D}(\exp(-\beta H/2))$ (in particular, $\limsup \phi(\exp(-\beta k_n)) < +\infty$), where H is a self-adjoint operator in $\mathcal{H}_\phi$, then*

$$\psi(a) = \left(\pi_\phi(a)\exp\left(\frac{-\beta H}{2}\right)1_\phi, \exp\left(\frac{-\beta H}{2}\right)1_\phi\right)\Big/ \left(\exp\left(\frac{-\beta H}{2}\right)1_\phi, \exp\left(-\frac{\beta H}{2}\right)1_\phi\right) \qquad (a \in A)$$

is a KMS state for $\{A, \exp(t(\widetilde{\delta_1 + \delta_2}))(t \in \mathbb{R})\}$ at β, *where $(\widetilde{\delta_1 + \delta_2})(a) = \mathrm{i}[h_n + k_n, a]$ for $a \in \mathcal{D}_n$ ($\mathcal{D}_n$ is the C*-subalgebra of A generated by A_n* and *$h_1, h_2, \ldots, h_n, k_1, k_2, \ldots, k_n$).*

Proof Since $(1 - \mathrm{i}H)^{-1} \in \mathcal{M}$, $H\eta\mathcal{M}$ and so

$$\exp\left(\frac{-\beta(H_\phi + H)}{2}\right)1_\phi = \exp\left(\frac{-\beta H}{2}\right)1_\phi.$$

Let $H = \int_{-\infty}^{\infty} \lambda\, \mathrm{d}E(\lambda)$ and $H_n = \int_{-n}^{n} \lambda\, \mathrm{d}E(\lambda)$ and put

$$g(x) = \left(x\exp\left(\frac{-\beta H}{2}\right)1_\phi, \exp\left(\frac{-\beta H}{2}\right)1_\phi\right) \qquad (x \in \mathcal{M});$$

then $\| f(\mathrm{i}\beta, H_n) - g \| \to 0$ $(n \to \infty)$.

$$f(\mathrm{i}\beta, H_n)(\pi_\phi(a)\exp(-\beta(H_\phi + H_n))\pi_\phi(b)\exp(\beta(H_\phi + H_n)))$$
$$= f(\mathrm{i}\beta, H_n)(\pi_\phi(b)\pi_\phi(a)) \qquad \left(a, b \in \bigcup_{n=1}^{\infty} A_n\right).$$

Hence

$$g(\pi_\phi(a)\exp(-\beta(H_\phi + H))\pi_\phi(b)\exp(\beta(H_\phi + H))) = g(\pi_\phi(b)\pi_\phi(a)). \qquad \square$$

4.6.6 Notes and remarks The study of commutative normal *-derivations is important, because they include classical lattice systems and Ising models.

Moreover they are much more manageable than non-commutative ones, because they can be treated purely algebraically (cf. 4.1.10).

References [167], [168], [169].

4.7 Phase transitions

In this section, we shall prove the uniqueness theorem for general normal *-derivations with bounded surface energy. For such derivations we establish the absence of phase transitions, i.e. the uniqueness of β-KMS states for all β.

4.7.1 Definition *Let $\{A,\alpha\}$ be a C*-dynamics and let β be a real number. Suppose that $\{A,\alpha\}$ has at least one KMS state at the inverse temperature β. Then $\{A,\alpha\}$ is said to have phase transition at β if it has at least two KMS states at β. If $\{A,\alpha\}$ has only one KMS state at β, then it is said to have no phase transition at β.*

In mathematical physics, we are often concerned with a C*-algebra A containing the identity and an increasing sequence of C*-subalgebras $\{A_n\}$ of A such that $1 \in A_n$ and the uniform closure of $\bigcup_{n=1}^{\infty} A_n$ is A. In addition, we are given a *-derivation δ in $\mathscr{A}$ satisfying the following conditions:

(1) $\mathscr{D}(\delta) = \bigcup_{n=1}^{\infty} A_n$;
(2) there is a sequence of self-adjoint elements $\{h_n\}$ in A such that $\delta(a) = \mathrm{i}[h_n, a]$ $(a \in A_n; n = 1, 2, \ldots)$.

Such a *-derivation is called a general normal *-derivation in A.

4.7.2 Definition *A general normal *-derivation δ is said to have bounded surface energy if there is a sequence $\{k_n\}$ of self-adjoint elements in A such that $k_n \in A_n$ $(n = 1, 2, \ldots)$ and $\| h_n - k_n \| = \mathrm{O}(1)$ $(n = 1, 2, \ldots)$.*

If δ has bounded surface energy, then by 4.1.9, δ is a pre-generator and $\exp(t\bar{\delta}) = \text{strong lim} \exp(t\delta_{\mathrm{i}h_n})$.

Now we shall prove the following theorem.

4.7.3 Theorem *Suppose that A_n $(n = 1, 2, \ldots,)$ has a unique tracial state τ_n (consequently, A has a unique tracial state τ). If a general normal *-derivation δ in A has bounded surface energy, then the C*-dynamics $\{A, \exp(t\bar{\delta})(t \in \mathbb{R})\}$ has a unique KMS state at β for each real number β (namely it has no phase transition at β for each real number β).*

Proof Since $\exp(t\bar{\delta}) = \text{strong}\lim \exp(t\delta_{ih_n})$ and A has a tracial state, by 4.3.17, it has a KMS state for each β. Let ϕ_1, ϕ_2 be two factorial KMS states for $\{A, \exp(t\bar{\delta})(t\in\mathbb{R})\}$ at β, and let $\{\pi_{\phi_1}, U_{\phi_1}, \mathscr{H}_{\phi_1}\}$ be the covariant representation of $\{A, \exp(t\bar{\delta})(t\in\mathbb{R})\}$ constructed via ϕ_1. Let $\delta_n = \bar{\delta} + \delta_{i(k_n - h_n)}$; then $\exp(t\delta_n)(a) = \exp(t\delta_{ik_n})(a)$ for $a\in A_n$. Since A_n has a unique tracial state, by 4.4.7

$$\frac{f(\mathrm{i}\beta, \pi_{\phi_1}(k_n - h_n))(\pi_\phi(a))}{f(\mathrm{i}\beta, \pi_{\phi_1}(k_n - h_n))(1_{\mathscr{H}_\phi})} = \frac{\tau(a\exp(-\beta k_n))}{\tau(\exp(-\beta k_n))} \qquad (a\in A_n).$$

Since $\|\pi_{\phi_1}(k_n - h_n)\| = \mathrm{O}(1)$, by 4.4.7, $\{f(\mathrm{i}\beta, \pi_\phi(k_n - h_n))\}$ is relatively $\sigma(\mathscr{M}_*, \mathscr{M})$-compact in $\mathscr{M}_*(\mathscr{M} = \pi_{\phi_1}(A)'')$, so that by Eberlein's theorem there is a subsequence $\{f(\mathrm{i}\beta, \pi_\phi(k_{n_j} - h_{n_j}))\}$ of $\{f(\mathrm{i}\beta, \pi_\phi(k_n - h_n))\}$ which converges to a normal faithful state ψ in $\sigma(\mathscr{M}_*, \mathscr{M})$.

Hence

$$\frac{\psi(\pi_{\phi_1}(a))}{\psi(1_{\mathscr{H}_{\phi_1}})} = \lim_{nj} \frac{\tau(a\exp(-\beta k_{nj}))}{\tau(\exp(-\beta k_{nj}))} \qquad (a\in A).$$

Quite similarly, we start with ϕ_2; then there is a normal state ξ on $\pi_{\phi_2}(A)''$ such that

$$\frac{\xi(\pi_{\phi_2}(a))}{\xi(1_{\mathscr{H}_{\phi_2}})} = \lim_{jm} \frac{\tau(a\exp(-\beta k_{hj_m}))}{\tau(\exp(-\beta k_{nj_m}))} \qquad (a\in A).$$

Hence

$$\frac{\psi(\pi_{\phi_1}(a))}{\psi(1_{\mathscr{H}_{\phi_1}})} = \frac{\xi(\pi_{\phi_2}(a))}{\xi(1_{\mathscr{H}_{\phi_1}})} \qquad (a\in A)$$

and so π_{ϕ_1} is quasi-equivalent to π_{ϕ_2} so that $\phi_1 = \phi_2$ (4.3.12). □

4.7.4 Proposition *Let $\{A, \exp(t\delta)(t\in\mathbb{R})\}$ be a C*-dynamics. Suppose that $\{A, \exp(t\delta)(t\in\mathbb{R})\}$ has a KMS state ϕ at β, and there is a sequence $\{A_n\}$ of C*-subalgebras in A and a sequence $\{\delta_n\}$ of *-derivations in A such that $\delta_n|\mathscr{D}(\delta_n)\cap\mathscr{D}(\delta) \supset \delta_n$ and $\mathscr{D}(\delta_n)\cap\mathscr{D}(\delta)$ is dense in A, $\|\delta_n - \delta\| = \mathrm{O}(1)$ on $\mathscr{D}(\delta_n)\cap\mathscr{D}(\delta)$, $\delta_n(A_n) \subset A_n$, A_n has a unique tracial state τ_n and $\bigcup_{n=1}^{\infty} A_n$ is dense in A; then $\{A, \exp(t\delta)(t\in\mathbb{R})\}$ has no phase transition at β.*

Proof Without loss of generality, we may assume that ϕ is factorial. Let $\{\pi_\phi, U_\phi, \mathscr{H}_\phi\}$ *be the covariant representation of* $\{A, \exp(t\delta)(t\in R)\}$ constructed via ϕ. $\delta_n - \delta$ can be uniquely extended to a bounded *-derivation on A, so that there is a self-adjoint element l_n in $\pi_\phi(A)''$ such that

$$\pi_\phi((\delta_n - \delta)(a)) = \mathrm{i}[l_n, \pi_\phi(a)] \qquad (a\in\mathscr{D}(\delta)\cap\mathscr{D}(\delta_n)).$$

$f(\mathrm{i}\beta, l_n)/f(\mathrm{i}\beta, l_n)(1_{\mathscr{H}_\phi})$ is a KMS state at β for the dynamics $\{A, \exp(t(\delta + \delta_{il_n}))\}$.

Put $\tilde{\delta}_n = \delta + \delta_{il_n}$; then $\delta_n \subset \tilde{\delta}_n$ and so $\tilde{\delta}_n(A_n) \subset A_n$; hence there is an self-adjoint element k_n in $\pi_\phi(A_n)''$ such that $\pi_\phi(\delta_n(a)) = \mathrm{i}[k_n, \pi_\phi(a)]$ $(a \in A_n)$. Since A_n has a unique tracial state τ_n,

$$\frac{f(\mathrm{i}\beta, l_n)(\pi_\phi(a))}{f(\mathrm{i}\beta, l_n)(1_{\mathscr{H}_\phi})} = \frac{\tau_n(a\exp(-\beta k_n))}{\tau_n(\exp(-\beta k_n))} \qquad (a \in A_n).$$

By 4.4.7, there is a normal faithful positive linear functional ψ on $\pi_\phi(A)'' = \mathscr{M}$ and a subsequence (n_j) of (n) such that $f(\mathrm{i}\beta, l_{n_j}) \to \psi$ in $\sigma(\mathscr{M}_*, \mathscr{M})$. Hence

$$\frac{\psi(\pi_\phi(a))}{\psi(1_{\mathscr{H}_\phi})} = \lim_{n_j} \frac{\tau_{n_j}(a\exp(-\beta k_{n_j}))}{\tau_{n_j}(\exp(-\beta k_{n_j}))} \qquad (a \in A).$$

If ϕ_1 is another factorial KMS state at β for $\{A, \exp(t\delta)(t \in R)\}$, then, by a similar discussion, there is a state ψ_1 on A which is quasi-equivalent to ϕ_1 and

$$\psi_1(a) = \lim_{m_j} \frac{\tau_{m_j}(a\exp(-\beta k_{m_j}))}{\tau_{m_j}(\exp(-\beta k_{m_j}))} \qquad (a \in A),$$

where (m_j) is a subsequence of (n_j). Hence ϕ and ϕ_1 are quasi-equivalent and so $\phi = \phi_1$. □

4.7.5 Proposition *Let* $\{A, \exp(t\delta_1)(t \in \mathbb{R})\}$ (resp. $\{A, \exp(t\delta_2)(t \in \mathbb{R})\}$) *be a C*-dynamics and suppose that there is a sequence* $\{\delta_n(n = 3, 4, \ldots)\}$ *of bounded *-derivations on A such that* $\{1 - (\delta_2 + \delta_n)\}^{-1} \to (1 - \delta_1)^{-1}$ *strongly and* $\|\delta_n\| = \mathrm{O}(1)\,(n = 3, 4, \ldots)$. *Then if* $\{A, \exp(t\delta_2)(t \in \mathbb{R})\}$ *has a KMS state at* β *and has no phase transition, then* $\{A, \exp(t\delta_1)(t \in \mathbb{R})\}$ *has also no phase transition at* β.

This proposition is a corollary of Proposition 4.4.11. Let ϕ be a KMS state for a C*-dynamics $\{A, \exp(t\delta)(t \in \mathbb{R})\}$ at β and let $\{\pi_\phi, U_\phi, \mathscr{H}_\phi\}$ be the covariant representation of $\{A, \exp(t\delta)(t \in \mathbb{R})\}$ constructed via ϕ. Let $\mathscr{M}$ be the weak closure of $\pi_\phi(A)$ in $\mathscr{H}_\phi$.

4.7.6 Proposition *Let* $\{h_n\}$ *be a sequence of self-adjoint elements in A such that there is a dense *-subalgebra* $\mathscr{D}_0$ *of* $\mathscr{D}(\delta)$ *with* $\overline{(1 - \delta)\mathscr{D}_0} = A$ *and* $\|[h_n, a]\| \to 0$ *for each* $a \in \mathscr{D}_0$. *Define* $\psi_n(a) = f(\mathrm{i}\beta, \pi_\phi(h_n))(\pi_\phi(a))/f(\mathrm{i}\beta, \pi_\phi(h_n))(1_{\mathscr{H}_\phi})$; *then any accumulation point* ψ *of* $\{\psi_n\}$ *in the state space of A is again a KMS state for* $\{A, \exp(t\delta)(t \in \mathbb{R})\}$ *at* β.

Proof Let $\delta_n = \delta + \delta_{\mathrm{i}h_n}$; then for $a \in \mathscr{D}_0$, $\delta_n(a) = \delta(a) + \mathrm{i}[h_n, a] \to \delta(a)$. Hence $(1 - \delta_n)^{-1} \to (1 - \delta)^{-1}$ (strongly). By 4.4.7, ψ_n is a KMS state for $\{A, \exp(t\delta_n)(t \in \mathbb{R})\}$ at β and by the proof of 4.3.17, ψ is a KMS state for $\{A, \exp(t\delta)(t \in \mathbb{R})\}$ at β. □

Let $k_n = h_n + \beta^{-1} \log f(\mathrm{i}\beta, \pi_\phi(h_n))(1_{\mathscr{H}_\phi})1$; then

$$f(\mathrm{i}\beta, \pi_\phi(k_n))(1_{\mathscr{H}_\phi}) = \left(\exp\left(-\frac{\beta(H_\phi + \pi_\phi(k_n))}{2} \right) 1_\phi, \exp\left(-\frac{\beta(H_\phi + \pi_\phi(k_n))}{2} \right) 1_\phi \right) = 1.$$

Moreover $\| [k_n, a] \| \to 0$ for $a \in \mathscr{D}_0$. Therefore without loss of generality, we may assume that $f(\mathrm{i}\beta, \pi_\phi(h_n))(1_{\mathscr{H}_\phi}) = 1$. Put $\pi_\phi(h_n) = l_n$; then $\exp(\mathrm{i}t(H_\phi + l_n)) \exp(-\mathrm{i}tH_\phi) \in \pi_\phi(A)$. For simplicity, we shall assume that ϕ is factorial. First we shall note the following fact.

$$\begin{aligned} &\| [\exp(\mathrm{i}t(H_\phi + l_n)) \exp(-\mathrm{i}tH_\phi), \pi_\phi(a)] \| \\ &\quad = \| \exp(-\mathrm{i}tH_\phi) \pi_\phi(a) \exp(\mathrm{i}tH_\phi) - \exp(-\mathrm{i}t(H_\phi + l_n)) \pi_\phi(a) \exp(\mathrm{i}t(H_\phi + l_n)) \| \\ &\quad = \| \pi_\phi(\exp(-t\delta(a))) - \pi_\phi(\exp(-t(\delta + \delta_{\mathrm{i}h_n}))(a)) \| \to 0 \end{aligned}$$

for $a \in A$.

Now we shall assume that A is separable. Let $\pi_\phi(A)'' = \mathscr{M}$ and let $\mathscr{M}_*$ be the pre-dual of $\mathscr{M}$; then $\mathscr{M}_*$ is a separable Banach space and so $L^1(\mathbb{R}, \mathscr{M}_*)^* = L^\infty(\mathbb{R}, \mathscr{M})$ (cf. [165]).

Let $\xi_n(t) = \exp(\mathrm{i}t(H_\phi + l_n)) \exp(-\mathrm{i}tH_\phi)$ $(t \in \mathbb{R})$; then $\xi_n \in L^\infty(\mathbb{R}, \mathscr{M})$. Since $\| \xi_n \| = 1$, there is a subsequence (n_j) of (n) and an element ξ in $L^\infty(\mathbb{R}, \mathscr{M})$ such that $\xi_{n_j} \to \xi$ with respect to $\sigma(L^\infty(\mathbb{R}, \mathscr{M}), L^1(\mathbb{R}, \mathscr{M}_*))$. For $\eta \in L^1(\mathbb{R}, \mathscr{M}_*)$, $\langle \xi_n, \eta \rangle = \int_{-\infty}^{\infty} \langle \xi_n(t), \eta(t) \rangle \,\mathrm{d}t$. Let $\xi_a(t) = \pi_\phi(a)$ $(a \in A, t \in \mathbb{R})$; then $\xi_a \in L^\infty(\mathbb{R}, \mathscr{M})$ and

$$\langle [\xi_n, \xi_a], \eta \rangle = \int_{-\infty}^{\infty} \langle [\xi_n(t), \pi_\phi(a)], \eta(t) \rangle \,\mathrm{d}t$$

$$\begin{aligned} |\langle [\xi_n, \xi_a], \eta \rangle| &\leqslant \int_{-\infty}^{\infty} \| [\xi_n(t), \pi_\phi(a)] \| \, \| \eta(t) \| \,\mathrm{d}t \\ &= \int_{-\infty}^{\infty} \| [\exp(\mathrm{i}t(H_\phi + l_n)) \exp(-\mathrm{i}tH_\phi), \pi_\phi(a)] \| \, \| \eta(t) \| \,\mathrm{d}t. \end{aligned}$$

Since

$$\| [\exp(\mathrm{i}t(H_\phi + l_n)) \exp(-\mathrm{i}tH_\phi), \pi_\phi(a)] \| \leqslant 2 \| \pi_\phi(a) \|$$

and

$$\| [\exp(\mathrm{i}t(H_\phi + l_n)) \exp(-\mathrm{i}tH_\phi), \pi_\phi(a)] \| \to 0,$$

$\langle [\xi, \xi_a], \eta \rangle = 0$ for all $\eta \in L^1(\mathbb{R}, \mathscr{M}_*)$. Hence $[\xi, \xi_a](t) = [\xi(t), \pi_\phi(a)] = 0$ a.e.

Since A is separable, $\xi(t) \in \mathscr{M}' \cap \mathscr{M}$ a.e. and so there is a bounded measurable function λ on $\mathbb{R}$ such that $\xi(t) = \lambda(t) 1_{\mathscr{H}_\phi}$ $(t \in \mathbb{R})$. Now let Λ_x $(x \in \mathscr{M})$ be the set of elements (p, q) in $L^\infty(\mathbb{R}) \oplus L^\infty(\mathbb{R})$ such that $\| p \|, \| q \| \leqslant \| x \|$. Let $\Lambda = \prod_{x \in \mathscr{M}} \Lambda_x$ be the weak infinite product space of $\{\Lambda_x | x \in \mathscr{M}\}$; then Λ is a compact space.

Let $p_{n_j}, x(t) = f(t, h_{n_j})(x)$ and $q_{n_j}, x(t) = f(t + \mathrm{i}\beta, h_{n_j})(x)$ $(t \in \mathbb{R})$; then

$$\begin{aligned} |f(t, h_{n_j})(x)| &= |\tilde{\phi}(x \exp(\mathrm{i}t(H_\phi + l_{n_j})) \exp(-\mathrm{i}t(H_\phi))| \\ &\leqslant \tilde{\phi}(xx^*)^{1/2} \tilde{\phi}(1_{\mathscr{H}_\phi})^{1/2} \leqslant \| x \| \end{aligned}$$

and

$$|f(t+\mathrm{i}\beta,h_{n_j})(x)| = |f(\mathrm{i}\beta,h_{n_j}))(\exp(\mathrm{i}t(H_\phi+l_{n_j}))(\exp(-\mathrm{i}tH_\phi)x)|$$
$$\leqslant f(\mathrm{i}\beta,h_{n_j})(1_{\mathscr{H}_\phi})^{1/2} f(\mathrm{i}\beta,h_{n_j})(x^*x)^{1/2} \leqslant \|x\|.$$

Hence $\{(p_{n_j,x},q_{n_j,x})|x\in\mathscr{M}\}\in\Lambda$:

Let $\mathscr{F}=[\{(p_{n_j,x},q_{n_j,x})|x\in\mathscr{M}\}|j=1,2,\ldots]$ and let $\{(p_x,q_x)|x\in\mathscr{M}\}$ be an accumulation point of $\mathscr{F}$ in Λ. Then for each $x\in\mathscr{M}$, there is a subsequence (n_k) of (n_j) such that $p_{n_k,x}\to p_x$ and $q_{n_k,x}\to q_x$ in $\sigma(L^\infty(\mathbb{R}),L^1(\mathbb{R}))$; hence

$$f(z,h_{n_k})(x) = \int_{-\infty}^{\infty} K_1(t,z) f(t,h_{n_k})(x)\,\mathrm{d}t + \int_{-\infty}^{\infty} K_2(t,z) f(t+\mathrm{i}\beta,h_{n_k})(x)\,\mathrm{d}t$$
$$\to \int_{-\infty}^{\infty} K_1(t,z)p_x(t)\,\mathrm{d}t + \int_{-\infty}^{\infty} K_2(t,z)q_x(t)\,\mathrm{d}t \quad (\text{say } F(z)(x))$$
$$\text{for } z\in S_\beta{}^0.$$

Therefore, $F(z)(x)$ is a bounded analytic function on $S_\beta{}^0$. Since $f(z,h_{n_k})$ is a linear functional on $\mathscr{M}$, $x\mapsto F(z)(x)$ is a linear functional on $\mathscr{M}$ for each $z\in S_\beta{}^0$. For $g\in L^1(R)$, let $\eta_{g,x}(t)=g(t)L_x\tilde{\phi}$, where $L_x\tilde{\phi}(y)=\tilde{\phi}(xy)$; then $\eta_{g,x}\in L^1(\mathbb{R},\mathscr{M}_*)$ and so

$$\langle \xi_{n_j},\eta_{g,x}\rangle = \int_{-\infty}^{\infty} \langle \xi_{n_j}(t), L_x\tilde{\phi}\rangle g(t)\,\mathrm{d}t$$
$$= \int_{-\infty}^{\infty} (x\exp(\mathrm{i}t(H_\phi+l_{n_j})\exp(-\mathrm{i}tH_\phi)1_\phi,1_\phi)g(t)\,\mathrm{d}t$$
$$\to \int_{-\infty}^{\infty} \langle \lambda(t)1_{\mathscr{H}_\phi}, L_x\tilde{\phi}\rangle g(t)\,\mathrm{d}t = \int_{-\infty}^{\infty} \lambda(t)(x1_\phi,1_\phi)g(t)\,\mathrm{d}t.$$

On the other hand,

$$\int_{-\infty}^{\infty} (x\exp(\mathrm{i}t(H_\phi+l_{n_j}))\exp(-\mathrm{i}tH_\phi)1_\phi,1_\phi)g(t)\,\mathrm{d}t = \int_{-\infty}^{\infty} f(t,h_{n_j})(x)g(t)\,\mathrm{d}t.$$

Hence

$$\int_{-\infty}^{\infty} p_x(t)g(t)\,\mathrm{d}t = \int_{-\infty}^{\infty} \lambda(t)(x1_\phi,1_\phi)g(t)\,\mathrm{d}t$$

and so $p_x(t)=\lambda(t)(x1_\phi,1_\phi)$ a.e.

Now we shall define $F(t)(x)=p_x(t)$ and $F(t+\mathrm{i}\beta)(x)=q_x(t)$ $(t\in\mathbb{R})$; then $F(t)(x)=\lambda(t)\tilde{\phi}(x)$ a.e., and

$$F(z)(x) = \int_{-\infty}^{\infty} K_1(t,z)F(t)(x)\,\mathrm{d}t + \int_{-\infty}^{\infty} K_2(t,z)F(t+\mathrm{i}\beta)(x)\,\mathrm{d}t$$
$$= \tilde{\phi}(x)\int_{-\infty}^{\infty} K_1(t,z)\lambda(t)\,\mathrm{d}t + \int_{-\infty}^{\infty} K_2(t,z)F(t+\mathrm{i}\beta)(x)\,\mathrm{d}t \qquad \text{for } z\in S_\beta{}^0.$$

If $\tilde{\phi}(x)=0$, then $F(t)(x)=0$ and so it is continuous. Hence $F(z)(x)=0$, so that there is a function $\mu(z)$ on $S_\beta{}^0$ such that $F(z)(x)=\mu(z)\tilde{\phi}(x)$ for $x\in\mathcal{M}$.

Since

$$(\exp(it(H_\phi+l_{n_j}))\exp(-itH_\phi)1_\phi,1_\phi)=(\exp(it(H_\phi+l_{n_j}))1_\phi,1_\phi)=f(t,h_{n_j})(1_{\mathscr{H}_\phi})$$

is a positive definite function on $\mathbb{R}$ and since $f(t,h_{n_j})(1_{\mathscr{H}_\phi})\to\lambda(t)$ with respect to $\sigma(L^\infty(\mathbb{R}),L^1(\mathbb{R}))$, λ is a measurable positive definite function on $\mathbb{R}$. By the well-known theorem, any measurable positive definite function on $\mathbb{R}$ is equivalent to a continuous positive definite function; hence one may assume that $\lambda(t)$ is continuous on $\mathbb{R}$. Therefore if we define $\mu(t)=\lambda(t)$, μ is continuous on $\{z|0\leqslant \mathrm{Im}(z)<\beta\}$. For $h\in A_2(\delta)$,

$$\begin{aligned}
f(t+i\beta,\pi_\phi(h))(1_{\mathscr{H}_\phi})&=(\exp(iz(H_\phi+\pi_\phi(h)))1_\phi,1_\phi)\\
&=(\exp(it(H_\phi+\pi_\phi(h)))\exp(-\beta(H_\phi+\pi_\phi(h)))1_\phi,1_\phi)\\
&=\left(\exp(it(H_\phi+\pi_\phi(h)))\exp\left(\frac{-\beta(H_\phi+\pi_\phi(h))}{2}\right)1_\phi,\exp\left(\frac{-(H_\phi+\pi_\phi(h))}{2}\right)1_\phi\right)
\end{aligned}$$

and so generally

$$\begin{aligned}
&f(t+i\beta,h_{n_j})(1_{\mathscr{H}_\phi})\\
&=\left(\exp(it(H_\phi+h_{n_j}))\exp\left(-\frac{\beta(H_\phi+h_{n_j})}{2}\right)1_\phi,\exp\left(-\frac{\beta(H_\phi+h_{n_j})}{2}\right)1_\phi\right).
\end{aligned}$$

Therefore $f(t+i\beta,h_{n_j})\,(1_{\mathscr{H}_\phi})$ is a positive definite function with respect to t. Since $f(t+i\beta,h_{n_j})(1_{\mathscr{H}_\phi})\to F(t+i\beta)(1_{\mathscr{H}_\phi})$ in $\sigma(L^\infty(\mathbb{R}),L^1(\mathbb{R}))$, $F(t+i\beta)(1_{\mathscr{H}_\phi})$ is a measurable positive definite function and so it is equivalent to a continuous positive definite function. Therefore we may assume that $t\mapsto F(t+i\beta)(1_{\mathscr{H}_\phi})$ is continuous. Put $\mu(t+i\beta)=F(t+i\beta)(1_{\mathscr{H}_\phi})$; then μ is a bounded continuous function on S_β and is analytic in the interior $S_\beta{}^0$.

Moreover we have the following

$$F(z)(x)=\left(\int_{-\infty}^{\infty}K_1(t,z)\mu(t)\,\mathrm{d}t\right)\tilde{\phi}(x)+\left(\int_{-\infty}^{\infty}K_2(t,z)\mu(t+i\beta)\,\mathrm{d}t\right)\tilde{\phi}(x)$$

for $z\in S_\beta{}^0$.

4.7.7 Theorem *Let $\{A,\exp(t\delta)(t\in\mathbb{R})\}$ be a C^*-dynamics with a separable C^*-algebra A and let ϕ be a factorial KMS state for $\{A,\exp(t\delta)(t\in\mathbb{R})\}$ at β. Let $\{h_n\}$ be a sequence of self-adjoint elements in A such that there is a dense *-subalgebra $\mathscr{D}_0$ of $\mathscr{D}(\delta)$ with $\overline{(1-\delta)\mathscr{D}_0}=A$ and $\|[h_n,a]\|\to 0$ $(a\in\mathscr{D}_0)$, and put $\psi_n(a)=f(i\beta,\pi_\phi(h_n))(\pi_\phi(a))/f(i\beta,\pi_\phi(h_n))(1_{\mathscr{H}_\phi})$ $(a\in A)$. Then any accumulation point ψ of $\{\psi_n\}$ in the state space of A is again a KMS state for $\{A,\exp(t\delta)(t\in\mathbb{R})\}$ at β. Moreover if ψ is disjoint with ϕ (in particular if ψ is a different factorial*

KMS state for $\{A, \exp(t\delta)(t\in\mathbb{R})\}$ *at* β), *then the sequence*

$$\{\exp(\mathrm{i}t\{H_\phi + \pi_\phi(h_n) + \beta^{-1}\log f(\mathrm{i}\beta, \pi_\phi(h_n))(1_{\mathscr{H}_\phi})\})\exp(-\mathrm{i}tH_\phi)\}$$

of unitary elements in $\pi_\phi(A)$ *converges to zero with respect to* $\sigma(L^\infty(\mathbb{R}, \mathscr{M}), L^1(\mathbb{R}, \mathscr{M}))$.

Proof It is enough to prove that

$$\{\exp(\mathrm{i}t\{(H_\phi + \pi_\phi(h_n)) + \beta^{-1}\log f(\mathrm{i}\beta, \pi_\phi(h_n))(1_{\mathscr{H}_\phi})\})\exp(-\mathrm{i}tH_\phi)\}$$

converges to zero with respect to $\sigma(L^\infty(\mathbb{R}, \mathscr{M}), L^1(\mathbb{R}, \mathscr{M}))$.

If this is not true, then one can choose a λ such that $\lambda \not\equiv 0$. Then $\mu \not\equiv 0$, and so $|\mu(t+\mathrm{i}\beta)| \leqslant \mu(\mathrm{i}\beta) \neq 0$. On the other hand, for $a(\geqslant 0)\in A$,

$$\begin{aligned}
&|f(t+\mathrm{i}\beta, l_n)(\pi_\phi(a))| \\
&\quad = |f(\mathrm{i}\beta, l_n)(\exp(\mathrm{i}t(H_\phi + l_n))\exp(-\mathrm{i}tH_\phi)\pi_\phi(a))| \\
&\quad \leqslant f(\mathrm{i}\beta, l_n)(\exp(\mathrm{i}t(H_\phi + l_n))\exp(-\mathrm{i}tH_\phi)\pi_\phi(a)\exp(\mathrm{i}tH_\phi)\exp(-\mathrm{i}t(H_\phi + l_n))^{1/2} \\
&\qquad \times f(\mathrm{i}\beta, l_n))(\pi_\phi(a))^{1/2} \\
&\quad = f(\mathrm{i}\beta, l_n)(\pi_\phi(\exp(-t\delta)(a)))^{1/2} f(\mathrm{i}\beta, l_n)(\pi_\phi(a))^{1/2}.
\end{aligned}$$

Hence,

$$F(t+\mathrm{i}\beta)(\pi_\phi(a)) = |\mu(t+\mathrm{i}\beta)\phi(a)| \leqslant \psi(\exp(-t\delta)(a))^{1/2}\psi(a)^{1/2}.$$

Since ψ is a KMS state for $\{A, \exp(t\delta)(t\in R)\}$ at β, $|\mu(t+\mathrm{i}\beta)\phi(a)| \leqslant \psi(a)^{1/2}\psi(a)^{1/2} = \psi(a)$. Since $\mu(\mathrm{i}\beta) > 0$, ϕ is not disjoint with ψ, a contradiction. □

4.7.8 Proposition *Let* $\{A, \exp(t\delta)(t\in\mathbb{R})\}$ *be a* C*-*dynamics with a separable* C*-*algebra* A *and let* ϕ *be a factorial KMS state for* $\{A, \exp(t\delta)(t\in\mathbb{R})\}$ *at* β. *Let* $\{h_n\}$ *be a sequence of self-adjoint elements in* A *such that there is a dense* *-*subalgebra* $\mathscr{D}_0$ *of* $\mathscr{D}(\delta)$ *with* $\overline{(1-\delta)\mathscr{D}_0} = A$ *and* $\|[h_n, a]\| \to 0$ $(a\in\mathscr{D}_0)$, *and let* $\psi_n(a) = f(\mathrm{i}\beta, \pi_\phi(h_n))(\pi_\phi(a))/f(\mathrm{i}\beta, \pi_\phi(h_n))(1_{\mathscr{H}_\phi})$ $(a\in A)$. *Then if a sequence* $\{f(\mathrm{is}, \pi_\phi(h_n))(1_{\mathscr{H}_\phi})\}$ *of continuous functions on* $0 \leqslant s \leqslant \beta$ *converges pointwise to a function* g *which is not zero at* β *and is continuous at* β, *then* $\{\psi_n\}$ *converges to* ϕ *in the state space of* A.

Proof Let $f(\mathrm{i}\beta, \pi_\phi(h_n))(1_{\mathscr{H}_\phi}) \to g(\mathrm{i}\beta)$. Let $k_n = h_n + \beta^{-1}\log f(\mathrm{i}\beta, \pi_\phi(h_n))(1_{\mathscr{H}_\phi})$; then $f(\mathrm{i}\beta, \pi_\phi(k_n))(1_{\mathscr{H}_\phi}) = 1$. Moreover

$$\begin{aligned}
&\lim f(\mathrm{is}, \pi_\phi(k_n))(1_{\mathscr{H}_\phi}) \\
&\quad = \lim\left(\exp\left(-\frac{s(H_\phi + \pi_\phi(k_n))}{2}\right)1_\phi, \exp\left(-\frac{s(H_\phi + \pi_\phi(k_n))}{2}\right)1_\phi\right) \\
&\quad = \lim f(\mathrm{is}, \pi_\phi(h_n))(1_{\mathscr{H}_\phi})\exp\left(-\frac{s\log f(\mathrm{i}\beta, \pi_\phi(h_n))(1_{\mathscr{H}_\phi})}{\beta}\right).
\end{aligned}$$

Hence without loss of generality, we may assume that $f(\mathrm{i}\beta, \pi_\phi(h_n))(1_{\mathscr{H}_\phi}) = 1$. By the previous considerations, $F(z)(x) = \mu(z)\tilde{\phi}(x)(x \in \mathscr{M}, z \in S_\beta)$, $\mu(\mathrm{i}\beta)\phi(a) \leqslant \psi(a)(a \geqslant 0, a \in A)$, where ψ is any accumulation point of $\{\psi_n\}$ in the state space of A. On the other hand,

$$\mu(\mathrm{i}\beta) = \lim_{s\uparrow\beta} \mu(\mathrm{i}s) = \lim_{s\uparrow\beta} \lim_j f(\mathrm{i}s, h_{n_j})(1_{\mathscr{H}_\phi})$$

$$= \lim_{s\uparrow\beta} g(\mathrm{i}s) = g(\mathrm{i}\beta) = \lim_j f(\mathrm{i}\beta, h_{n_j})(1_{\mathscr{H}_\phi}) = 1.$$

Hence $\phi(a) \leqslant \psi(a)$, and $\phi = \psi$. □

4.7.9 Proposition *Let ϕ be a KMS state for a C^*-dynamics $\{A, \exp(t\delta)(t \in \mathbb{R})\}$ at β, and let $\{l_n\}$ be a sequence of self-adjoint elements of $\pi_\phi(A)'' = \mathscr{M}$ such that $\sup_{\substack{0 \leqslant s \leqslant \beta \\ 1 \leqslant n < +\infty}} |f(\mathrm{i}s, l_n)(l_n)| < +\infty$. Then there is a subsequence $\{f(\mathrm{i}s, l_{n_j})(1_{\mathscr{H}_\phi})\}$ of a sequence $\{f(\mathrm{i}s, l_n)(1_{\mathscr{H}_\phi})\}$ of continuous functions on $0 \leqslant s \leqslant \beta$ which converges to a continuous function on $0 \leqslant s \leqslant \beta$ uniformly.*

Proof For $h \in A_2(\delta)$,

$$f(\mathrm{i}s, \pi_\phi(h))(1_{\mathscr{H}_\phi}) = (\exp(-s(H_\phi + \pi_\phi(h)))1_\phi, 1_\phi),$$

and so

$$\frac{\mathrm{d}}{\mathrm{d}s} f(\mathrm{i}s, \pi_\phi(h))(1_{\mathscr{H}_\phi}) = (-(H_\phi + \pi_\phi(h))\exp(-s(H_\phi + \pi_\phi(h)))1_\phi, 1_\phi)$$

$$= -(\pi_\phi(h)\exp(-s(H_\phi + \pi_\phi(h)))1_\phi, 1_\phi);$$

hence

$$\frac{\mathrm{d}}{\mathrm{d}s} f(\mathrm{i}s, \pi_\phi(h))(1_{\mathscr{H}_\phi}) = -f(\mathrm{i}s, \pi_\phi(h))(\pi_\phi(h)).$$

For $l \in \mathscr{M}$, take $\{\pi_\phi(h_\alpha)\}$ such that $h_\alpha \in A_2(\delta)$, $\|\pi_\phi(h_\alpha)\| \leqslant \|l\|$ and $\pi_\phi(h_\alpha) \to l$ (strongly); then $f(\mathrm{i}s, \pi_\phi(h_\alpha))(\pi_\phi(h_\alpha)) \to f(\mathrm{i}s, l)(l)$ uniformly on $0 \leqslant s \leqslant \beta$. Hence

$$\int_0^s -f(\mathrm{i}p, \pi_\phi(h_\alpha))(\pi_\phi(h_\alpha))\mathrm{d}p = f(\mathrm{i}s, \pi_\phi(h_\alpha))(1_{\mathscr{H}_\phi}) - f(0, \pi_\phi(h_\alpha))(1_{\mathscr{H}_\phi})$$

$$\to f(\mathrm{i}s, l)(1_{\mathscr{H}_\phi}) - f(0, l)(1_{\mathscr{H}_\phi}).$$

Hence

$$f(\mathrm{i}s, l)(1_{\mathscr{H}_\phi}) - f(0, l)(1_{\mathscr{H}_\phi}) = \int_0^s -f(\mathrm{i}p, l)(l)\mathrm{d}p$$

and so

$$\frac{\mathrm{d}f(\mathrm{i}s, l)}{\mathrm{d}s}(1_{\mathscr{H}_\phi}) = -f(\mathrm{i}s, l)(l).$$

Therefore if $\sup_{\substack{0\leqslant s\leqslant\beta \\ n=1,2,\ldots}} |f(\mathrm{i}s, l_n)(l_n)| < +\infty$, then take a subsequence $\{f(\mathrm{i}s, l_{n_j})(l_{n_j})\}$ such that $f(\mathrm{i}s, l_{n_j})(l_{n_j}) \to g(\mathrm{i}s)$ in $\sigma(L^\infty(0,\beta), L^1(0,\beta))$; then

$$\int_0^s f(\mathrm{i}p, l_{n_j})(l_{n_j})\mathrm{d}p \to \int_0^s g(\mathrm{i}p)\mathrm{d}p$$

and so

$$f(\mathrm{i}s, l_{n_j})(1_{\mathscr{H}_\phi}) - f(0, l_{n_j})(1_{\mathscr{H}_\phi}) \to \int_0^s g(\mathrm{i}p)\mathrm{d}p.$$

Since $f(0, l_{n_j})(1_{\mathscr{H}_\phi}) = \tilde{\phi}(1_{\mathscr{H}_\phi}) = 1$,

$$f(\mathrm{i}s, l_{n_j})(1_{\mathscr{H}_\phi}) \to f(0, l_{n_j})(1_{\mathscr{H}_\phi}) + \int_0^s g(\mathrm{i}p)\mathrm{d}p = 1 + \int_0^s g(\mathrm{i}p)\mathrm{d}p. \qquad \square$$

4.7.10 Corollary *Let $\{A, \exp(t\delta)(t\in\mathbb{R})\}$ be a C*-dynamics with a separable C*-algebra A and let ϕ be a factorial KMS state for $\{A, \exp(t\delta)(t\in R)\}$ at β. Let $\{h_n\}$ be a sequence of self-adjoint elements in A such that there is a dense *-subalgebra $\mathscr{D}_0$ of $\mathscr{D}(\delta)$ with $\overline{(1-\delta)\mathscr{D}_0} = A$ and $\|[h_n, a]\| \to 0\ (a\in\mathscr{D}_0)$, and let*

$$\psi_n(a) = f(\mathrm{i}\beta, \pi_\phi(h_n))(\pi_\phi(a))/f(\mathrm{i}\beta, \pi_\phi(h_n))(1_{\mathscr{H}_\phi}) \qquad (a\in A).$$

If $\sup_{\substack{0\leqslant s\leqslant\beta \\ 1\leqslant n<+\infty}} |f(\mathrm{i}s, \pi_\phi(h_n))(\pi_\phi(h_n))| < +\infty$, then $\{\psi_n\}$ converges to ϕ in the state space of A.

In particular if $\{\|h_n\|\}$ is bounded, then $\{\psi_n\}$ converges to ϕ.

Remark If $\{\|h_n\|\}$ is bounded, then by 4.4.11, we can eliminate the assumption of separability from the above theorem.

4.7.11 Notes and Remarks Theorem 4.7.3 was first proved by the author [168] for commutative normal *-derivations in 1975. Later in the same year, Araki [5] proved it for normal *-derivations in UHF algebra. Kishimoto [104] also gave a simplified proof of Araki's theorem.

References [5], [104], [168], [169], [170].

4.8 Continuous quantum systems

The study of continuous quantum systems is so far incomplete. In fact, for general interacting models, the time evolution has not yet been constructed. This means that the corresponding *-derivation has not been constructed globally for interacting models. In this section, we shall generalize the notion of a C*-dynamical system to include a fairly wide class of interacting models in continuous quantum systems.

We shall formulate a system of four axioms (1)–(4) which is satisfied by a fairly wide class of interacting models in continuous quantum systems, and show that a globally defined *-derivation exists under these four axioms. Next we shall add a further three axioms (5)–(7) to study the existence of time evolution and KMS states, and to show that the combined seven axioms (1)–(7) assure the existence of time evolution and KMS states. Unfortunately it is not so easy to check axioms (6) and (7) for interacting models in continuous quantum systems. These two axioms are more or less automatic if the model has time evolution and a KMS state. It is a problem for the future to check them for general interacting models. We shall give some applications of these two axioms in some concrete models. For continuous quantum systems, another hopeful method might be to consider dynamical systems on unbounded operator algebras.

In the last part of this section, we shall present the systems of five axioms (1′)–(5′) to formulate dynamical systems on unbounded operator algebras, which might be applicable to a wide class of interacting models. We shall show that these five axioms (1′)–(5′) ensure again the existence of time evolution and KMS states.

Here again Axiom (5′) is difficult to verify for general interacting models. Both Axioms (7) and (5′) comprise the so-called core problem in mathematics which appears in various branches of mathematical physics and is a traditionally difficult problem.

In the previous sections, we have formulated the theory of quantum lattice systems within the framework of unbounded *-derivations in C*-algebras. In particular, time evolution has been constructed as a strongly continuous one-parameter group of *-automorphisms on a C*-algebra by integrating a given unbounded *-derivation in the algebra. Consequently, C*-dynamics is well fitted to be an abstraction of quantum lattice systems. On the other hand, for continuous quantum systems it is impossible to construct time evolution as a strongly continuous one-parameter group of *-automorphisms on an appropriate C*-algebra. At best, one may hope to construct time evolution as a σ-weakly continuous one-parameter group of *- automorphisms on the weak closure of a C*-algebra on an appropriate Hilbert space.

This is the case of the ideal Bose gas. However for general interacting models, time evolution has not been constructed even in this weak sense. As a matter of fact, even the *-derivation has not been constructed globally for interacting models.

In this section, we shall generalize the notion of C*-dynamical systems so that our system can include a wide class of interacting models in continuous quantum systems and discuss the existence of time evolution and KMS states.

In mathematical physics, we are often concerned with a C*-algebra $\mathscr{A}$ containing an identity and an increasing sequence $\{\mathscr{A}_n\}$ of C*-subalgebras of $\mathscr{A}$ such that $1\in\mathscr{A}_1\subset\mathscr{A}_2\subset\cdots\subset\mathscr{A}_n\subset\cdots$ and the uniform closure of $\bigcup_{n=1}^{\infty}\mathscr{A}_n$ is $\mathscr{A}$.

To deal with continuous quantum systems, we shall assume that $\mathscr{A}_n = B(\mathscr{H}_n)$, where $B(\mathscr{H}_n)$ is the W*-algebra of all bounded operators on a separable Hilbert space $\mathscr{H}_n$. By the separability of $\mathscr{H}_n$, $B(\mathscr{H}_n)$ is weakly closed in $B(\mathscr{H}_m)$ (cf. [165]) and the σ-weak topology of $B(\mathscr{H}_n)$ coincides with the σ-weak topology of $B(\mathscr{H}_m)$ on $B(\mathscr{H}_n)$ for $m \geqslant n$.

A linear mapping δ in $\mathscr{A}$ is said to be a *-derivation if it satisfies the following conditions:

(1) There is an increasing sequence $\{\mathscr{D}_n\}$ of *-subalgebras of $\mathscr{A}$ containing the identity such that $1\in\mathscr{D}_1\subset\mathscr{D}_2\subset\cdots\subset\mathscr{D}_n\subset\cdots$.
(2) There is an increasing subsequence $\{m(n)\}$ of $\{n\}$ such that $\mathscr{D}_n\subset\mathscr{A}_{m(n)}$ and the σ-weak closure of $\mathscr{D}_n$ in $\mathscr{A}_{m(n)}$ contains $\mathscr{A}_n$ for each n.
(3) $\mathscr{D}(\delta)=\bigcup_{n=1}^{\infty}\mathscr{D}_n$, where $\mathscr{D}(\delta)$ is the domain of δ, and $\delta(ab)=\delta(a)b+a\delta(b)$ and $\delta(a^*)=\delta(a)^*$ for $a,b\in\mathscr{D}(\delta)$.
(4) For each $\mathscr{D}_n$ there is a self-adjoint operator h_n in $\mathscr{H}_{m(n)}$ such that $\delta(a)=\overline{\mathrm{i}[h_n,a]}$ for $a\in\mathscr{D}_n$, where $\overline{(\cdot)}$ is the closure of $(\cdot)$ on the space $\mathscr{H}_{m(n)}$.

Now we shall give some examples of the above system. For details, we refer the reader to Bratteli and Robinson [42]. Let $L^2(\mathbb{R}^\nu)$ be the Hilbert space of all square integrable functions on the ν-dimensional euclidean space $\mathbb{R}^\nu$, and let $\mathscr{F}_\pm(L^2(\mathbb{R}^\nu))$ be the Bose–Fock and Fermi–Fock spaces. Let $\Lambda_n=\{x\in\mathbb{R}^\nu \mid \|x\|<n\}$, where $\|\cdot\|$ is the euclidean norm of $\mathbb{R}^\nu$ and let $L^2(\Lambda_n)_\pm{}^m$ denote the subspace of $L^2(\Lambda_n{}^m)$ formed by the totally symmetric (plus sign) and totally anti-symmetric (minus sign) functions of m-variables $x\in\Lambda_n$.

Consider the associated Fock space $\mathscr{F}_\pm(L^2(\Lambda_n))=\sum_{m\geqslant 0}L^2(\Lambda_n)_\pm{}^m$ with $L^2(\Lambda_n)$, and let $\mathscr{A}_\pm(\Lambda_n)$ denote the CCR and the CAR algebras respectively. Take the σ-weak closure of $\mathscr{A}_\pm(\Lambda_n)$ in $B(\mathscr{F}_\pm(L^2(\Lambda_n))$; then it is $B(\mathscr{F}_\pm(L^2(\Lambda_n))$. Now let Λ be a bounded open subset of the configuration space $\mathbb{R}^\nu$ such that $0\in\Lambda$ and $-\Lambda=\Lambda$. Consider a Bose or Fermi gas interaction through a two-body potential Φ; then

$$U_\Phi=\frac{1}{2}\iint \mathrm{d}x\,\mathrm{d}y\,\Phi(x-y)a^*(x)a^*(y)a(y)a(x).$$

Suppose that $\Phi\in L^2(\Lambda)$ and $f\in L^2(\Lambda_1)$, where Λ_1 is a bounded open subset of R^ν. Then

$$([U_\Phi,a(f)]\psi)^{(m)}(x_1,x_2,\ldots,x_m)$$
$$=\int \mathrm{d}x\bar{f}(x)\left\{\sum_{i=1}^{m}\Phi(x-x_i)\right\}\psi^{(m+1)}(x,x_1,x_2,\ldots,x_m)\qquad \text{for } \psi\in\mathscr{F}_\pm(L^2(\mathbb{R}^\nu)).$$

If $x_i \notin \Lambda_1 - \Lambda$ $(i = 1, 2, \ldots, m)$, then $([U_\Phi, a(f)]\psi)^{(m)} = 0$. Hence

$$[U_\Phi, a(f)] = [U_{\Phi_{\Lambda_1 - \Lambda}}, a(f)] = [U_{\Phi_{\Lambda_1 + \Lambda}}, a(f)].$$

We can also easily see that

$$[U_\Phi, a^*(f)] = [U_{\Phi_{\Lambda_1 + \Lambda}}, a^*(f)].$$

Let $W(f)$ be the Weyl operator $f \in L^2(\Lambda_1)$; then

$$\|[U_\Phi, W(f)]\psi\| \leqslant \|\Phi\|_2 \|f\|_2 (a_0 + a_1 \|f\|_2) \|(1 + N_{\Lambda_1 + \Lambda})^{3/2}\psi\| \quad \text{for } \psi \in \mathscr{F}_\pm(L^2(\mathbb{R}^\nu)),$$

where $N_{\Lambda_1 + \Lambda}$ is the number operator of $\mathscr{F}_\pm(L^2(\Lambda_1 + \Lambda))$ and a_0, a_1 are positive constants independent of Φ, f and ψ (cf. Lemma 6.3.34 in [42]). Therefore $[U_\Phi, W(f)] = [U_{\Phi_{\Lambda_1 + \Lambda}}, W(f)]$, and $[U_\Phi, W(f)](1 + N_{\Lambda_1 + \Lambda})^{-3/2}$ is bounded.

4.8.1 Lemma $[U_{\Phi_{\Lambda_1 + \Lambda}}, N_{\Lambda_2}] = 0$ *and* $[U_\Phi, N_{\Lambda_2}] = 0$, *where* Λ_2 *is a bounded open subset of* $\Lambda_1 + \Lambda$.

Proof For $\phi^{(m)} \in L^2(\Lambda_2)_\pm{}^m$ and $\psi^{(l)} \in L^2((\Lambda_1 + \Lambda) \cap \Lambda_2{}^c)_\pm{}^l$,

$$\begin{aligned} &U_{\Phi_{\Lambda_1 + \Lambda}} N_{\Lambda_2} P_\pm(\phi^{(m)} \otimes \psi^{(l)})(x_1, x_2, \ldots, x_m, x_{m+1}, \ldots, x_{m+l}) \\ &= U_{\Phi_{\Lambda_1 + \Lambda}} m P_\pm(\phi^{(m)} \otimes \psi^{(l)})(x_1, x_2, \ldots, x_m, x_{m+1}, \ldots, x_{m+l}) \\ &= m U_{\Phi_{\Lambda_1 + \Lambda}}(x_1, x_2, \ldots, x_m, x_{m+1}, \ldots, x_{m+l})(P_\pm(\phi^{(m)} \otimes \psi^{(l)})) \\ &(x_1, x_2, \ldots, x_m, x_{m+1}, \ldots, x_{m+l}). \end{aligned}$$

On the other hand,

$$\begin{aligned} &\{N_{\Lambda_2} U_{\Phi_{\Lambda_1 + \Lambda}} P_\pm(\phi^{(m)} \otimes \psi^{(l)})\}(x_1, x_2, \ldots, x_m, x_{m+1}, \ldots, x_{m+l}) \\ &= m U_{\Phi_{\Lambda_1 + \Lambda}}(x_1, x_2, \ldots, x_m, x_{m+1}, \ldots, x_{m+l}) \\ &\quad \cdot P_\pm(\phi^{(m)} \otimes \psi^{(l)})(x_1, \ldots, x_m, x_{m+1}, \ldots, x_{m+l}). \end{aligned}$$

Hence $[U_{\Phi_{\Lambda_1 + \Lambda}}, N_{\Lambda_2}] = 0$ and analogously $[U_\Phi, N_{\Lambda_2}] = 0$. □

Since $[U_\Phi, W(f)](1 + N_{\Lambda_1 + \Lambda})^{-3/2}$ is bounded on $\mathscr{F}_\pm(L^2(\Lambda_1 + \Lambda))$ it is also bounded on $\mathscr{F}_\pm(L^2(\mathbb{R}^\nu))$. Since

$$\begin{aligned} 1 + N_{\Lambda_1 + \Lambda} = 1 + N_{\Lambda_1} + N_{(\Lambda_1 + \Lambda) \cap \Lambda_1{}^c}, \quad &(1 + N_{\Lambda_1})(1 + N_{(\Lambda_1 + \Lambda) \cap \Lambda_1{}^c}) \\ &\geqslant (1 + N_{\Lambda_1 + \Lambda}) \end{aligned}$$

and so

$$(1 + N_{\Lambda_1})^{-3/2}(1 + N_{(\Lambda_1 + \Lambda) \cap \Lambda_1{}^c})^{-3/2} \leqslant (1 + N_{\Lambda_1 + \Lambda})^{-3/2}.$$

By 4.8.1,

$$\begin{aligned} &[U_\Phi, W(f)(1 + N_{\Lambda_1})^{-3/2}(1 + N_{(\Lambda_1 + \Lambda) \cap \Lambda_1{}^c})^{-3/2}] \\ &= [U_\Phi, W(f)](1 + N_{\Lambda_1})^{-3/2}(1 + N_{(\Lambda_1 + \Lambda) \cap \Lambda_1{}^c})^{-3/2} \end{aligned}$$

and

$$\begin{aligned}&[U_\Phi, W(f)(1+N_{\Lambda_1})^{-3/2}(1+N_{(\Lambda_1+\Lambda)\cap\Lambda_1{}^c})^{-3/2}]\\&\quad=[U_{\Phi_{\Lambda_1+\Lambda}}, W(f)(1+N_{\Lambda_1})^{-3/2}(1+N_{(\Lambda_1+\Lambda)\cap\Lambda_1{}^c})^{-3/2}]\\&\quad=[U_{\Phi_{\Lambda_1+\Lambda}}, W(f)](1+N_{\Lambda_1})^{-3/2}(1+N_{(\Lambda_1+\Lambda)\cap\Lambda_1{}^c})^{-3/2}\end{aligned}$$

Analogously,

$$\begin{aligned}&[U_\Phi, (1+N_{(\Lambda_1+\Lambda)\cap\Lambda_1{}^c})^{-3/2}(1+N_{\Lambda_1})^{-3/2}W(f)(1+N_{\Lambda_1})^{-3/2}(1+N_{(\Lambda_1+\Lambda)\cap\Lambda_1{}^c})^{-3/2}]\\&\quad=[U_{\Phi_{\Lambda_1+\Lambda}}, (1+N_{(\Lambda_1+\Lambda)\cap\Lambda_1{}^c})^{-3/2}(1+N_{\Lambda_1})^{-3/2}W(f)\\&\qquad\times(1+N_{\Lambda_1})^{-3/2}(1+N_{(\Lambda_1+\Lambda)\cap\Lambda_1{}^c})^{-3/2}]\\&\quad=(1+N_{(\Lambda_1+\Lambda)\cap\Lambda_1{}^c})^{-3/2}(1+N_{\Lambda_1})^{-3/2}[U_{\Phi_{\Lambda_1+\Lambda}}, W(f)]\\&\qquad\times(1+N_{\Lambda_1})^{-3/2}(1+N_{(\Lambda_1+\Lambda)\cap\Lambda_1{}^c})^{-3/2}.\end{aligned}$$

Now let $\mathscr{A}_1 = B(\mathscr{F}_\pm(L^2(\Lambda_1))$ and let $\mathscr{D}_1$ be a *-subalgebra of $B(\mathscr{F}_\pm(L^2(\Lambda_1+\Lambda))$ generated by $\{(1+N_{(\Lambda_1+\Lambda)\cap\Lambda_1{}^c})^{-3/2}(1+N_{\Lambda_1})^{-3/2}W(f)(1+N_{\Lambda_1})^{-3/2}(1+N_{(\Lambda_1+\Lambda)\cap\Lambda_1{}^c})^{-3/2} \mid f \in C_0{}^\infty(\Lambda_1)\cap L^2(\Lambda_1)\}$ and the identity. Let $\mathscr{A}_2 = B\in\mathscr{F}_\pm(L^2(\Lambda_1+\Lambda)$; then the σ-weak closure of $\mathscr{D}_1$ contains $(1+N_{(\Lambda_1+\Lambda)\cap\Lambda_1{}^c})^{-3/2}\mathscr{A}_1(1+N_{(\Lambda_1+\Lambda)\cap\Lambda_1{}^c})^{-3/2}$, because $\mathscr{A}_\pm(\Lambda_1)$ is weakly dense in $B(\mathscr{F}_\pm(L^2(\Lambda_1)))$, and so it contains $(1+N_{(\Lambda_1+\Lambda)\cap\Lambda_1{}^c})^{-3}$; hence it contains $\mathscr{A}_1$. Next let $-\nabla^2$ be the Laplacian and let $T = \mathrm{d}\Gamma(-\nabla^2)$ be the kinetic energy operator. Then for $f\in C_0{}^\infty(\Lambda_2)\cap L^2(\Lambda_2)$,

$$\|[T, W(f)]\chi\| \leqslant \|\nabla f\|_2^2\|\psi\| + 2\|\nabla^2 f\|_2\|(1+N_{\Lambda_2})^{1/2}\psi\|$$

for $\psi\in\mathscr{F}_\pm(L^2(\mathbb{R}^\nu))$ (cf. Lemma 6.3.37 in [42]), where Λ_2 is a bounded open subset of $\mathbb{R}^\nu$.

Since $[T, W(f)] = [T_{\Lambda_2}, W(f)]$ and $[T, N_{\Lambda_2}] = [T_{\Lambda_2}, N_{\Lambda_2}] = 0$,

$$[U_\Phi + T - \mu N, a] = [U_{\Phi_{\Lambda_1+\Lambda}} + T_{\Lambda_1+\Lambda} - \mu N_{\Lambda_1+\Lambda}, a] \qquad \text{for } a\in\mathscr{D}_1,$$

where N is the number operator on $\mathscr{F}_\pm(L^2(\mathbb{R}^\nu))$ and μ is a real number. Moreover $[U_\Phi + T - \mu N, a]$ is bounded on $\mathscr{F}_\pm(L^2(\mathbb{R}^\nu))$. Next, starting from $\mathscr{A}_2$, we construct a *-subalgebra $\mathscr{D}'_2$ using a method similar to the one above, and let $\mathscr{D}_2$ be a *-subalgebra of $\mathscr{A}_3(= B(\mathscr{F}_\pm(L^2(\Lambda_1+\Lambda+\Lambda))))$ generated by $\mathscr{D}_1$ and $\mathscr{D}'_2$; then clearly,

$$[U_\Phi + T - \mu N, b] = [U_{\Phi_{\Lambda_1+\Lambda}} + T_{\Lambda_1+\Lambda+\Lambda} - \mu N_{\Lambda_1+\Lambda+\Lambda}, b]$$

for $b\in\mathscr{D}_2$ and $[U_\Phi + T - \mu N, b]$ is bounded. Moreover the σ-weak closure of $\mathscr{D}_2$ contains $\mathscr{A}_2$.

Continuing this process, we can construct increasing sequences $\{\mathscr{A}_n\}$ and $\{\mathscr{D}_n\}$ such that $\mathscr{D}_n \subset \mathscr{A}_{n+1}$ and the σ-weak closure of $\mathscr{D}_n$ in $\mathscr{A}_{n+1}$ contains $\mathscr{A}_n$.

Now define a *-derivation δ on $\overline{\bigcup_{n=1}^\infty \mathscr{D}_n}$ as follows.

$$\delta(a) = \mathrm{i}[U_{\Phi_{\Lambda_1+\underbrace{\Lambda+\cdots+\Lambda}_{m}}} + T_{\Lambda_1+\underbrace{\Lambda+\cdots+\Lambda}_{m}} - N_{\Lambda_1+\underbrace{\Lambda+\cdots+\Lambda}_{m}}, a]$$

for $a \in \mathscr{D}_m$, where $\overline{[\cdot]}$ is the closure of $[\cdot]$. Then δ satisfies the conditions (1)–(3).

If $U_{\Phi_{\underbrace{\Lambda_1+\Lambda+\cdots+\Lambda}_{m}}} + T_{\underbrace{\Lambda_1+\Lambda+\cdots+\Lambda}_{m}} - N_{\underbrace{\Lambda_1+\Lambda+\cdots+\Lambda}_{m}}$ has a self-adjoint extension for each m, then it satisfies Condition (4). This may be done under suitable conditions on the function Φ, and a suitable boundary condition on $-\nabla^2$.

By the above considerations, we have seen that if an interacting continuous quantum model is defined through a two-body interaction potential with $\Phi \in L^2(\Lambda)$ (Λ, a bounded open subset of $\mathbb{R}^\nu$), there is always a globally defined *-derivation δ. Next we shall formulate the conditions which ensure the existence of time evolution and the KMS state.

For continuous quantum systems, we may further assume the following conditions:

(5) $\{\xi \in \mathscr{H}_{m(n)} \mid a \exp(-\beta h_n)\xi \in \mathscr{D}(h_n)\}$ is dense in $\mathscr{H}_{m(n)}$ for $\beta > 0$ and $a \in \mathscr{D}_n$, where $\mathscr{D}(h_n)$ is the domain of h_n, and $\exp(-\beta h_n)$ is from the trace class in $B(\mathscr{H}_{m(n)})$.

Now by Condition (5), we can define a state on $\mathscr{A}_{m(n)}$ for each n as follows: $\phi_n(a) = \mathrm{Tr}(a \exp(-\beta h_n))/\mathrm{Tr}(\exp(-\beta h_n))$ for $a \in \mathscr{A}_{m(n)}$.

The following two conditions are difficult to prove for interacting models. However, to assure the existence of time evolution and KMS state, they are more or less necessary, though they might be weakened.

(6) There is a locally normal state ϕ on $\mathscr{A}$ (the uniform closure of $\bigcup_{n=1}^\infty \mathscr{A}_n$ on $\mathscr{F}_\pm(L^2(\mathbb{R}^\nu)))$ such that $\phi(a) = \lim_{n\to\infty} \phi_n(a)$ for $a \in \mathscr{D}(\delta)$ and $\phi(a^*\delta(a)) = \lim_{n\to\infty} \phi_n(a^*\delta(a))$ for $a \in \mathscr{D}(\delta)$.

By the local normality, we mean that the restriction of ϕ to $\mathscr{A}_n$ is normal for each n.

Let ϕ be the state on $\mathscr{A}$ satisfying Condition (6), and let $\{\pi_\phi, \mathscr{H}_\phi\}$ be the GNS representation of $\mathscr{A}$ constructed via ϕ on a Hilbert space $\mathscr{H}_\phi$.

Since ϕ is locally normal, the restriction of π_ϕ to $\mathscr{A}_n$ is faithful for each n; hence π_ϕ is faithful on $\mathscr{A}$. Now we shall identify $\mathscr{A}$ with $\pi_\phi(\mathscr{A})$.

(7) $(1 \pm \delta)\mathscr{D}(\delta)$ is σ-weakly dense in $\mathscr{M}$, where $\mathscr{M}$ is the σ-weak closure of $\pi_\phi(\mathscr{A})(= \mathscr{A})$ in $B(\mathscr{H}_\phi)$.

Now we shall prove the following theorem.

4.8.2 Theorem *Suppose that δ satisfies conditions* (1)–(7). *Then under the identification of $\mathscr{A}$ with $\pi_\phi(\mathscr{A})$ in $\mathscr{M}$, δ is a σ-weakly closable *-derivation in $\mathscr{M}$, and its closure $\bar{\delta}$ is the generator of a σ-weakly continuous one-parameter group $\{\alpha_t \mid t \in \mathbb{R}\}$ of *-automorphisms on $\mathscr{M}$. Moreover let $\tilde{\phi}(x) = (x1_\phi, 1_\phi)$ for $x \in \mathscr{M}$; then $\tilde{\phi}$ is a KMS state at β for the W^*-dynamical system $\{\mathscr{M}, \alpha\}$.*

To prove the theorem, we shall provide some considerations. Since the σ-weak closure of $\mathscr{D}(\delta)\cap\mathscr{A}_{m(n)}$ in $\mathscr{A}_{m(n)}$ contains $\mathscr{A}_n$ and ϕ is locally normal, the σ-weak closure of $\mathscr{D}(\delta)\cap\mathscr{A}_{m(n)}$ in $\mathscr{M}$ contains $\mathscr{A}_n$, for $\pi_\phi(\mathscr{A}_{m(n)})$ is σ-weakly closed in $B(\mathscr{H}_\phi)$; hence the σ-weak closure of $\mathscr{D}(\delta)$ in $\mathscr{M}$ is $\mathscr{M}$. For $a\in\mathscr{D}_n$,

$$\phi(\delta(a)) = \lim_{p\to\infty}\phi_p(\delta(a)) = \lim_{p\to\infty}\frac{\operatorname{Tr}(\mathrm{i}\overline{[h_p,a]}(\exp(-\beta h_p))}{\operatorname{Tr}(\exp-\beta h_p)}.$$

Let $h_p=\sum_{j=1}^\infty\lambda_j e_j$ with $\operatorname{Tr}(e_j)=1$, where $\{e_j\}$ is a family of mutually orthogonal one-dimensional projections; then $\exp(-\beta h_p)=\sum_{j=1}^\infty\exp(-\beta\lambda_j)e_j$; hence $\operatorname{Tr}(\exp(-\beta h_p))=\sum_{j=1}^\infty\exp(-\beta\lambda_j)<+\infty$ for all positive numbers β. Therefore except for a finite number of $\{\lambda_j\}$, λ_j must be positive; hence h_p is lower bounded. For a positive β_0 with $\beta_0<\beta$, $|\lambda_j|\leqslant\exp(\beta_0\lambda_j)$ for $j\geqslant j_0$, where j_0 is some number. Since $\exp(-(\beta-\beta_0)h_p)\in T(\mathscr{H}_{m(p)})$, $h_p\exp(-\beta h_p)\in T(\mathscr{H}_{m(p)})$, where $T(\mathscr{H}_{m(p)})$ is the Banach space of all trace class operators on $\mathscr{H}_{m(p)}$. Let $V=\{\xi\in\mathscr{H}_{m(p)}\,|\,a\exp(-\beta h_p)\xi\in\mathscr{D}(h_p)\}$ for $a\in\mathscr{D}_p$. For $\xi\in V$,

$$(\mathrm{i}h_p a\exp(-\beta h_p)-\mathrm{i}ah_p\exp(-\beta h_p))\xi=\delta(a)\exp(-\beta h_p)\xi.$$

Since V is dense in $\mathscr{H}_{m(p)}$ and $\mathrm{i}ah_p\exp(-\beta h_p)+\delta(a)\exp(-\beta h_p)$ is a bounded operator, $\overline{\mathrm{i}h_p a\exp(-\beta h_p)|V}=\mathrm{i}ah_p\exp(-\beta h_p)+\delta(a)\exp(-\beta h_p)$. Since $\mathrm{i}h_p a\exp(-\beta h_p)$ is a closed operator,

$$\mathrm{i}h_p a\exp(-\beta h_p)=\mathrm{i}ah_p\exp(-\beta h_p)+\delta(a)\exp(-\beta h_p).$$

Hence,

$$\operatorname{Tr}(\delta(a)\exp(-\beta h_p))=\operatorname{Tr}(\mathrm{i}h_p a\exp(-\beta h_p))-\operatorname{Tr}(\mathrm{i}ah_p\exp(-\beta h_p)).$$

$$\operatorname{Tr}(\mathrm{i}h_p a\exp(-\beta h_p))=\operatorname{Tr}(\mathrm{i}h_p a\exp(-\beta_1 h_p)\exp(-\beta_2 h_p)),$$

where $\beta_1,\beta_2>0$ and $\beta_1+\beta_2=\beta$.

$$\operatorname{Tr}(\mathrm{i}h_p a\exp(-\beta_1 h_p)\exp(-\beta_2 h_p))=\operatorname{Tr}(\mathrm{i}\exp(-\beta_2 h_p)\cdot h_p a\exp(-\beta_1 h_p)),$$

for $h_p a\exp(-\beta_1 h_p)$ is bounded.

$$\begin{aligned}&\operatorname{Tr}(\mathrm{i}\exp(-\beta_2 h_p)\cdot h_p a\exp(-\beta_1 h_p))\\&\quad=\operatorname{Tr}(\overline{(\mathrm{i}\exp(-\beta_2 h_p)h_p)}a\exp(-\beta_1 h_p),\\&\quad=\operatorname{Tr}(\mathrm{i}h_p\exp(-\beta_2 h_p)a\exp(-\beta_1 h_p))=\operatorname{Tr}(\mathrm{i}a\exp(-\beta_1 h_p)h_p\exp(-\beta_2 h_p))\\&\quad=\operatorname{Tr}(\mathrm{i}ah_p\exp(-(\beta_1+\beta_2)h_p))=\operatorname{Tr}(\mathrm{i}ah_p\exp(-\beta h_p)).\end{aligned}$$

Hence

$$\operatorname{Tr}(\delta(a)\exp(-\beta h_p))=\operatorname{Tr}(\mathrm{i}ah_p\exp(-\beta h_p))-\operatorname{Tr}(\mathrm{i}ah_p\exp(-\beta h_p))=0.$$

Therefore for $a\in\mathscr{D}_n$, $\phi_n(\delta(a))=0$. If $m\geqslant n$, $a\in\mathscr{D}_n$ implies $a\in\mathscr{D}_m$; hence $\phi_m(\delta(a))=0$. Therefore $\phi(\delta(a))=0$ for $a\in\bigcup_{n=1}^\infty\mathscr{D}_n$.

Hence we have the following lemma.

4.8.3 Lemma $\phi(\delta(a)) = 0$ *for all* $a \in \mathscr{D}(\delta)$.

Now define $\mathrm{i}Ha_\phi = \delta(a)_\phi$ for $a \in \mathscr{D}(\delta)$; then H is well-defined. In fact, for $a, b \in \mathscr{D}(\delta)$,

$$|\phi(\delta(a)b)| = |\phi(\delta(ab)) - \phi(a\delta(b))| = |\phi(a\delta(b))| \\ \leqslant \phi(a^*a)^{1/2}\phi(\delta(b)^*\delta(b))^{1/2} = 0,$$

if $\phi(a^*a) = 0$. For $a, b \in \mathscr{D}(\delta)$,

$$(Ha_\phi, b_\phi) = -\mathrm{i}(\delta(a)_\phi, b_\phi) = -\mathrm{i}\phi(b^*\delta(a)) = \mathrm{i}\phi(\delta(b^*)a) \\ = \mathrm{i}(a_\phi, iHb_\phi) = (a_\phi, Hb_\phi);$$

hence H is symmetric.

4.8.4 Lemma *H is essentially self-adjoint.*

Proof It is enough to show that $(1 + \mathrm{i}H)\mathscr{D}(H)$ is dense in $\mathscr{H}_\phi$. Since $(1 + \mathrm{i}H)\mathscr{D}(H) = \{(1 \pm \delta)\mathscr{D}(\delta)\}_\phi$, it is enough to show that $(1 \pm \delta)\mathscr{D}(\delta)$ is σ-weakly dense in $\mathscr{M}$. By Condition (7), $(1 \pm \delta)\mathscr{D}(\delta)$ is σ-weakly dense in $\mathscr{M}$.

□

For $a, b \in \mathscr{D}(\delta)$ with $a^* = a$,

$$\mathrm{i}[H, a]b_\phi = \mathrm{i}(Ha - aH)b_\phi = \delta(ab)_\phi - (a\delta(b))_\phi = \delta(a)b_\phi;$$

hence $\delta(a) = \overline{\mathrm{i}[H, a]}$. Since $\mathrm{i}[\bar{H}, a]$ is symmetric and $\mathrm{i}[H, a] \subset \mathrm{i}[\bar{H}, a]$, $\mathrm{i}[\bar{H}, a] \subset \{\mathrm{i}[\bar{H}, a]\}^* \subset \{\mathrm{i}[H, a]\}^* = \overline{i[H, a]}$; hence $\mathrm{i}[\bar{H}, a] = \overline{\mathrm{i}[H, a]}$ for $a \in \mathscr{D}(\delta)$. By Corollary 3.2.5b in [41], δ is $\sigma(\mathscr{M}, \mathscr{M}_*)$-closable. For $x \in B(\mathscr{H}_\phi)$, define $\rho_t(x) = \exp(\mathrm{i}t\bar{H})x\exp(-\mathrm{i}t\bar{H})$; then $t \mapsto \rho_t$ is a σ-weakly continuous one-parameter group of *-automorphisms on $B(\mathscr{H}_\phi)$. Let $\tilde{\delta}$ be the generator of ρ; then $\delta \subset \tilde{\delta}$ and so $\bar{\delta} \subset \tilde{\delta}$, where $\bar{\delta}$ is the weak closure of δ. Again by Corollary 3.2.5b in [41], $\|(1 - \lambda\bar{\delta})(a)\| \geqslant \|a\|$ for $a \in \mathscr{D}(\bar{\delta})$ and $\lambda \in R$. Now we shall show that $(1 \pm \bar{\delta})\mathscr{D}(\bar{\delta}) = \mathscr{M}$. Since $(1 \pm \delta)\mathscr{D}(\delta)$ are σ-weakly dense in $\mathscr{M}$, for $x \in \mathscr{M}$ there is a directed set $\{a_\alpha\}$ in $\mathscr{D}(\delta)$ such that $(1 + \delta)(a_\alpha) \to x$ (σ-weakly). $(1 + \bar{\delta})^{-1}(1 + \delta)(a_\alpha) = (1 + \tilde{\delta})^{-1}(1 + \delta)(a_\alpha) = a_\alpha \to (1 + \tilde{\delta})^{-1}x$ (σ-weakly), because $\tilde{\delta}$ is the generator. Hence $(1 + \tilde{\delta})^{-1}x \in \mathscr{D}(\bar{\delta})$ and $(1 + \bar{\delta})(1 + \tilde{\delta})^{-1}x = x$ and so $(1 + \bar{\delta})\mathscr{D}(\bar{\delta}) = \mathscr{M}$.

Analogously $(1 - \bar{\delta})\mathscr{D}(\bar{\delta}) = \mathscr{M}$. Now by the Hille–Yoshida theorem (cf. Theorem 3.1.10 in [41]), $\bar{\delta}$ is the generator of a $\sigma(\mathscr{M}, \mathscr{M}_*)$-continuous one-parameter group of *-automorphisms on $\mathscr{M}$. Let $\alpha_t = \exp(t\bar{\delta})(t \in R)$; then $\phi(\delta(a)) = 0$ for $a \in \mathscr{D}(\delta)$ and so $\phi(\bar{\delta}(a)) = 0$ for $x \in \mathscr{D}(\bar{\delta})$; hence $\phi(\alpha_t(a)) = \phi(a)$ for $a \in \mathscr{M}$. Define $u_t a_\phi = \alpha_t(a)_\phi$; then u_t can be uniquely extended to a unitary operator on $\mathscr{H}_\phi$ (denoted again by u_t). One can easily see that $t \mapsto u_t$ is a strongly continuous one-parameter group of unitary operators on $\mathscr{H}_\phi$. Let

$u_t = \exp(\mathrm{i}tK)$ (one may take $\bar{H}$ as K); then $\alpha_t(x) = \exp(\mathrm{i}tK)x\exp(-\mathrm{i}tK)$ for $x \in \mathscr{M}$.

4.8.5 Lemma *$\tilde{\phi}$ is a KMS state for $\{\mathscr{M}, \alpha\}$ at β.*

Proof It is enough to show that by 4.3.16,

$$-\mathrm{i}\beta\tilde{\phi}(a^*\bar{\delta}(a)) \geqslant \tilde{\phi}(a^*a)\log(\tilde{\phi}(a^*a)/\tilde{\phi}(aa^*)) \qquad \text{for } a \in \mathscr{D}(\bar{\delta}).$$

For $a \in \mathscr{D}_n$,

$$\begin{aligned} &-\mathrm{i}\phi_p(a^*\delta(a)) \\ &= -\mathrm{i}\beta \frac{\mathrm{Tr}(a^*\overline{[\mathrm{i}h_p, a]}\exp(-\beta h_p))}{\mathrm{Tr}(\exp(-\beta h_p))} \\ &= -\mathrm{i}\beta \frac{\mathrm{Tr}(a^*\mathrm{i}\overline{[h_p, a]}\exp(-\beta h_p))}{\mathrm{Tr}(\exp(-\beta h_p))} \\ &\geqslant \frac{\mathrm{Tr}(a^*a\exp(-\beta h_p))}{\mathrm{Tr}(\exp(-\beta h_p))} \cdot \log\left\{\frac{\mathrm{Tr}(a^*a\exp(-\beta h_p))}{\mathrm{Tr}(\exp(-\beta h_p))} \Bigg/ \frac{\mathrm{Tr}(aa^*\exp(-\beta h_p))}{\mathrm{Tr}(\exp(-\beta h_p))}\right\} \end{aligned}$$

for $p \geqslant n$, because $\mathrm{Tr}(x\exp(-\beta h_p))/\mathrm{Tr}(\exp(-\beta h_p))(x \in \mathscr{A}_{m(p)})$ is a KMS state for $\{\mathscr{A}_{m(p)}, \gamma\}$ at β, where $\gamma_t(x) = \exp(\mathrm{i}th_p)x\exp(-\mathrm{i}th_p)(x \in \mathscr{A}_{m(p)})$.

Taking the limit, we have

$$-\mathrm{i}\beta\phi(a^*\delta(a)) \geqslant \phi(a^*a)\log(\phi(a^*a)/\phi(aa^*)),$$

if $\phi(a^*a) \neq 0$ and $\phi(aa^*) \neq 0$. If $\phi(a^*a) = 0$, then

$$|\phi(a^*\delta(a))| \leqslant \phi(a^*a)^{1/2}\phi(\delta(a)^*)\delta(a))^{1/2} = 0.$$

$u\log(u/v)$ is defined to be zero if $u = 0, v \geqslant 0$; hence

$$-\mathrm{i}\beta\phi(a^*\delta(a)) \geqslant \phi(a^*a)\log(\phi(a^*a)/\phi(aa^*))$$

if $\phi(a^*a) = 0$. Next suppose that $\phi(a^*a) \neq 0$ and $\phi(aa^*) = 0$; then

$$-\mathrm{i}\beta\phi_p(a^*\delta(a)) \geqslant \phi_p(a^*a)\log(\phi_p(a^*a)/\phi_p(aa^*))$$

implies

$$-\mathrm{i}\beta\phi(a^*\delta(a)) \geqslant \phi(a^*a)\{\log\phi(a^*a) - \lim\log\phi_p(aa^*)\}.$$

Hence $-\mathrm{i}\phi(a^*\delta(a)) \geqslant +\infty$. This is impossible; hence $\phi(aa^*) = 0$ implies $\phi(a^*a) = 0$. Next for $a \in \mathscr{D}(\bar{\delta})$, there is a directed set $\{a_\alpha\}$ in $\mathscr{D}(\delta)$ such that $a_\alpha \to a$ (strongly) and $\delta(a_\alpha) \to \delta(a)$ (strongly), for $\bar{\delta}$ is the σ-weak closure of δ. Hence,

$$-\mathrm{i}\beta\tilde{\phi}(a^*\bar{\delta}(a)) \geqslant \tilde{\phi}(a^*a)\log(\tilde{\phi}(a^*a)/\tilde{\phi}(aa^*))$$

if $\tilde{\phi}(a^*a) \neq 0$ and $\tilde{\phi}(aa^*) \neq 0$. If $\tilde{\phi}(a^*a) = 0$, then $\tilde{\phi}(a^*\bar{\delta}(a)) = 0$ by a discussion similar to one above. If $\tilde{\phi}(a^*a) \neq 0$ and $\tilde{\phi}(aa^*) = 0$, then

$$-\mathrm{i}\beta\tilde{\phi}(a^*\delta(a)) \geqslant \tilde{\phi}(a^*a)(\log \tilde{\phi}(a^*a) - \lim_{\alpha} \log \tilde{\phi}(a_\alpha a_\alpha^*))$$

$$\geqslant +\infty.$$

This is impossible; hence $\tilde{\phi}(aa^*) = 0$ implies $\tilde{\phi}(aa^*) = 0$.

By the above lemma, we have finished the proof of Theorem 4.8.2.

Now we shall discuss some concrete cases.

4.8.6 The ideal Fermi gas In this case, we can discuss our subject quite abstractly. Let $\mathscr{F}_-(\mathscr{H})$ be the anti-symmetric Fock space built over the one-particle Hilbert space $\mathscr{H}$, and let H be a self-adjoint operator in $\mathscr{H}$. Then the time evolution α is given as follows.

$$\alpha_t(x) = \Gamma(\exp(\mathrm{i}tH))x\Gamma(\exp(-\mathrm{i}tH)) \qquad (x \in B(\mathscr{F}_-(\mathscr{H}))),$$

where Γ is the second quantization of unitary operators on $\mathscr{H}$. The action on the annihilation and creation operators are $\alpha_t(a(f)) = a(\exp(\mathrm{i}tH)f)$ and $\alpha_t(a^*(f)) = a^*(\exp(\mathrm{i}tH)f)$. Therefore the time evolution can be expressed as a one-parameter group of *-automorphisms on the canonical anti-commutation relation algebra $\mathscr{A}_-(\mathscr{H})$. We shall use the notation from §3.1, Example (6), except for using $\mathscr{A}_-(\mathscr{H})$ instead of $\mathscr{A}(\mathscr{H})$. It is strongly continuous, because

$$\|\alpha_t(a(f)) - a(f)\| = \|a(\exp(\mathrm{i}tH)f - f)\| = \|\exp(\mathrm{i}tH)f - f\|.$$

Now suppose that H is a finite-rank operator with the range space $R(H)$, and let $f_1, f_2, \ldots, f_n$ be an orthonormal basis of $R(H)$. Let $f_1, f_2, \ldots, f_n, f_{n+1}, \ldots f_{n+p}, \ldots$ be an orthonormal basis of $\mathscr{H}$. Then $\alpha_t(a(f_{n+p})) = a(f_{n+p})$ and $\alpha_t(a^*(f_{n+p})) = a^*(f_{n+p})(p = 1, 2, \ldots)$. Since $\mathscr{A}_-(R(H))$ is finite-dimensional, α_t is uniformly continuous on $\mathscr{A}_-(R(H))$ so that it is uniformly continuous on $\mathscr{A}_-(\mathscr{H})$. Let $\alpha_t = \exp(t\delta)$; then δ is a bounded *-derivation. Since $\mathscr{A}_-(\mathscr{H})$ is simple, by 2.5.7 there is a self-adjoint element h in $\mathscr{A}_-(\mathscr{H})$ such that $\delta(a) = \mathrm{i}[h, a] (a \in \mathscr{A}_-(\mathscr{H}))$.

On the other hand, $\mathscr{A}_-(\mathscr{H})$ is a UHF algebra; therefore there is a unique tracial state τ on $\mathscr{A}_-(\mathscr{H})$. Clearly, a KMS state for $\{\mathscr{A}_-(\mathscr{H}), \alpha\}$ at β is

$$\phi(a) = \tau(a \exp(-\beta h))/\tau(\exp(-\beta h)) \qquad (a \in \mathscr{A}_-(\mathscr{H})).$$

Moreover by the considerations at the beginning of §4.3, such a KMS state at β is unique. Next, let H be a general self-adjoint operator in $\mathscr{H}$; then by Weyl's theorem for arbitrary $\varepsilon(>0)$, H can be expressed as $H = H_\varepsilon + K_\varepsilon$, where H_ε is a diagonal self-adjoint operator and K_ε is of Hilbert–Schmidt class with $\|K_\varepsilon\| < \varepsilon$. Hence there exists a sequence $\{H_n\}$ of finite-rank self-adjoint operators such that $\|\exp(\mathrm{i}tH_n)f - \exp(\mathrm{i}tH)f\| \to 0$ $(n \to \infty)$ for each $f \in \mathscr{H}$, and so α is approximately inner. By 4.3.17, $\{\mathscr{A}_-(\mathscr{H}), \alpha\}$ has a KMS state ϕ at β. Now we shall show that the ϕ is unique for $\{\mathscr{A}_-(\mathscr{H}), \alpha\}$ at β. Suppose that ψ be another KMS state for $\{\mathscr{A}_-(\mathscr{H}), \alpha\}$ at β. Then by 4.3.4 for $f \in \mathscr{D}(\exp(-\beta H))$, and $g \in \mathscr{H}$,

$$\begin{aligned}\psi(a^*(f)a(g)) &= \psi(a(g)\alpha_{i\beta}(a^*(f))) = \psi(a(g)a^*(\exp(-\beta H)f)) \\ &= \psi(\langle g, \exp(-\beta H)f\rangle 1 - a^*(\exp(-\beta H)f)a(g)) \\ &\qquad\qquad \text{(use the anti-commutation relation)} \\ &= \langle g, \exp(-\beta H)f\rangle - \psi(a^*(\exp(-\beta H)f)a(g)).\end{aligned}$$

Hence,

$$\psi(a^*((1 + \exp(-\beta H)f)a(g)) = \langle g, \exp(-\beta H)f\rangle.$$

Let $h \in \mathscr{H}$ and put $f = (1 + \exp(-\beta H))^{-1}h$; then $f \in \mathscr{D}(\exp(-\beta H))$ and so $\psi(a^*(h)a(g)) = \langle g, \exp(-\beta H)(1 + \beta\exp(-\beta H))^{-1}h\rangle$ for $f, g \in \mathscr{H}$. Analogously,

$$\begin{aligned}\psi\left(\prod_{i=1}^n a^*(f_i)\prod_{j=1}^n a(g_j)\right) &= \psi\left(\prod_{i=2}^n a^*(f_i)\prod_{j=1}^n a(g_j)a^*(\exp(-\beta H)f_1)\right) \\ &= \sum_{p=1}^n (-1)^{n-p}\langle g_p, \exp(-\beta H)f_1\rangle\psi\left(\prod_{i=2}^n a^*(f_i)\prod_{\substack{j=1\\ j\neq p}}^n a(g_j)\right) \\ &\quad - \psi\left(a^*(\exp(-\beta H)f_1)\prod_{i=2}^n a^*(f_i)\prod_{j=1}^n a(g_j)\right).\end{aligned}$$

Therefore by linearity and the replacement of f_1 by $(1 + \exp(-\beta H))^{-1}f_1$,

$$\begin{aligned}\psi\left(\prod_{i=1}^n a^*(f_i)\prod_{j=1}^n a(g_i)\right) &= \sum_{p=1}^n (-1)^{n-p}\langle g_p, \exp(-\beta H)(1 + \exp(-\beta H))^{-1}f_1\rangle \\ &\quad \times \psi\left(\prod_{i=2}^n a^*(f_i)\prod_{\substack{j=1\\ j\neq p}}^n a(g_j)\right).\end{aligned}$$

Hence by induction, the value of ψ on the product of $a^*(f_i)a(g_j)$ can be expressed by the sum of products of two-point functions $\psi(a^*(f_i)a(g_j))$.

If the numbers of $a^*(f_i)$ and $a(g_j)$ differ, then the corresponding value of ψ is zero. In fact, by the previous discussion, if the numbers are different, then $\psi(\prod_{i=1}^m a^*(f_i)\prod_{j=1}^n a(g_j))$ is expressed by the sum of products of numbers with factors $\psi(\prod_{p=1}^{m-n} a(f_{i_p}))$ or $\psi(\prod_{q=1}^{n-m} a(g_{j_q}))$; therefore it is enough to show that $\psi(\prod_{j=1}^n a(f_j)) = 0$ $(n = 1, 2, \ldots,)$, because $\psi(\prod_{j=1}^n a^*(f_j)) = (-1)^{n-1}\overline{\psi(\prod_{j=1}^n a(f_j))}$.

$$\psi\left(\prod_{j=1}^{n} a(f_j)\right)=\psi\left(a(f_1)\prod_{j=2}^{n} a(f_j)\right)=\psi\left(\prod_{j=2}^{n} a(f_j)a(\exp(-\beta H)f_1)\right)$$
$$=(-1)^{n-1}\psi\left(a(\exp(-\beta H)f_1)\prod_{j=2}^{n} a(f_j)\right)$$
$$=(-1)^{2(n-1)}\psi\left(a(\exp(-2\beta H)f_1)\prod_{j=2}^{n} a(f_j)\right)$$
$$=\cdots=(-1)^{k(n-1)}\psi\left(a(\exp(-k\beta H)f_1)\prod_{j=2}^{n} a(f_j)\right).$$

Let $H=\int_{-\infty}^{\infty}\lambda\,\mathrm{d}E_\lambda$ be the spectral decomposition of H. If $f_1\in E([\varepsilon,\infty))$ $(\varepsilon>0)$, then

$$\|\exp(-k\beta H)f_1\|^2=\int_{\varepsilon}^{\infty}\exp(-2k\beta\lambda)\,\mathrm{d}\|E_\lambda f_1\|^2\to 0 \qquad (k\to\infty)$$

(assume $\beta>0$). Hence $\psi(\prod_{j=1}^{n}a(f_j))=0$. If $f_1\in E((-\infty,-\varepsilon])$ $(\varepsilon>0)$, then $\exp(k\beta H)f_1$ exists. Replace f_1 by $\exp(k\beta H)f_1$; then

$$\psi\left(a(\exp(k\beta H)f_1)\prod_{j=2}^{n} a(f_j)\right)=(-1)^{k(n-1)}\psi\left(\prod_{j=1}^{n} a(f_j)\right).$$

Since $\exp(k\beta H)f_1\to 0$ (strongly), $\psi(\prod_{j=1}^{n}a(f_j))=0$.

$$a(f_1)=\lim_{\varepsilon\to 0}\{a(E((-\infty,-\varepsilon])f_1+E([\varepsilon,+\infty))f_1\}+\lim_{\varepsilon\to 0}a(E((-\varepsilon,\varepsilon))f_1).$$

Hence

$$\psi\left(\prod_{j=1}^{n} a(f_j)\right)=\lim_{\varepsilon\to 0}\psi\left(a(E((-\varepsilon,\varepsilon))f_1)\prod_{j=2}^{n} a(f_j)\right)$$
$$=\psi\left(a((E_{+0}-E_0)f_1)\prod_{j=2}^{n} a(f_j)\right).$$

Therefore it is enough to show that $\psi(\prod_{j=1}^{n}a(f_j))=0$ under the assumption $Hf_1=0$.

Suppose that $\|f_1\|=1$; then $a(f_1)a^*(f_1)a(f_1)=a(f_1)$. Hence,

$$\psi\left(\prod_{j=1}^{n} a(f_j)\right)=\psi\left(a(f_1)a^*(f_1)a(f_1)\prod_{j=2}^{n} a(f_j)\right)$$
$$=\psi\left(a^*(f_1)a(f_1)\prod_{j=2}^{n} a(f_j)a(\exp(-\beta H)f_1)\right)$$
$$=\psi\left(a^*(f)a(f)\prod_{j=2}^{n} a(f_j)a(f_1)\right)$$

$$= (-1)^{n-1}\psi\left(a^*(f_1)a(f_1)a(f_1)\prod_{j=2}^{n} a(f_j) \right) = 0,$$

because $a(f_1)^2 = 0$.

Hence we have proved the following theorem.

4.8.7 Theorem *Let H be a self-adjoint operator in a Hilbert space and let $\mathscr{A}_-(\mathscr{H})$ be the anti-commutation relation algebra over $\mathscr{H}$. Let $\{\mathscr{A}_-(\mathscr{H}), \alpha\}$ be the C^*-dynamics defined by $\alpha_t = \exp(t\bar{\delta}_{iH})$; then $\{\mathscr{A}_-(\mathscr{H}), \alpha\}$ has a unique KMS state ϕ_β for each $\beta \in \mathbb{R}$ and*

$$\phi_\beta(a^*(f)a(g)) = \langle g, \exp(-\beta H)(1 + \exp(-\beta H))^{-1} f \rangle \qquad \text{for } f, g \in \mathscr{H}.$$

Now we shall discuss Gibbs thermodynamic limit or, more generally, KMS states for non-interacting Fermi systems. Let Λ be a bounded open subset of the configuration space $\mathbb{R}^\nu$ and define a self-adjoint operator H_Λ in $L^2(\Lambda)$ as a self-adjoint extension of the Laplacian $-\nabla^2$, where $-\nabla^2$ is defined on all infinitely differentiable functions with support in Λ. The number of the possible extensions is partially governed by the smoothness of the boundary of Λ.

4.8.8 Theorem *For each bounded open subset $\Lambda \subset \mathbb{R}^\nu$, let H_Λ be a self-adjoint extension of $-\nabla^2$ in $L^2(\Lambda)$ and let H be the unique self-adjoint extension in $L^2(\mathbb{R}^\nu)$ of $-\nabla^2$. Let $\mathscr{A}_-(L^2(\Lambda))$ be the anti-commutation relation algebra over $L^2(\Lambda)$ and $\mathscr{A}_-(L^2(\mathbb{R}^\nu))$ be the corresponding algebra over $L^2(\mathbb{R}^\nu)$, and let α^Λ and α be the one-parameter groups of *-automorphisms of $\mathscr{A}_-(L^2(\Lambda))$ and $\mathscr{A}_-(L^2(R^\nu))$ such that $\alpha_t^\Lambda(a(f)) = a(\exp(itH_\Lambda)f)$ and $\alpha_t(a(f)) = a(\exp(itH)f)$. Then we have the following.*

(1) $\lim_{\Lambda' \to \infty} \| \alpha_t^{\Lambda'}(x) - \alpha_t(x) \| = 0$ *for all $x \in \mathscr{A}_-(L^2(\Lambda))$ and all $\Lambda \subset \mathbb{R}^\nu$ uniformly for t in finite intervals of $\mathbb{R}$, where $\Lambda' \to \infty$ in the sense that Λ' eventually contains any given $\Lambda \subset R^\nu$.*

(2) *Let $\phi_{\Lambda,\mu}$ be the unique KMS state for $\{\mathscr{A}_-(L^2(\Lambda)), \alpha^{\Lambda,\mu}\}$ with $\alpha_t^{\Lambda,\mu} = \exp(t\bar{\delta}_{iH_\Lambda - i\mu 1})$ $(\mu \in \mathbb{R})$ at $\beta \in \mathbb{R}$ such that*

$$\phi_{\Lambda,\mu}(a^*(f)a(g)) = \langle g, z\exp(-\beta H_\Lambda)(1 + z\exp(-\beta H_\Lambda))^{-1} f \rangle,$$

where $z = \exp(\beta\mu)$; then $\lim_{\Lambda' \to \infty} \phi_{\Lambda'\mu}(x) = \phi_\mu(x)$ for $x \in \mathscr{A}_-(L^2(\Lambda))$ and all $\Lambda \subset \mathbb{R}^\nu$, where ϕ_μ is the unique KMS state for $\{\mathscr{A}_-(L^2(\mathbb{R}^\nu)), \alpha^\mu\}$ at β with $\alpha_t^\mu = \exp(t\bar{\delta}_{iH - i\mu 1})$ such that

$$\phi_\mu(a^*(f)a(g)) = \langle g, z\exp(-\beta H)(1 + z\exp(-\beta H))^{-1} f \rangle.$$

(3) *$\phi_{\Lambda,\mu}$ and ϕ_μ are gauge-invariant – i.e. $\phi_{\Lambda,\mu}(\gamma_t(x)) = \phi_{\Lambda,\mu}(x)$ for $x \in \mathscr{A}_-(L^2(\Lambda))$ and $\phi_\mu(\gamma_t(x)) = \phi_\mu(x)$ for $x \in \mathscr{A}_-(L^2(\mathbb{R}^\nu))$, where $\gamma_t = \exp(t\bar{\delta}_{i1})$.*

Proof Clearly $\|H_{\Lambda'}f - Hf\| \to 0$ $(\Lambda' \to \infty)$ for $f \in L^2(\Lambda) \cap C_0^{\infty}(\Lambda)$ and $\bigcup_{\Lambda \subset \mathbb{R}^\nu}(1 \pm \mathrm{i}H)\{L^2(\Lambda) \cap C_0^{\infty}(\Lambda)\}$ is dense in $L^2(\mathbb{R}^\nu)$. Therefore $\exp(\mathrm{i}tH_{\Lambda'}) \to \exp(\mathrm{i}tH)$ (strongly), so that $\|\alpha_t{}^{\Lambda}(a(f)) - \alpha_t(a(f))\| \to 0$. From this, one can easily conclude that $\|\alpha_t{}^{\Lambda'}(x) - \alpha_t(x)\| \to 0$ for $x \in \mathscr{A}_-(L^2(\Lambda))$ and $\Lambda \subset \mathbb{R}^\nu$. Define $\phi_{\Lambda,\mu}{}^s(x) = \phi_{\Lambda,\mu}(\gamma_s(x))$ for $x \in \mathscr{A}_-(L^2(\Lambda))$; then $\alpha_t{}^{\Lambda}\gamma_s = \gamma_s\alpha_t{}^{\Lambda}$ implies:

$$\begin{aligned}\phi_{\Lambda,\mu}{}^s(x\alpha_{\mathrm{i}\beta}{}^{\Lambda,\mu}(y)) &= \phi_{\Lambda,\mu}(\gamma_s(x\alpha_{\mathrm{i}\beta}{}^{\Lambda,\mu}(y))) = \phi_{\Lambda,\mu}(\gamma_s(x)\alpha_{\mathrm{i}\beta}{}^{\Lambda,\mu}(\gamma_s(y))) \\ &= \phi_{\Lambda,\mu}(\gamma_s(y)\gamma_s(x)) = \phi_{\Lambda,\mu}{}^s(yx).\end{aligned}$$

Hence $\phi_{\Lambda,\mu}{}^s$ is given a KMS state for $\{\mathscr{A}_-(L^2(\Lambda)), \alpha^{\Lambda,\mu}\}$ at β, so that by the unicity, $\phi_{\Lambda,\mu}{}^s = \phi_{\Lambda,\mu}$. Analogously, ϕ_μ is gauge-invariant. Moreover

$$\begin{aligned}\phi_\mu(a^*(f)a(g)) &= \langle g, \exp(-\beta(H-\mu 1))(1+\exp(-\beta(H-\mu 1)))^{-1}f\rangle \\ &= \langle g, \exp(\beta\mu)\exp(-\beta H)(1+\exp(-\beta\mu)\exp(-\beta H))^{-1}f\rangle \\ &= \langle g, z\exp(-\beta H)(1+z\exp(-\beta H))^{-1}f\rangle \qquad \text{for } f, g \in L^2(\mathbb{R}^\nu).\end{aligned}$$

The rest is clear. □

Now we shall discuss the relation between the ideal Fermi gas and our system.

Let $\mathscr{F}_-(L^2(\mathbb{R}^\nu))$ (resp. $\mathscr{F}_-(L^2(\Lambda))$) be the associated Fermi–Fock space with $L^2(\mathbb{R}^\nu)$ (resp. $L^2(\Lambda)$).

Let $\mathscr{A}_-(L^2(\mathbb{R}^\nu))$ (resp. $\mathscr{A}_-(L^2(\Lambda))$) be the CAR algebra corresponding to $\mathscr{F}_-(L^2(\mathbb{R}^\nu))$ (resp. $\mathscr{F}_-(L^2(\Lambda))$). Let $\Lambda_n = \{p \in \mathbb{R}^\nu \mid \|p\| < n\}$; then $\mathscr{A}_-(L^2(\Lambda_n))$ is σ-weakly dense in $B(\mathscr{F}_-(L^2(\Lambda_n))) = \mathscr{A}_n$ and $\mathscr{A} =$ the uniform closure of $\bigcup_{n=1}^{\infty}\mathscr{A}_n$.

Let $\mathscr{D}_n$ be the *-subalgebra of $\mathscr{A}_n$ generated by $\{a(f) \mid f \in C_c{}^{\infty}(\Lambda)\}$ and the identity operator on $\mathscr{F}_-(L^2(\Lambda))$, where $C_c{}^{\infty}(\Lambda)$ is the space of all infinitely differentiable functions with compact support in Λ_n. Let H_{Λ_n} be a self-adjoint extension of the Laplacian $-\nabla^2$ in $L^2(\Lambda_n)$, and let T_{Λ_n} be the second quantization of H_{Λ_n} on $\mathscr{F}_-(L^2(\Lambda_n))$. Let $K_{\mu,\Lambda_n} = T_{\Lambda_n} - \mu N_{\Lambda_n}$, where $\mu \in \mathbb{R}$ and N_{Λ_n} is the number operator on $\mathscr{F}_-(L^2(\Lambda_n))$.

Then

$$\mathrm{i}\overline{[K_{\mu,\Lambda_n}, a(f)]} = a(\mathrm{i}(H_{\Lambda_n} - \mu 1)f)$$

and

$$\mathrm{i}\overline{[K_{\mu,\Lambda_n}, a^*(f)]} = a^*(\mathrm{i}(H_{\Lambda_n} - \mu 1)f) \qquad \text{for } f \in C_c{}^{\infty}(\Lambda_n)$$

and so

$$\mathrm{i}\overline{[K_{\mu,\Lambda_n}, x]} = \delta_{\mathrm{i}(H-\mu 1)}(x) \qquad \text{for } x \in \mathscr{D}_n.$$

Let $\mathscr{D}(\delta) = \bigcup_{n=1}^{\infty}\mathscr{D}_n$ and define $\delta_\mu(x) = \lim_{n\to\infty}\mathrm{i}\overline{[K_{\mu,\Lambda_n}, x]} = \mathrm{i}\overline{[K_{\mu,\Lambda_m}, x]}$ for $x \in \bigcup_{n=1}^{\infty}\mathscr{D}_n$ with some m; then δ_μ is a *-derivation in $\mathscr{A}$. It is clear that the above models satisfy Conditions (1)–(4). If we put a classical boundary condition on Λ_n, then H_{Λ_n} is lower bounded and $\exp(-\beta(H_{\Lambda_n} - \mu 1))$ has a finite trace value for $\beta > 0$ and $\mu \in \mathbb{R}$ (cf. Proposition 5.2.22 in [42]). Then by

Lemma 4.8.4, one can easily see that the models satisfy all of the Conditions (1)–(7).

4.8.9 Ideal Bose gas The Bose annihilation and creation operators $a(f)$, $a^*(f)$ are no longer bounded; therefore we have to replace them with Weyl operators

$$\{W(f) \mid f \in L^2(\mathbb{R}^\nu)\}.$$

$W(f)$ is a unitary operator on $\mathscr{F}_+(L^2(\mathbb{R}^\nu))$ and satisfies the commutation relation:

$$W(f)W(g) = \exp(-\operatorname{Im}\langle f, g\rangle/2)W(f+g) = \exp(-\mathrm{i}\operatorname{Im}\langle f, g\rangle)W(g)W(f).$$

Let $\mathscr{A}_+(\Lambda)$ be a C*-subalgebra of $B(\mathscr{F}_+(L^2(\Lambda)))$ generated by $\{W(f) \mid f \in L^2(\Lambda)\}$; then the σ-weak closure of $\mathscr{A}_+(\Lambda)$ in $B(\mathscr{F}_+(L^2(\Lambda)))$ is $B(\mathscr{F}_+(L^2(\Lambda)))$. Let Λ_n $(n = 1, 2, \ldots,)$ be an increasing sequence of bounded open subsets in $\mathbb{R}^\nu$ such that for any bounded open subset Λ of $\mathbb{R}^\nu$, there is a Λ_{n_0} such that $\Lambda \subset \Lambda_{n_0}$. Let H_Λ be a self-adjoint extension of the Laplacian $-\nabla^2$ on $L^2(\Lambda)$ corresponding to a classical boundary condition, and let H be the unique self-adjoint extension of $-\nabla^2$ in $L^2(\mathbb{R}^\nu)$. Put $H_{\mu,\Lambda} = H_\Lambda - \mu 1$ and $H_\mu = H - \mu 1$.

Let $K_{\mu,\Lambda}$ (resp. K_μ) be the second quantization of $H_\Lambda - \mu 1$ (resp. $H - \mu 1$) on $\mathscr{F}_+(L^2(\Lambda))$ (resp. $\mathscr{F}_+(L^2(\mathbb{R}^\nu))$).

Since H_Λ is lower bounded, $K_{\mu,\Lambda}$ is also lower bounded. At this time, it is difficult to choose $\mathscr{D}_n$ inside of $\mathscr{A}_+(\Lambda_n)$. Let $\mathscr{D}'_n$ be the *-subalgebra of $B(\mathscr{F}_+(L^2(\Lambda_n)))$ generated by $\{(1 + N_{\Lambda_n})^{-1}W(f)(1 + N_{\Lambda_n})^{-1} \mid f \in C_c^\infty(\Lambda_n)\}$ and the identity operator on $\mathscr{F}_+(L^2(\Lambda_n))$. Let $\mathscr{D}_n$ be the *-subalgebra of $B(\mathscr{F}_+)(L^2(\Lambda_n))$ generated by $\{\mathscr{D}'_1, \mathscr{D}'_2, \ldots, \mathscr{D}'_n\}$ and let $\mathscr{D} = \bigcup_{n=1}^\infty \mathscr{D}_n$. Define δ_μ on $\mathscr{D}$ as follows.

$$\delta_\mu(x) = \lim_{n\to\infty} \overline{\mathrm{i}[K_{\mu,\Lambda_n}, x]} = \overline{\mathrm{i}[K_{\mu,\Lambda_{n_0}}, x]} \qquad \text{for } x \in \mathscr{D},$$

where n_0 is a number depending on x; then δ_μ is a *-derivation in $\mathscr{A}$ (= the uniform closure of $\bigcup_{n=1}^\infty \mathscr{A}_+(\Lambda_n)$).

Then, the model satisfies the Conditions (1)–(4).

It is known that $\exp(-\beta H_{\Lambda_n})$ is of trace class, and if $\beta(H_{\Lambda_n} - \mu 1) > 0$ (strictly positive) then $\exp(-\beta K_{\mu,\Lambda_n})$ is of trace class (cf. Proposition 5.2.27 in [42]). Let $\Phi(f) = \overline{a(f) + a^*(f)/(2)^{1/2}}$; then $\Phi(f)$ is self-adjoint and all elements in $L^2(\Lambda_n)_+{}^p$ $(p = 0, 1, 2, \ldots,)$ are analytic vectors for $\Phi(f)$ and

$$\sum_{m\geqslant 0} \frac{|t|^m}{m!} \|\Phi(f)^m \psi^{(p)}\| \leqslant \sum_{m\geqslant 0} \frac{2^{1/2}|t|^m}{m!} \left(\frac{(p+m)!}{p!}\right)^{1/2} \|f\|^p \|\psi^{(p)}\|$$

(cf. Proposition 5.2.3 in [42]).

Moreover

$$W(f)=\exp(\mathrm{i}\Phi(f)), \qquad \mathrm{i}[K_{\mu,\Lambda_n},\Phi(f)]=\Phi(\mathrm{i}(H_{\Lambda_n}-\mu 1)f).$$

Therefore for $f\in C_c^{\infty}(\Lambda_n)$,

$$\sum_{m\geqslant 0}\frac{|t|^m}{m!}\|[K_{\mu,\Lambda_n},\Phi(f)^m]\psi^{(p)}\|$$

$$\leqslant \sum_{m\geqslant 1}\frac{|t|^m}{m!}m(p+m)^{1/2}(p+m-1)^{1/2}\cdots$$

$$\times(p+1)^{1/2}\|\psi^{(p)}\|\,\|f\|^{m-1}\|\mathrm{i}(H_{\Lambda_n}-\mu 1)f\|$$

$$=|t|\,\|\mathrm{i}(H_{\Lambda_n}-\mu 1)f\|\sum_{m\geqslant 0}\frac{|t|^m}{m!}(p+m)^{1/2}(p+m-1)^{1/2}\cdots$$

$$\times(p+1)^{1/2}\|\psi^{(p)}\|\,\|f\|^{m-1}<+\infty \qquad \text{for } \psi^{(p)}\in\mathscr{D}(K_{\mu,\Lambda_n}).$$

Let $\sum_{m=0}^{r}(\mathrm{i})^m/m!\,\Phi(f)^m=W_r$; then

$$\mathrm{i}[K_{\mu,\Lambda_n},W_r]\psi^{(p)}=(\mathrm{i}K_{\mu,\Lambda_n}W_r-\mathrm{i}W_rK_{\mu\Lambda_n})\psi^{(p)}$$

$$=\sum_{m=0}^{r}\frac{(\mathrm{i})^m}{m!}\mathrm{i}[K_{\mu,\Lambda_n},\Phi(f)^m]\psi^{(p)}\to\xi,$$

where ξ is an element in $\mathscr{F}_+(L^2(\Lambda))$. Since $K_{\mu,\Lambda_n}\psi^{(p)}\in L^2(\Lambda_n)_+{}^p$, $\mathrm{i}W_rK_{\mu,\Lambda_n}\psi^{(p)}\to \mathrm{i}W(f)K_{\mu,\Lambda_n}\psi^{(p)}$; hence $\mathrm{i}K_{\mu,\Lambda_n}W_r\psi^{(p)}\to\xi-\mathrm{i}W(f)K_{\mu,\Lambda_n}\psi^{(p)}$.

Since $W_r\psi^{(p)}\to W(f)\psi^{(p)}$ and K_{μ,Λ_n} is closed, $W(f)\psi^{(p)}\in\mathscr{D}(K_{\mu,\Lambda_n})$ and $\mathrm{i}K_{\mu,\Lambda_n}W(f)\psi^{(p)}=\xi-\mathrm{i}W(f)K_{\mu,\Lambda_n}\psi^{(p)}$. Therefore if $\psi^{(p)}\in\mathscr{D}(K_{\mu,\Lambda_n})$, $W(f)\psi^{(p)}\in\mathscr{D}(K_{\mu,\Lambda_n})$. Since $\exp(-\beta K_{\mu,\Lambda_n})$ preserves $L^2(\Lambda_n)_+{}^p$ invariant, $\exp(-\beta K_{\mu,\Lambda_n})L^2(\Lambda_n)_+{}^p\subset L^2(\Lambda_n)_+{}^p\cap\mathscr{D}(K_{\mu,\Lambda_n})$; hence $\{\xi\in\mathscr{F}_+(L^2(\Lambda_n))|\,W(f)\xi\in\mathscr{D}(K_{\mu,\Lambda_n})\}$ is dense in $\mathscr{F}_+(L^2(\Lambda_n))$. Since $(1+N_{\Lambda_n})^{-1}L^2(\Lambda_n)_+{}^p\subset L^2(\Lambda_n)_+{}^p$, $(1+N_\Lambda)^{-1}\exp(-\beta K_{\mu,\Lambda_n})L^2(\Lambda_n)_+{}^p\subset L^2(\Lambda_n)_+{}^p$. Since $[K_{\mu,\Lambda_n},(1+N_{\Lambda_n})^{-1}]=0$, $K_{\mu,\Lambda_n}(1+N_{\Lambda_n})^{-1}W(f)(1+N_{\Lambda_n})^{-1}\exp(-\beta K_{\mu,\Lambda_n})$ has a dense domain, so that Condition (5) is satisfied.

Conditions (6) and (7) may need more careful discussion than for the ideal Fermi gas. However the most important point is to show the existence of time evolution and the KMS state for the ideal Bose gas. This can be done by circumventing complicated discussions.

Put $\phi_n(x)=\mathrm{Tr}(x\exp(-\beta K_{\mu,\Lambda_n}))/\mathrm{Tr}(\exp(-\beta K_{\mu,\Lambda_n}))$, whenever $\phi_n(x)$ is defined. By a discussion similar to the Fermi case, we can show that for $f_1, f_2,\ldots,f_m,g_1,g_2,\ldots,g_l\in C_c^{\infty}(\Lambda_n)$, $\phi_n(a^*(f_1)a^*(f_2)\cdots a^*(f_m)a(g_1)a(g_2)\cdots a(g_l))=0$ if $m\neq l$, and $\phi_n(a^*(f_1)a^*(f_2)\cdots a^*(f_m)a(g_1)a(g_2)\cdots a(g_l))$ is a sum of products of two-point functions $\phi_n(a^*(f_i)a(g_j))$ if $m=l$. Moreover we can easily see that

$$\phi_n(a^*(f)a(g))=\langle z\exp(-\beta H_{\Lambda_n})(1-z\exp(-\beta H_{\Lambda_n})),f,g\rangle$$

where $z = \exp(\beta\mu)$ (cf. page 60 in [42]), so that

$$\phi_n(W(f)) = \exp\left\{ -\frac{\langle (1 + z\exp(-\beta H_{\Lambda_n}))(1 - z\exp(-\beta H_{\Lambda_n}))^{-1} f, f \rangle}{4} \right\}.$$

Now let H be the unique self-adjoint extension of the Laplacian $-\nabla^2$ in $L^2(\mathbb{R}^\nu)$; then $H_{\Lambda_n} f = Hf$ for $f \in C_c^\infty(\Lambda_n)$. Hence $\exp(itH_{\Lambda_n}) \to \exp(itH)$ (strongly). From this, one can easily show that

$$\phi_n(W(f)) \to \phi(W(f)) = \exp\left\{ -\frac{\langle (1 + z\exp(-\beta H))(1 - z\exp(-\beta H))^{-1} f, f \rangle}{4} \right\}$$

if $z < 1$. If $z = 1$ and Λ_n satisfies the Dirichlet boundary condition, then we can show again that $\phi_n(W(f)) \to \phi(W(f)$.

It is known that there is a locally normal state $\tilde{\phi}$ on $\mathscr{A}$ such that $\tilde{\phi} = \phi$ on the uniform closure of $\bigcup_{n=1}^\infty \mathscr{A}_+(\Lambda_n)$.

Define $\alpha_{n,t}(x) = \exp(itK_{\mu,\Lambda_n})x\exp(-itK_{\mu,\Lambda_n})$ and

$$\alpha_t(x) = \exp(itK_\mu)x\exp(-itK_\mu) \text{ for } x \in B(\mathscr{F}_+(L^2(\mathbb{R}^\nu))),$$

where K_μ is the second quantization of $H - \mu 1$. $\alpha_{n,t}(x) \to \alpha_t(x)$ in the strong operator topology. Moreover,

$$\alpha_{n,t}(W(f)) = \exp(itK_{\mu,\Lambda_n})W(f)\exp(-itK_{\mu,\Lambda_n}) = W(\exp(it(H_{\Lambda_n} - \mu 1))f) \quad \text{for } f \in L^2(\Lambda_n)$$

and

$$\alpha_t(W(f)) = \exp(itK_\mu)W(f)\exp(-itK_\mu) = W(\exp(it(H - \mu 1)f) \quad \text{for } f \in L^2(\mathbb{R}^\nu).$$

$$\begin{aligned}
\phi_n(W(g)\alpha_{n,t}W(f)) &= \phi_n(W(g)W(\exp(itH_n)f)) \\
&= \phi_n(\exp(-\operatorname{Im}\langle g, \exp(itH_n)f\rangle/2)W(g + \exp(itH_n)f)) \\
&= \exp(-\operatorname{Im}\langle g, \exp(itH_n)f\rangle/2) \\
&\quad \times \exp\left\{ -\frac{\langle (1 + z\exp(-\beta H_n))(1 - z\exp(-\beta H_n))^{-1}(g + \exp(itH_n)f), (g + \exp(itH_n)f)\rangle}{4} \right\} \\
&\to \exp(-\operatorname{Im}\langle g, \exp(itH)f\rangle/2) \\
&\quad \times \exp\left\{ -\frac{\langle (1 + \exp(-\beta H))(1 - z\exp(-\beta H))^{-1}(g + \exp(itH)f), (g + \exp(itH)f)\rangle}{4} \right\} \\
&= \phi(W(g)\alpha_t(W(f)).
\end{aligned}$$

Therefore by a discussion similar to that for the proof of 4.3.17, we can easily see that ϕ is a KMS state at β for the dynamical system $\{\mathscr{A}, \alpha\}$.

Now we shall consider Conditions (6) and (7). Put

$$\phi_n(x) = \frac{\mathrm{Tr}(x\exp(-\beta K_{\mu,\Lambda_n}))}{\mathrm{Tr}(\exp(-\beta K_{\mu,\Lambda_n}))},$$

whenever $\phi_n(x)$ is defined. Let $\mathscr{L}_n$ be a *-algebra generated by $\{a^*(f), a(f) \mid f \in C_c^\infty(\Lambda_n)\}$; then it is known that $\{\phi_n(x)\}$ converges for each $x \in \mathscr{L}_m$ $(m = 1, 2, \ldots,)$. Denote its limit by $\phi(x)$. Then

$$[\mathrm{i}K_{\mu,\Lambda_n}, a(f)] = a(\mathrm{i}H_{\mu,\Lambda_n} f) \qquad \text{and} \qquad [\mathrm{i}K_{\mu,\Lambda_n}, a^*(f)] = a^*(\mathrm{i}H_{\mu,\Lambda_n} f)$$

for $f \in C_c^\infty(\Lambda_n)$, so that

$$[\mathrm{i}K_{\mu,\Lambda_n}, a^*(f)] = a^*(\mathrm{i}H_\mu f) \qquad \text{and} \qquad [\mathrm{i}K_{\mu,\Lambda_n}, a(f)] = a(\mathrm{i}H_\mu f).$$

Therefore

$$[\mathrm{i}K_{\mu,\Lambda_n}, x] = [\mathrm{i}K_\mu, x] \qquad \text{for } x \in \mathscr{L}_n.$$

Define $\delta(x) = \lim\,[\mathrm{i}K_{\mu,\Lambda_n}, x]$ for $x \in \bigcup_{m=1}^\infty \mathscr{L}_m$; then δ is a *-derivation defined on $\bigcup_{m=1}^\infty \mathscr{L}_m$ (denoted by $\mathscr{L}$)

Clearly $\phi_n(\delta(x)) = 0$ for $x \in \mathscr{L}_n$ and so $\phi(\delta(x)) = 0$ for $x \in \bigcup_{m=1}^\infty \mathscr{L}_m$. This implies that Condition (6) is satisfied for the *-algebra $\bigcup_{m=1}^\infty \mathscr{L}_m$. ϕ_n is also a KMS state at β for $\{B(\mathscr{F}_+(L^2(\Lambda_n))), \alpha_n\}$ with

$$\alpha_{n,t}(a) = \exp(\mathrm{i}tK_{\mu,\Lambda_n}) a \exp(-\mathrm{i}tK_{\mu,\Lambda_n}) \ (a \in B(\mathscr{F}_+(L^2(\Lambda_n))));$$

hence

$$-\mathrm{i}\beta\phi_n(x^*)\delta(x)) \geqslant \phi_n(x^*x)\log\frac{\phi_n(x^*x)}{\phi_n(xx^*)} \qquad \text{for } x \in \mathscr{L}_n.$$

By taking a limit, we have

$$-\mathrm{i}\beta\phi(x^*\delta(x)) \geqslant \phi(x^*x)\log\frac{\phi(x^*x)}{\phi(xx^*)} \qquad \text{for } x \in \mathscr{L}$$

Consider the GNS representation $\{\pi_\phi, \mathscr{H}_\phi\}$ of the *-algebra $\mathscr{L}$ constructed via ϕ. Define $\mathrm{i}H_\phi x_\phi = (\delta(x))\phi$ for $x \in \mathscr{L}$; then, H_ϕ is a symmetric operator in $\mathscr{H}_\phi$. Let $\bar{H}_\phi$ be the closure of H_ϕ in $\mathscr{H}_\phi$. Let $\alpha_t = \exp(t\bar{\delta})$; then $\bar{\delta}(x) = \mathrm{i}\overline{[K_\mu, x]}$ for $x \in \mathscr{L}$ in $\mathscr{F}_+(L^2(R^\nu))$. $(1-\bar{\delta})^{-1} a(f_1)a(f_2)\cdots a(f_n) = \int_0^\infty a(\exp(\mathrm{i}tH_\mu)f_1)$ $a(\exp(\mathrm{i}tH_\mu)f_2)\cdots a(\exp(\mathrm{i}tH_\mu)f_n)\exp(-t)\,\mathrm{d}t$.

Since

$$\varphi\Bigg(\left(\int_0^\infty a(\exp(\mathrm{i}tH_\mu)f_1)a(\exp(\mathrm{i}tH_\mu)f_2)\cdots a(\exp(\mathrm{i}tH_\mu)f_n)\exp(-t)\,\mathrm{d}t\right)^*$$
$$\times\left(\int_0^\infty a(\exp(\mathrm{i}tH_\mu)f_1)a(\exp(\mathrm{i}tH_\mu)f_2)\cdots a(\exp(\mathrm{i}tH_\mu)f_n)\exp(-t)\,\mathrm{d}t\right)\Bigg)$$

$$\leqslant \int_0^\infty \int_0^\infty C^2 \|f_1\|^2 \|f_2\|^2 \cdots \|f_n\|^2 \exp(-(s+t))\,\mathrm{d}s\,\mathrm{d}t \qquad < +\infty,$$

where C is a suitable constant,

$$\bar{\delta}(1-\bar{\delta})^{-1} a(f_1)a(f_2)\cdots a(f_n) = (1-\bar{\delta})^{-1}\bar{\delta}(a(f_1)a(f_2)\cdots a(f_n))$$

$$= \int_0^\infty \delta(a)(\exp(itH_\mu)f_1)a(\exp(itH_\mu)f_2)\cdots a(\exp(itH_\mu)f_n))\exp(-t)\,\mathrm{d}t.$$

Hence $\{(1-\bar{\delta})^{-1}a(f_1)a(f_2)\cdots a(f_n)\}_\phi \in \mathscr{D}(\bar{H}_\phi)$ and so $(1-\mathrm{i}\bar{H}_\phi)\mathscr{D}(\bar{H}_\phi)$ is dense in $\mathscr{H}_\phi$. Analogously $(1+\mathrm{i}\bar{H}_\phi)\mathscr{D}(\bar{H}_\phi)$ is dense in $\mathscr{H}_\phi$ so that $\bar{H}_\phi$ is selfadjoint. Therefore $\exp(\mathrm{i}t\bar{H}_\phi)W(f)\exp(-\mathrm{i}t\bar{H}_\phi) = \exp(\mathrm{i}tK_\mu)W(f)\exp(-\mathrm{i}tK_\mu)$ on $\mathscr{H}_\phi$ for $f \in L^2(R^\nu)$. From this one can easily conclude that the linear span of $\{((1\pm\delta)W(f))_\phi \mid f \in C_c{}^\infty(R^\nu)\}$ is dense in $\mathscr{H}_\phi$ and so $\{(1\pm\delta)\bigcup_{n=1}^\infty \mathscr{D}_n)_\phi\}$ is dense in $\mathscr{H}_\phi$, so that Condition (7) is satisfied.

4.8.10 Interacting models To discuss more general interacting models we shall reformulate our axioms for continuous quantum systems by using unbounded operator *-algebras. Let $L^2(\mathbb{R}^\nu)$ be the Hilbert space of all square integrable functions on the ν-dimensional euclidean space R^ν, and let $\mathscr{F}_\pm(L^2(\mathbb{R}^\nu))$ be the Bose–Fock space and Fermi–Fock space over $L^2(\mathbb{R}^\nu)$. Let $\Lambda_n = \{x \in \mathbb{R}^\nu \mid \|x\| < n\}$, where $\|\cdot\|$ is the norm of the euclidean space $\mathbb{R}^\nu$, and let $L^2(\Lambda_n)_\pm{}^m$ denote the subspace of $L^2(\Lambda_n{}^m)$ formed by the totally symmetric (plus sign) and totally anti-symmetric (minus sign) functions of m-variables $x \in \Lambda_n$. Consider the associated Fock space $\mathscr{F}_\pm(L^2(\Lambda_n)) = \sum_{m \geqslant 0} L^2(\Lambda_n)_\pm{}^m$ with $L^2(\Lambda_n)$, and let $\mathscr{B}_\pm(\Lambda_n)$ denote a *-algebra generated by $\{a_\pm(f), a_\pm(f^*) \mid f \in L^2(\Lambda_n)\}$ respectively. Then $\mathscr{B}_+(\Lambda_n)$ is a *-algebra of unbounded operators in $\mathscr{F}_+(L^2(\Lambda_n))$ and $\mathscr{B}_-(\Lambda_n)$ is a *-algebra of bounded operators in $\mathscr{F}_-(L^2(\Lambda_n))$, and $\mathscr{B}_\pm(\Lambda_n) \subset \mathscr{B}_\pm(\Lambda_{n+1})$. Let $\mathscr{B}_\pm = \bigcup_{n=1}^\infty \mathscr{B}_\pm(\Lambda_n)$. Let $\mathscr{D}_\pm(\Lambda_n)$ be a *-subalgebra of $\mathscr{B}_\pm(\Lambda_n)$ generated by $\{a_\pm(f), a_\pm{}^*(f) \mid f \in C_c{}^\infty(\Lambda_n)\}$ and let $\mathscr{D}_\pm = \bigcup_{n=1}^\infty \mathscr{D}_\pm(\Lambda_n)$. Let $\Omega = (1, 0, 0, \ldots,)$ be the vector in $\mathscr{F}_\pm(L^2(R^\nu))$ corresponding to the zero-particle state and let $V_\pm = \mathscr{D}_\pm\Omega$.

Now we shall consider the following conditions.

(1′) There is a mapping δ (called a *-derivation) of $\mathscr{D}_\pm$ into densely defined closable operators in the Hilbert space $\mathscr{F}_\pm(L^2(R^\nu))$ such that

(i) $\overline{\delta(a)} = \overline{\delta(a) \mid V_\pm}$ for $a \in \mathscr{D}_\pm$;
(ii) $\delta(\lambda a) \mid V_\pm = \lambda\delta(a) \mid V_\pm$ for $a \in \mathscr{D}_\pm$;
(iii) $\delta(a+b) \mid V_\pm = \delta(a) \mid V_\pm + \delta(b) \mid V_\pm$;
(iv) $\delta(ab) \mid V_\pm = \delta(a)b \mid V_\pm + a\delta(b) \mid V_\pm$ for $a, b \in \mathscr{D}_\pm$;
(v) $\overline{\delta(a^*) \mid V_\pm} = \delta(a)^*$ for $a \in \mathscr{D}_\pm$, where $(\cdot) \mid V_\pm$ implies the restriction of $(\cdot)$ to $V_\pm$ respectively.

(2′) For each n, there is an $\mathscr{F}_{\pm}(L^2(\Lambda_{m(n)}))$ with $n \leqslant m(n)$ and self-adjoint operator h_n in $\mathscr{F}_{\pm}(L^2(\Lambda_{m(n)}))$ such that $\overline{\delta(a)} = \overline{\mathrm{i}[h_n, a]}$ for $a \in \mathscr{D}_{\pm}(\Lambda_n)$.

(3′) For $\beta > 0$, $\bar{a}\exp(-\beta h_n)$ and $\overline{a\delta(b)}\exp(-\beta h_n)$ are of trace class for each n and $a, b \in \mathscr{D}_{\pm}(\Lambda_n)$, where $\overline{(\cdot)}$ is the closure of $(\cdot)$ in $\mathscr{F}_{\pm}(L^2(\mathbb{R}^\nu))$.

Now let ϕ be a locally normal state on the C*-algebra $\mathscr{A}_{\pm}$ (the uniform closure of $\bigcup_{n=1}^{\infty} B(\mathscr{F}_{\pm}(L^2(\Lambda_n)))$; then there is a unique trace class positive operator $T_{m(n)}$ on $\mathscr{F}_{\pm}(L^2(\Lambda_{m(n)}))$ such that $\phi(x) = \mathrm{Tr}(xT_{m(n)})$ for $x \in B(\mathscr{F}_{\pm}(L^2(\Lambda_{m(n)})))$.

Since $\bar{a}$ and $\overline{a\delta(a)}$ are closed operators in $\mathscr{F}_{\pm}(L^2(\Lambda_{m(n)}))$, we can consider the closed operators $\bar{a}T_{m(n)}$ and $\overline{a\delta(b)}T_{m(n)}$. Assume that they are of trace class; then we can define $\phi(a)$ and $\phi(a\delta(a))$ by $\mathrm{Tr}(\bar{a}T_{m(n)})$ and $\mathrm{Tr}(\overline{a\delta(a)}T_{m(n)})$.

(4′) There is a locally normal state ϕ on $\mathscr{A}_{\pm}$ such that $\lim_{n\to\infty}\phi_n(a) = \phi(a)$ and $\lim_{n\to\infty}\phi_n(a^*\delta(b)) = \phi(a^*\delta(b))$ for $a, b \in \mathscr{D}_{\pm}$, where $\phi_n(a) = \mathrm{Tr}(\bar{a}\exp(-\beta h_n))/\mathrm{Tr}(\exp(-\beta h_n))$ and $\phi_n(a^*\delta(a)) = \mathrm{Tr}(\overline{\alpha^*\delta(b)}\exp(-\beta h_n))/\mathrm{Tr}(\exp(-\beta h_n))$ for $a, b \in \mathscr{D}_{\pm}$.

Now suppose that a *-derivation δ satisfies Conditions (1′)–(4′). By (3′),

$$\bar{\delta}(a)\exp(-\beta h_n) = \mathrm{i}\overline{[h_n, a]}\exp(-\beta h_n) = \mathrm{i}\overline{(h_n a - a h_n)}\exp(-\beta h_n);$$

hence

$$\overline{\delta(a)}\exp(-\beta h_n) \supset \mathrm{i}h_n a\exp(-\beta h_n) - \mathrm{i}ah_n\exp(-\beta h_n).$$

By a discussion similar to that in the proof of Theorem 4.8.2, $\bar{a}h_n\exp(-\beta h_n)$ is of trace class and so

$$\overline{\delta(a)}\exp(-\beta h_n) + \mathrm{i}\bar{a}h_n\exp(-\beta h_n) \supset \mathrm{i}h_n a\exp(-\beta h_n);$$

hence $\overline{\delta(a)}\exp(-\beta h_n) + \mathrm{i}\bar{a}h_n\exp(-\beta h_n) = \mathrm{i}h_n\bar{a}\exp(-\beta h_n)$. Therefore,

$$\begin{aligned}\phi_n(\delta(a)) &= \mathrm{Tr}(\overline{\delta(a)}\exp(-\beta h_n))/\mathrm{Tr}(\exp(-\beta h_n)) \\ &= \mathrm{Tr}(\mathrm{i}h_n\bar{a}\exp(-\beta h_n) - \mathrm{i}\bar{a}h_n\exp(-\beta h_n))/\mathrm{Tr}(\exp(-\beta h_n)).\end{aligned}$$

Let $h_n = \sum_{j=1}^{\infty}\lambda_j e_j$, where $\dim(e_j) = 1$ and $\{e_j\}$ is a mutually orthogonal family of projections with $\sum_{j=1}^{\infty} e_j = 1$. Then,

$$\begin{aligned}\mathrm{Tr}(\mathrm{i}h_n\bar{a}\exp(-\beta h_n)) &= \sum_{j=1}^{\infty}(\mathrm{i}h_n\bar{a}\exp(-\beta h_n)\xi_j, \xi_j) = \sum_{j=1}^{\infty}(\mathrm{i}\bar{a}\exp(-\beta h_n)\xi_j, \lambda_j\xi_j) \\ &= \sum_{j=1}^{\infty}(\mathrm{i}\bar{a}\exp(-\beta h_n)h_n\xi_j, \xi_j),\end{aligned}$$

where $e_j\xi_j = \xi_j$ with $\|\xi_j\| = 1$. Hence $\mathrm{Tr}(\mathrm{i}h_n\bar{a}\exp(-\beta h_n)) = \mathrm{Tr}(\overline{\bar{a}\exp(-\beta h_n)\mathrm{i}h_n})$, also $\mathrm{Tr}(\mathrm{i}\bar{a}h_n\exp(-\beta h_n))=\mathrm{Tr}(\mathrm{i}\overline{\bar{a}\exp(-\beta h_n)h_n})$ and so $\phi_n(\delta(a))=0$ for $a\in\mathscr{D}_\pm$. Therefore we have $\phi(\delta(a))=0$ for $a\in\mathscr{D}_\pm$. For $x\in B(\mathscr{F}_\pm(L^2(\Lambda_{m(n)})))$, define $\psi_n(x)=\mathrm{Tr}(x\exp(-\beta h_n))/\mathrm{Tr}(\exp(-\beta h_n))$; then ψ_n is a KMS state at β for the W*-dynamical system $\{B(\mathscr{F}_\pm(L^2(\Lambda_{m(n)}))),\gamma_n\}$, where $\gamma_{n,t}(x)=\exp(\mathrm{i}th_n)x\exp(-\mathrm{i}th_n)$ for $x\in B(\mathscr{F}_\pm(L^2(\Lambda_{m(n)})))$. Therefore,

$$-\mathrm{i}\beta\psi_n(x^*\tilde{\delta}_n(x))\geqslant\psi_n(x^*x)\log\frac{\psi_n(x^*x)}{\psi_n(xx^*)}\qquad\text{for } x\in\mathscr{D}_\pm,$$

where $\gamma_{n,t}=\exp(t\tilde{\delta}_n)$.

Consider the GNS representation $\{\pi_{\psi_n},\mathscr{H}_{\psi_n}\}$ of $B(\mathscr{F}_\pm(L^2(\Lambda_{m(n)})))$ on a Hilbert space $\mathscr{H}_{\psi_n}$ constructed via ψ_n. Define $\mathrm{i}H_nx_{\psi_n}=\tilde{\delta}(x)_{\psi_n}$ for $x\in\mathscr{D}_\pm$; then H_n is an essentially self-adjoint operator in $\mathscr{H}_{\psi_n}$ for $x\in\mathscr{D}_\pm$. Let $\bar{H}_n$ be the closure of H_n; then

$$\beta(\bar{H}_n\xi,\xi)\geqslant(\xi,\xi)\log(\xi,\xi)/(\exp(-\beta\bar{H})\xi,\xi)\qquad\text{for } \xi\in\mathscr{D}(\bar{H}_n)\cap\mathscr{D}(\exp(-\beta\bar{H}_n)).$$

For x, $y\in B(\mathscr{F}_\pm(L^2(\Lambda_{m(n)})))$,

$$(x1_{\psi_n},y1_{\psi_n})=\frac{\mathrm{Tr}(y^*x\exp(-\beta h_n))}{\mathrm{Tr}(\exp(-\beta h_n))}=\left(x\exp\left(\frac{-\beta h_n}{2}\right),y\exp\left(\frac{-\beta h_n}{2}\right)\right).$$

By Condition (3′), for $a\in\mathscr{D}_\pm(\Lambda_{m(n)})$, $\bar{a}\exp(-\beta h_n/2)$ is of trace class; hence we can define the image $\bar{a}_{\psi_n}$ of $\bar{a}$ in $\mathscr{H}_{\psi_n}$ uniquely. Moreover,

$$(\exp(\mathrm{i}th_n)\bar{a}\exp(\mathrm{i}th_n))_{\psi_n}=\exp(\mathrm{i}t\bar{H}_n)\bar{a}_{\psi_n}$$

and so we have

$$\overline{\delta(a)}_{\psi_n}=\mathrm{i}\overline{[h_n,a]}_{\psi_n}=\mathrm{i}\bar{H}_n\bar{a}_{\psi_n}\qquad\text{for } a\in\mathscr{D}_\pm(\Lambda_{m(n)}).$$

Therefore,

$$-\mathrm{i}\beta\phi_n(a^*\delta(a))\geqslant\phi_n(a^*a)\log\frac{\phi_n(a^*a)}{\phi_n(aa^*)}\qquad\text{for } a\in\mathscr{D}_\pm(\Lambda_{m(n)}).$$

By taking a limit, we have

$$-\mathrm{i}\beta\phi(a^*\delta(a))\geqslant\phi(a^*a)\log\frac{\phi(a^*a)}{\phi(aa^*)}\qquad\text{for } a\in\mathscr{D}_\pm.$$

Let $\{\pi_\phi,\mathscr{H}_\phi\}$ be a GNS representation of $\mathscr{A}_\pm$ on a Hilbert space $\mathscr{H}_\phi$ constructed via ϕ. For x, $y\in B(\mathscr{F}_\pm(L^2(\Lambda_{m(n)})))$,

$$\begin{aligned}\phi(y^*x)&=(x1_\phi,y1_\phi)=\mathrm{Tr}(y^*xT_{m(n)})=\mathrm{Tr}({T_{m(n)}}^{1/2}y^*x{T_{m(n)}}^{1/2})\\&=\langle x{T_{m(n)}}^{1/2},y{T_{m(n)}}^{1/2}\rangle.\end{aligned}$$

Therefore we can define the images $\bar{a}_\phi$ and $\overline{\delta(a)}_\phi$ of $\bar{a}$ and $\overline{\delta(a)}$ $(a\in\mathscr{D}_\pm)$ in $\mathscr{H}_\phi$.

Define $\mathrm{i}H_\phi\bar{a}_\phi=\overline{\delta(a)}\phi$; then one can easily see that H_ϕ is a densely defined symmetric operator in $\mathscr{H}_\phi$.

Now we shall consider the following condition

(5′) $\{[\overline{(1 \pm \delta)a}]_\phi | a \in \mathscr{D}_\pm\}$ is dense in $\mathscr{H}_\phi$. Then the closure of H_ϕ is self-adjoint. Let $\pi_\phi(\alpha_t(x)) = \exp(it\bar{H}_\phi)\pi_\phi(x)\exp(-it\bar{H}_\phi)$ for $x \in \mathscr{A}_\pm$; then the state φ on $\mathscr{A}_\pm$ is a KMS state at β for the dynamics $\{\mathscr{A}_\pm, \alpha\}$.

Therefore we have the following theorem.

4.8.11 Theorem *If a dynamical system* $\{\mathscr{A}_\pm, \delta\}$ *satisfies Conditions* (1′)–(5′), *then it has a time evolution* $t \mapsto \alpha_t$ $(t \in \mathbb{R})$ *and a KMS state at* β $(\beta > 0)$.

Now we shall apply the above system of conditions to interacting Bose models.

Let Φ be an even positive function on $\mathbb{R}^\nu$ with compact support K which is continuous except possibly at the origin. Consider a two-body interaction

$$U_\Phi = \frac{1}{2}\iint \Phi(x-y)a^*(x)a^*(y)a(y)a(x)\,\mathrm{d}x\,\mathrm{d}y.$$

Then,

$$U_{\Phi,\Lambda_n} = \frac{1}{2}\int_{\Lambda_n}\int_{\Lambda_n} \Phi(x-y)a^*(x)a^*(y)a(y)a(x)\,\mathrm{d}x\,\mathrm{d}y.$$

Let $T_{\infty,\Lambda_n}{}^{(1)}$ denote the self-adjoint extension of $-\nabla^2$ corresponding to the Dirichlet boundary condition $\psi = 0$ on $\partial\Lambda_n$. Define $T_{\infty,\Lambda_n} = \mathrm{d}\Gamma(T_{\infty,\Lambda}{}^{(1)})$ (the second quantization of $T_{\infty,\Lambda_n}{}^{(1)}$). Let $H_{\infty,\Lambda_n} = T_{\infty,\Lambda_n} + U_{\Phi,\Lambda_n}$ (the sum is not necessarily the operator sum) (cf. page 356 in [42]); then H_{∞,Λ_n} is self-adjoint. Moreover $\exp\{-\beta(H_{\infty,\Lambda_n} - \mu N_{\Lambda_n})\}$ is of trace class for all $\beta > 0$ and $\mu \in \mathbb{R}$ for Fermi statistics and all $\mu < \infty$ for Bose statistics (cf. Proposition 6.3.5 in [42]). For $f \in C_c{}^\infty(\Lambda_n)$,

$$\begin{aligned}
2[U_\Phi, a(f)] &= \left[\iint \Phi(x-y)a^*(x)a^*(y)a(y)a(x)\,\mathrm{d}x\,\mathrm{d}y, \int \overline{f(z)}a(z)\,\mathrm{d}z\right] \\
&= \iiint \Phi(x-y)\overline{f(z)}[a^*(x)a^*(y)a(y)a(x), a(z)]\,\mathrm{d}x\,\mathrm{d}y\,\mathrm{d}z \\
&= \iiint \Phi(x-y)\overline{f(z)}(-\delta(x-z)a^*(y)a(y)a(y)a(x))\,\mathrm{d}x\,\mathrm{d}y\,\mathrm{d}z \\
&\quad + \iiint \Phi(x-y)\overline{f(z)}(-\delta(y-z)a^*(x)a(y)a(x))\,\mathrm{d}x\,\mathrm{d}y\,\mathrm{d}z \\
&= -\iint \Phi(x-y)\overline{f(x)}a^*(x)a(y)a(x)\,\mathrm{d}x\,\mathrm{d}y \\
&\quad - \iint \Phi(x-y)\overline{f(y)}a^*(x)a(y)a(x)\,\mathrm{d}x\,\mathrm{d}y.
\end{aligned}$$

Hence

$$[U_\Phi, a(f)] = [U_{\Phi,K+\Lambda_n}, a(f)].$$

Quite similarly,

$$[U_\Phi, a^*(f)] = [U_{\Phi,K+\Lambda_n}, a^*(f)].$$

Take a positive integer $m(n)$ such that $K + \Lambda_n \subset \Lambda_{m(n)}$; then for $a \in \mathscr{D}_+(\Lambda_n)$,

$$\mathrm{i}[T + U_\Phi - \mu N, a] = \mathrm{i}[T_{\infty,\Lambda_{m(n)}} + U_{\Phi,\Lambda_{m(n)}} - \mu N_{\Lambda_{m(n)}}, a],$$

where $T = \mathrm{d}\Gamma(-\nabla^2)$ in $\mathscr{F}_+(L^2(R^\nu))$.

Define $\delta(a) = \mathrm{i}[T + U_\Phi - \mu N, a]$ for $a \in \mathscr{D}_+$, and put

$$\omega_{\Lambda_n}(a) = \frac{\mathrm{Tr}(\bar{a}\exp(-\beta(T_{\infty,\Lambda_n} + U_{\Phi,\Lambda_n} - \mu N_{\Lambda_n})))}{\mathrm{Tr}(\exp(-\beta(T_{\infty,\Lambda_n} + U_{\Phi,\Lambda_n} - \mu N_{\Lambda_n})))} \qquad \text{for } a \in \mathscr{D}_+(\Lambda_n).$$

Then,

$$\omega_{\Lambda_n}(a^*(f_1)a^*(f_2)\cdots a^*(f_m)a(g_m)a(g_{m-1})\cdots a(g_1))$$
$$= \int \mathrm{d}x_1\,\mathrm{d}x_2\cdots\mathrm{d}x_m\,\mathrm{d}y_1\,\mathrm{d}y_2\cdots\mathrm{d}y_m \overline{g_1(y_1)}\,\overline{g_2(y_2)}\cdots\overline{g_m(y_m)}f_1(x_1)f_2(x_2)\cdots f_m(x_m)$$
$$\times \tilde{\rho}_{\Lambda_n}(y_1, y_2, \ldots, y_m; x_1, x_2, \ldots, x_m),$$

where $\tilde{\rho}_{\Lambda_n}(y_1, y_2, \ldots, y_m; x_1, x_2, \ldots, x_m)$ are the Dirichlet reduced density matrices, which are a positive bounded continuous function and a sequence $\{\hat{\rho}_{\Lambda_n}\}$ converges uniformly to a function $\tilde{\rho}$ on every compact subset in $\mathbb{R}^{\nu m} \times \mathbb{R}^{\nu m}$ (cf. Corollary 6.3.20 in [42]). Therefore,

$$\lim_m \omega_{\Lambda_n}(a^*(f_1)a^*(f_2)\cdots a^*(f_m)a(g_m)a(g_{m-1})\cdots a(g_1))$$
$$= \lim_n \int \mathrm{d}x_1\,\mathrm{d}x_2\cdots\mathrm{d}x_m\,\mathrm{d}y_2\cdots\mathrm{d}y_m \overline{g_1(y_1)}\,\overline{g_2(y_2)}\cdots\overline{g_m(y_m)}f_1(x_1)f_2(x_2)\cdots f_m(x_m)$$
$$\times \tilde{\rho}_{\Lambda_n}(y_1, y_2, \ldots, y_m; x_1, x_2, \ldots, x_m)$$
$$= \int \mathrm{d}x_1\,\mathrm{d}x_2\cdots\mathrm{d}x_m\,\mathrm{d}y_1\,\mathrm{d}y_2\cdots\mathrm{d}y_m \overline{g_1(y_1)}\,\overline{g_2(y_2)}\cdots\overline{g_m(y_m)}f_1(x_1)f_2(x_2)\cdots f_m(x_m)$$
$$\times \tilde{\rho}(y_1, y_2, \ldots, y_m; x_1, x_2, \ldots, x_m)$$
$$= \omega(a^*(f_1)a^*(f_2)\cdots a^*(f_m)a(g_m)a(g_{m-1})\cdots a(g_1))$$

for $f_1, f_2, \ldots, f_m, g_1, g_2, \ldots, g_m \in C_c^\infty(\mathbb{R}^\nu)$.

Moreover, it is known that ω will define a locally normal state on $\mathscr{A}_+$ (cf. Theorem 6.3.22 in [42]).

Now we shall examine Condition (4′). For $f \in C_c^\infty(\Lambda_n)$,

$$\delta(a(f)) = \mathrm{i}[T_{\infty,\Lambda_{m(n)}} + U_{\Phi,\Lambda_{m(n)}} - \mu N_{\Lambda_{m(n)}}, a(f)].$$
$$\mathrm{i}[T_{\infty,\Lambda_{m(n)}} - \mu N_{\Lambda_{m(n)}} a(f)] = a(\mathrm{i}(-\nabla^2 - \mu 1)f)$$

and

$$\mathrm{i}[T_{\infty,\Lambda_{m(n)}} - \mu N_{\Lambda_{m(n)}}, a^*(f)] = a^*(\mathrm{i}(-\nabla^2 - \mu 1)f).$$

On the other hand,

$$\mathrm{i}[U_{\Phi,\Lambda_{m(n)}}, a(f)] = \frac{\mathrm{i}}{2}\int_{\Lambda_{m(n)}}\int_{\Lambda_{m(n)}} \mathrm{d}x\,\mathrm{d}y\,\Phi(x-y)\left[a^*(x)a^*(y)a(y)a(x), \int \bar{f}(z)a(z)\mathrm{d}z\right]$$

$$= \frac{\mathrm{i}}{2}\int_{\Lambda_{m(n)}}\int_{\Lambda_{m(n)}}\int_{\Lambda_n} \mathrm{d}x\,\mathrm{d}y\,\mathrm{d}z\Phi(x-y)\overline{f(z)}[a^*(x)a^*(y)a(y)a(x), a(z)]$$

$$= \frac{\mathrm{i}}{2}\int_{\Lambda_{m(n)}}\int_{\Lambda_{m(n)}} \mathrm{d}x\,\mathrm{d}y\Phi(x-y)\overline{f(y)}(-a^*(x)a(y)a(x))$$

$$+ \frac{\mathrm{i}}{2}\int_{\Lambda_{m(n)}}\int_{\Lambda_{m(n)}} \mathrm{d}x\,\mathrm{d}y\Phi(x-y)\bar{f}(y)(-a^*(x)a(y)a(x)).$$

Similarly,

$$\mathrm{i}[U_{\Phi,\Lambda_{m(n)}}, a^*(f)] = \frac{\mathrm{i}}{2}\int_{\Lambda_{m(n)}}\int_{\Lambda_{m(n)}} \mathrm{d}x\,\mathrm{d}y\Phi(x-y)f(y)a^*(x)a^*(y)a(x)$$

$$+ \frac{\mathrm{i}}{2}\int_{\Lambda_{m(n)}}\int_{\Lambda_{m(n)}} \mathrm{d}x\,\mathrm{d}y\Phi(x-y)f(x)a^*(x)a^*(y)a(y).$$

On the other hand,

$$\begin{aligned}
&\omega_{\Lambda_{m(n)}}(a^*(f_1)a^*(f_2)\cdots a^*(f_m)\delta(a(g_m)a(g_{m-1})\cdots a(g_1)))\\
&\quad = \omega_{\Lambda_{m(n)}}(a^*(f_1)a^*(f_2)\cdots a^*(f_m)\delta(a(g_m))a(g_{m-1})\cdots a(g_1))\\
&\qquad + \omega_{\Lambda_{m(n)}}(a^*(f_1)a^*(f_2)\cdots a^*(f_m)a(g_m)\delta(a(g_{m-1}))a(g_{m-2})\cdots a(g_1))\\
&\qquad + \cdots + \omega_{\Lambda_{m(n)}}(a^*(f_1)a^*(f_2)\cdots a^*(f_m)a(g_m)a(g_{m-1})\cdots \delta(a(g_1)))
\end{aligned}$$

Moreover,

$$\begin{aligned}
&\omega_{\Lambda_{m(n)}}(a^*(f_1)a^*(f_2)\cdots a^*(f_m)\mathrm{i}[U_\Phi, a(g_m)]a(g_{m-1})\cdots a(g_1))\\
&\quad = \frac{\mathrm{i}}{2}\int \mathrm{d}x_1\,\mathrm{d}x_2\cdots \mathrm{d}x_m \mathrm{d}x\,\mathrm{d}y\,\mathrm{d}y_{m-1}\,\mathrm{d}y_{m-2}\cdots \mathrm{d}y_1 f_1(x_1)f_2(x_2)\cdots f_m(x_m)\\
&\qquad \times(-\Phi(x-y)\overline{g_{m(x)}}\overline{g_{m-1}(y_{m-1})}\cdots\overline{g_1(y_1)}\tilde{\rho}_{\Lambda_{m(n)}}\\
&\qquad \times(x_1, x_2, \ldots, x_m, y; y, x, y_{m-1}, \ldots, y_1) + \frac{\mathrm{i}}{2}\int \mathrm{d}x_1\,\mathrm{d}x_2\cdots \mathrm{d}x_m\,\mathrm{d}x\,\mathrm{d}y f_1(x_1)\\
&\qquad \times f_2(x_2)\cdots f_m(x_m)(-\Phi(x-y)\overline{g_m(y)})\overline{g_{m-1}(y_{m-1})}\cdots\overline{g_1(y_1)}\tilde{\rho}_{\Lambda_{m(n)}}\\
&\qquad \times(x_1, x_2, \ldots, x_m, x; x, y, y_{m-1}, \ldots, y_1)\\
&\quad \to \omega(a^*(f_1)a^*(f_2)\cdots a^*(f_m)\,\mathrm{i}[U_\Phi, a(g_m)]a(g_{m-1})\cdots a(g_1))
\end{aligned}$$

for $f_1, f_2, \ldots, f_m, g_1, g_2, \ldots, g_m \in C_c^{\infty}(\mathbb{R}^\nu)$.

Quite similarly, we can show that

$$\omega_{\Lambda_n}(a^*\delta(a)) \to \omega(a^*\delta(a)) \qquad (n \to \infty) \text{ for } a \in \bigcup_{n=1}^{\infty} \mathscr{D}_+(\Lambda_n) = \mathscr{D}_+$$

Therefore Condition (4′) is satisfied.

It is difficult to prove Condition (5′) for interacting models and this is one of the most important problems in the continuous quantum system. We shall leave this important problem for future research. In any case, it is one of the most important problems to construct time evolution and KMS states for interacting models.

References [41] [42] [177] [178].

REFERENCES

1. C.A. Akemann and G.K. Pedersen (1979), Central sequences and inner derivations, *Amer. J. Math.*, **101**, 1047–61.
2. H. Araki (1964), On the algebra of all local observables, *Prog. Theor. Phys.*, **32**, 844–54.
3. H. Araki (1973), *Relative Hamiltonian for Faithful Normal States*, Vol. 9 Research Institute for Mathematical Sciences, 165–209.
4. H. Araki (1974), Some properties of modular conjugation operators of von Neumann algebras and a non-commutative Radon–Nikodym theorem with a chain rule, *Pacific J. Math.*, **50**(2), 309–54.
5. H. Araki (1975), On the uniqueness of one-dimensional quantum lattice systems, *Commun. Math. Phys.*, **44**, 1–7.
6. W. Arveson (1974), On groups of automorphisms of operator algebras, *J. Funct. Anal.*, **15**(3), 217–43.
7. C.J.K. Batty (1978), Dissipative mappings and well-behaved derivations, *J. London Math. Soc.*, **18**, 527–33.
8. C.J.K. Batty (1978), Unbounded derivations of commutative C*-algebras, *Commun. Math. Phys.*, **61**, 261–68.
9. C.J.K. Batty (1981), Derivations on compact spaces, *Proc. London Math. Soc.*, **42**, 299–330.
10. C.J.K. Batty (1983), Connected spaces without derivations, *Bull. London Math. Soc.*, **15**, 349–52.
11. C.J.K. Batty (1985), Local operators and derivations on C*-algebras, *Trans. Amer. Math. Soc.*, **287**, 343–52.
12. C.J.K. Batty (1987), Derivations on the line and flows along orbits, *Pacific J. Math.*, **136**(2), 209–25.
13. C.J.K. Batty, A.L. Carey, D.E. Evans and D.W. Robinson (1984), *Extending Derivations*, Publ. R.I.M.S., Kyoto Univ., **20**, 119–30.
14. C.J.K. Batty and A. Kishimoto (1985), Derivations and one-parameter subgroups of C*-dynamical systems, *J. London Math. Soc.*, **31**, 526–36.
15. C.J.K. Batty and D.W. Robinson (1985), The characterization of differential operators by locality: abstract derivations, *Ergod. Th. Dyn. Syst.*, **5**, 171–83.
16. H.J. Borchers (1966), Energy and momentum as observables in quantum field theory, *Commun. Math. Phys.*, **2**, 49–54.

17. H.J. Brascomp (1970), Equilibrium states for a classical lattice gas, *Commun. Math. Phys.*, **18**, 82–96.
18. O. Bratteli (1976), Unbounded derivations and invariant trace state, *Commun. Math. Phys.*, **46**, 31–5.
19. O. Bratteli (1986), *Derivations, Dissipations and Groups Actions on* C*-algebras, Springer Lecture Notes in Mathematics, Vol. 1229, Springer, Berlin, Heidelberg, New York.
20. O. Bratteli, T. Digerness and G.A. Elliott (1985), Locality and differential operators on C*-algebras II, in H. Araki, C.C. Moore, S. Strălilă and D. Voiculescu (eds.), Springer Lecture Notes in Mathematics, Vol. 1132, pp. 46–83, Springer, Berlin, Heidelberg, New York.
21. O. Bratteli, T. Digerness and D.W. Robinson (1984), Relative locality of derivations, *J. Funct. Anal.*, **59**, 12–40.
22. O. Bratteli, T. Digerness, F. Goodman and D.W. Robinson (1985), Integration in abelian C*-dynamical systems, *Res. Inst. Math. Sci. Kyoto Univ.*, **2**, 1001–30.
23. O. Bratteli, G.A. Elliott and P. Jørgensen (1984), Decomposition of unbounded derivations into invariant and approximately inner parts, *J. Reine Agnew. Math.*, **346**, 166–93.
24. O. Bratteli, G.A. Elliott and D.W. Robinson (1985), The characterization of differential operators by locality; Dissipations and ellipticity, *Res. Inst. Math. Sci. Kyoto Univ.*, **21**, 1031–49.
25. O. Bratteli, G.A. Elliott and D.W. Robinson (1985), Strong topological transitivity and C*-dynamical systems, *J. Math. Soc. Japan*, **37**, 115–33.
26. O. Bratteli, G.A. Elliott and D.W. Robinson (1986), The characterization of differential operators by locality; classical flows, *Compos. Math.*, **58**, 279–319.
27. O. Bratteli, G.A. Elliott and D. Evans (1986), Locality and differential operators on C*-algebras, *J. Diff. Equations*, **64**, 221–73.
28. O. Bratteli, D. Evans (1986), Derivations tangential to compact group; the non abelian case, *Proc. London Math. Soc.*, **52**, 369–88.
29. O. Bratteli, D. Evans, F. Goodman and P. Jørgensen (1986), A dichotomy for derivations on O_n, *Res. Inst. Math. Sci. Kyoto Univ.*, **22**, 103–17.
30. O. Bratteli, D. Evans and A. Kishimoto, Covariant and extremely non-covariant representations of a C*-algebra, in preparation.
31. O. Bratteli and F. Goodman (1985), Derivations tangential to compact group actions: Spectral conditions in the weak closure, *Can. J. Math.*, **37**, 160–92.
32. O. Bratteli, F. Goodman and P. Jørgensen (1985), Unbounded derivations tangential to compact groups of automorphisms II, *J. Funct. Anal.*, **61**, 241–89.
33. O. Bratteli and P. Jørgensen (1982), Derivations commuting with abelian gauge actions on lattice systems, *Commun. Math. Phys.*, **87**, 353–64.
34. O. Bratteli, P. Jørgensen, A. Kishimoto and D.W. Robinson (1984), A C*-algebraic Schöenberg theorem, *Ann. Inst. Fourier*, **33**, 155–87.
35. O. Bratteli and A. Kishimoto (1980), Generations of semi-group, and two-dimensional quantum lattice systems, *J. Funct. Anal.*, **35**, 344–68.
36. O. Bratteli and A. Kishimoto (1985), Automatic continuity of derivations on eigen-space, Proceedings of the Conference on Operator Algebras and Mathematical Physics, Iowa, *Contemp. Math.*, **60**, 403–12.
37. O. Bratteli and A. Kishimoto (1986), Derivations and free group actions on C*-algebras, *J. Operator Theory*, **15**, 377–410.
38. O. Bratteli, A. Kishimoto and D.W. Robinson (1987), Imbedding product type actions into C*-dynamical systems, *J. Funct. Anal.*, **75**, 188–210.

39. O. Bratteli and D.W. Robinson (1975), Unbounded derivations of C*-algebras, *Commun. Math. Phys.*, **42**, 253–68.
40. O. Bratteli and D.W. Robinson (1976), Unbounded derivations, II, *Commun. Math. Phys.*, **46**, 11–50.
41. O. Bratteli and D.W. Robinson (1979), *Operator Algebras and Statistical Mechanics*, Vol. I, Springer, Berlin, Heidelberg, New York.
42. O. Bratteli and D.W. Robinson (1981), *Operator algebras and Statistical Mechanics*, Vol. II, Springer, Berlin, Heidelberg, New York.
43. G. Brink and M. Winnink (1976), Spectra of Liouville operators, *Commun. Math. Phys.*, **51**, 135–40.
44. G. Brink (1976), On a class of approximately inner automorphisms, *Commun. Math. Phys.*, **46**, 31–5.
45. A. Brown and C. Pearcy (1965), Structure theorem of commutators of operators, *Ann. Math.*, **82**(2), 112–27.
46. O. Buchholtz and J.E. Roberts (1976) Bounded perturbations of dynamics, *Commun. Math. Phys.*, **49**, 161–77.
47. D.P. Chi (1975), Derivations in C*-algebras, Dissertation, University of Pennsylvania.
48. G. Choquet (1956), Existence unicitè des representations integrales au moyen des points extrêmaux dans les cones convexes, *C.R. Acad. Sci., Paris*, **243**, 555–57.
49. I. Colojoara and C. Foìas (1968), *Theory of Generalized Spectral Theory*, Gordon and Breach, New York.
50. A. Connes (1985), Non commutative differential geometry, *Inst. Hautes Etudes Sci.*, **62**, 257–360.
51 J. Daughtry (1975), An invariant subspace theorem, *Proc. Amer. Math. Soc.*, **49**, 267–8.
52. G.F. Dell'Antonio (1966), On some groups of automorphisms of physical observables, *Commun. Math. Phys.*, **2**, 384–97.
53. J. Dixmier (1969), Sur les groupes d'automorphisms normiquement continus des C*-algèbre, *C.R. Acad. Sci. Paris*, **263**, 643–4.
54. J. Dixmier and P. Malliavin (1978), Factorizations de functions et vecteurs indefiniment differentiables, *Bull. Soc. Math. France*, 2nd series, **102**, 305–30.
55. S. Doplicher (1965), An algebraic spectrum condition, *Commun. Math. Phys.*, **1**, 1–5.
56. S. Doplicher, R.V. Kadison, D. Kastler and D.W. Robinson (1957), Asymptotically abelian systems, *Commun. Math. Phys.*, **6**, 101–20.
57. S. Doplicher, D. Kastler and E. Størmer (1969), Invariant states and asymptotically abelianness, *J. Funct. Anal.*, **3**, 21–5.
58. S. Doplicher (1979), A remark of a theorem of Powers and Sakai, *Commun. Math. Phys.*, **45**, 59.
59. E. Effros and F. Hahn (1967), Locally compact transformation groups and C*-algebras, *Memoirs Amer. Math. Soc.*, **76** (1967).
60. G.A. Elliott (1970), Derivations of matroid C*-algebras, *Inventions Math.*, **9**, 253–69.
61. G.A. Elliott (1977), Some C*-algebras with outer derivations, III, *Ann. Math.*, **106**(2), 121–43.
62. B. Fuglede (1950), A commutativity theorem for normal operators, *Proc. Nat. Acad., USA*, **36**, 35–40.

63. F. Goodman (1980), Closed derivations in commutative C*-algebras, *J. Funct. Anal.*, **39**, 308–14.
64. F. Goodman (1981), Translation invariant closed *-derivations, *Pacific J. Math.*, **97**. 403–13.
65. F. Goodman and P. Jørgensen (1981), Unbounded derivations commuting with compact group actions, *Commun. Math. Phys.*, **82**, 399–405.
66. F. Goodman and P. Jørgensen (1983), Lie algebra of unbounded derivations commuting with compact group actions, *J. Funct. Anal.*, **52**, 369–84.
67. F. Goodman, P. Jørgensen and C. Peligrad (1986), Smooth derivations commuting with Lie group actions, *Proc. Camb. Phil. Soc.*, **99**, 307–14.
68. O. Bratteli, G.A. Elliott, F. Goodman and P. Jørgensen (1989), Smooth Lie group actions on non-commutative tori, *Nonlinearity*, **2**, 271–86.
69. F. Goodman and A.J. Wassermann (1984), Unbounded derivations commuting with compact group actions II, *J. Funct. Anal.*, **55**, 389–97.
70. L. Gross (1972), Existence and uniqueness of physical ground states, *J. Funct. Anal.*, **10**, 52–109.
71. R. Haag and D. Kastler (1964), An algebraic approach to quantum field theory, *J. Math. Phys.*, **7**, 848–61.
72. R. Haag, N. Hugenholtz and M. Winnink (1967), On the equilibrium states in quantum statistical mechanics, *Commun. Math. Phys.*, **5**, 215–36.
73. R. Haag, D. Kastler and E.B. Trych-Pohlmeyer (1974), Stability and equilibrium states, *Commun. Math. Phys.*, **38**, 173–93.
74. A. Helemskii and Ya. Sinai (1973), A description of differentiations in algebra of the type of local observables, *Funct. Anal. Appl.*, **6**, 343–4.
75. J.W. Helton and R.E. Howe (1973), *Integral Operators; Commutators Traces, Index and Homology*, Springer Lecture Notes in Mathematics, Vol. 345, Springer, Berlin, Heidelberg, New York.
76. R. Herman (1975), Unbounded derivations, *J. Funct. Anal.*, **20**, 234–9.
77. E. Hille (1959), *Analytic Function Theory*, Vols I and II, Ginn, Boston, NA, reprinted 1962.
78. E. Hille and R. Phillips (1957), *Functional Analysis and Semi-groups, Proceedings of Symposia in Pure Mathematics*, Vol. 31, American Mathematical Society, Rhode Island.
79. A. Ikunishi (1983), Derivations in C*-algebras commuting with compact actions, *Res. Inst. Math. Sci. Kyoto Univ.*, **19**, 99–106.
80. A. Ikunishi (1986), Derivations in covariant representations of C*-algebras, *Res. Inst. Math. Sci. Kyoto Univ.*, **22**, 527–36.
81. A. Ikunishi (1988), The W*-dynamical system associated with a C*-dynamical system and unbounded derivations, *J. Funct. Anal.*, **79**, 1–8.
82. M.C. Irwin (1980), *Smooth Dynamical Systems*, Academic Press, London, New York.
83. B.E. Johnson (1969), Continuity of derivations on commutative algebras, *Amer. J. Math.*, **91**, 1–10.
84. B.E. Johnson and A.M. Sinclair (1968), Continuity of derivations and a problem of Kaplansky, *Amer. J. Math.*, **90**, 1067–73.
85. P.E.T. Jørgensen (1976), Approximately reducing subspaces for unbounded derivations, *J. Funct. Anal.*, **23**, 392–414.
86. P.E.T. Jørgensen (1977), Trace states and KMS states of approximately inner dynamical one-parameter groups of *-automorphisms, *Commun. Math. Phys.*, **53**, 135–42.

87. P.E.T. Jørgensen (1984), A structure theorem for Lie algebras of unbounded derivations in C*-algebras, *Compos. Math.*, **52**, 85–98.
88. P.E.T. Jørgensen (1984), New results on unbounded derivations and ergodic groups of automorphisms, *Exp. Math.*, **2**, 3–24.
89. P.E.T. Jørgensen and G.L. Price (1986), Extending quasi-free derivations on the CAR algebra, *J. Operator Theory*, **16**, 147–55.
90. R.V. Kadison (1966), Derivations on operator algebras, *Ann. Math.*, **83**, 280–93.
91. R.V. Kadison, (ed.) (1982), *Operator Algebras and Applications: Proceedings of symposia on Pure Mathematics*, Vol. 38, Parts 1 and 2, American Mathematical Society, Rhode Island.
92. R.V. Kadison, C.E. Lance and J. Ringrose (1967), Derivations and automorphisms of operator algebras, II, *J. Funct. Anal.*, **1**, 204–21.
93. R.V. Kadison and J.R. Ringrose (1967), Derivations and automorphisms of operator algebras, *Commun. Math. Phys.*, **4**, 37–63.
94. R.V. Kadison and I.M. Singer (1952), *Proc. Nat. Acad. Sci., USA*, **38**, 419–23.
95. R.R. Kallman (1969), Unitary groups and automorphisms of operator algebras, *Amer. J. Math.*, **91**, 785–806.
96. I. Kaplansky (1953), Modules over operator algebras, *Amer. J. Math.*, **75**, 839–53.
97. I. Kaplansky (1958), *Some Aspects of Analysis and Probability*, pp. 1–34, John Wiley & Sons, New York.
98. T. Kato (1966), *Perturbation Theory for Linear Operators*, Springer, Berlin, Heidelberg, New York.
99. T. Kato and O. Taussky (1956), Commutators of A and A*, *J. Washington Acad. Sci.*, **46**, 38–40.
100. S. Kantorovitz (1965), Classification of operators by means of their operators, II, *Trans. Amer. Math. Soc.*, **115**, 192–214.
101. H.W. Kim, C. Pearcy and A.L. Shields (1975), Rank-one commutators and hyper invariant subspaces, *Michigan Math. J.*, **22**, 193–4.
102. H.W. Kim, C. Pearcy and A.L. Shields (1976), Sufficient conditions for rank-one commutators and hyper invariant subspaces, *Michigan Math. J.*, **23**, 235–43.
103. A. Kishimoto (1976), Dissipations and derivations, *Commun. Math. Phys.*, **47**, 25–32.
104. A. Kishimoto (1976), On uniqueness of one-dimensional quantum lattice systems, *Commun. Math. Phys.*, **47**, 167–70.
105. A. Kishimoto, Equilibrium states of semi-quantum lattice systems, unpublished.
106. A. Kishimoto (1984), Derivation with a domain condition, *Yokohama Math. J.*, **32**, 215–23.
107. A. Kishimoto and H. Takai (1978), Some remarks on C*-dynamical systems with a compact group, *Res. Inst. Math. Sci. Kyoto Univ.*, **14**, 383–97.
108. A. Kishimoto and D.W. Robinson (1985), Dissipations, Derivations, dynamical systems, and asymptotic abelianness, *J. Operator Theory*, **13**, 237–53.
109. A. Kishimoto and D.W. Robinson (1985), Derivations, dynamical systems and spectral restrictions, *Math. Scand.*, **56**, 83–95.
110. C.D. Kleinecke (1957), On operator commutators, *Proc. Amer. Math. Soc.*, **8**, 535–6.

111. P. Kruszynski (1976), On existence of KMS states for invariant approximately inner dynamics, *Bull. Acad. Pol. Sci.*, **XXIV**(4), 299–301.
112. H. Kurose (1981), An example of non quasi-well behaved derivations in $C(I)$, *J. Funct. Anal.*, **43**, 193–201.
113. H. Kurose (1982), A closed derivations in $C(I)$, *Memoirs of the faculty of Science, Kyushu University*, Ser. A, **36**, 193–8.
114. H. Kurose (1983), Closed derivations in $C(I)$, *Tohoku Math. J.*, **35**, 341–7.
115. H. Kurose (1983), Unbounded *-derivations commuting with actions of R^n in C*-algebras, *Memoirs of the faculty of Science, Kyushu University*, Ser. A, **37**, 107–12.
116. H. Kurose (1986), Closed derivations on compact spaces, *J. London Math. Soc.*, **34**, 524–33.
117. C. Lance and A. Niknam (1976), Unbounded derivations of group C*-algebras, *Proc. Amer. Math. Soc.*, **6**, 310–14.
118. O. Lanford and D. Ruelle (1967), Integral representations of invariant states on a B*-algebra, *J. Math. Phys.*, **8**, 1460–3.
119. R. de Laubenfels (1984), Well-behaved derivations on $C(I)$, *Pacific J. Math.*, **115**, 73–80.
120. G. Lindblad (1976), On the generators of quantum dynamical semi-groups, *Commun. Math. Phys.*, **48**, 147.
121. V.I. Lomonosov (1973), Invariant subspaces of the family of operators that commutes with a completely continuous operator, *Anal. i prilozèn*, **7**, 55–6.
122. R. Longo (1977), On perturbed derivations of C*-algebras, *Rep. Math. Phys.*, **12**, 119–24.
123. R. Longo (1979), Automatic relative boundedness of derivations in C*-algebras, *J. Funct. Anal.*, **34**, 21–8.
124. R. Longo and C. Peligrad (1984), Non-commutative topological dynamics and compact actions on C*, *J. Funct. Anal.*, **58**, 157–74.
125. G. Lumer and R.S. Phillips (1961), Dissipative operators in a Banach space, *Pacific J. Math.*, **11** 679–98.
126. R. McGovern (1977), Quasi-free derivations on the canonical anticommutation relation algebra, *J. Funct. Anal.*, **26**, 89–101.
127. A. McIntosh (1978), Functions and derivations of C*-algebras, *J. Funct. Anal.*, **30**, 264–75.
128. D. Montgomery and L. Zippin (1965), *Topological Transformation Groups*, John Wiley & Sons, New York.
129. C.C. Moore, *Group Representations in Mathematics and Physics, Springer Lecture Notes in Physics*, Vol. 6, pp. 1–35, Berlin, New York, Springer, Heidelberg.
130. R.D. Mosak (1975), *Banach Algebras*, University of Chicago Press, Chicago.
131. B. Sz-Nagy (1947), On uniformly bounded linear transformations in Hilbert space, *Acta. Sci. Math.*, (Szeged), **11**, 152–7.
132. H. Nakazato (1982), Closed *-derivations on compact groups, *J. Math Soc. Japan*, **34**, 83–93.
133. H. Nakazato (1984), Extension of derivations in the algebra of compact operators, *J. Funct. Anal.*, **57**, 101–10.
134. E. Nelson (1958), Analytic vectors, *Ann. Math.*, **70**, 572–615.
135. A. Niknam (1983), Closable derivations of simple C*-algebras, *Glasgow Math. J.*, **24**, 181–3.

136. K. Nishio (1985), A local kernel property of closed derivations in $C(I \times I)$, *Proc. Amer. Math. Soc.*, **95**, 573–6.
137. D. Olesen and G. Pedersen (1976), Group of automorphisms with spectrum condition and the lifting theorem, *Commun. Math. Phys.*, **51**, 85–95.
138. S. Ota (1978), Certain operator algebras induced by *-derivations in C*-algebras on an indefinite inner product, *J. Funct. Anal.*, **30**, 238–44.
139. G.K. Pedersen (1979), *C*-algebras and their Automorphism Groups*, Academic Press, London, New York.
140. C. Peligrad (1981), *Derivations of C*-algebras which are invariant under an automorphism group*, Vol. 2, OT series, pp. 259–68, Birkhäuser, Basel, Boston.
141. C. Peligrad (1982), *Derivations of C*-algebra which are invariant under an automorphism group, II*, *Advanced Applications*, Vol. 6, Birkhäuser, Basel, Boston.
142. R.T. Powers (1967), Representations of the canonical anti-commutations relation, Thesis, Princeton University.
143. R.T. Powers (1975), A remark on the domain of an unbounded derivation of a C*-algebra, *J. Funct. Anal.*, **18**, 85–95.
144. R.T. Powers and G.L. Price (1982), Derivations vanishing on $S(\infty)$, *Commun. Math. Phys.*, **84**, 439–47.
145. R.T. Powers and S. Sakai (1975), Existence of ground states and KMS states for approximately inner dynamics, *Commun. Math. Phys.*, **39**, 273–88.
146. R.T. Powers and S. Sakai (1975), Unbounded derivations in operator algebras, *J. Funct. Anal.*, **19**, 81–95.
147. G.L. Price (1983), Extensions of quasi-free derivations on the CAR algebra, *Res. Inst. Math. Sci. Kyoto Univ.*, **19**, 345–54.
148. G.L. Price (1984), On some non-extendable derivations of the gauge invariant CAR algebras, *Trans. Amer. Math. Soc.*, **285**, 185–201.
149. C.R. Putnam (1950), On normal operators in Hilbert space, *Amer. J. Math.*, **73**, 357–62.
150. C.R. Putnam (1967), *Commutator Properties of Hilbert Space Operators and Related Topics*, Ergebnisse series, Vol. 36, Springer, Berlin, Heidelberg, New York.
151. C.E. Rickart (1960), *General Theory of Banach Algebras*, van Nostrand, New York.
152. J.R. Ringrose (1972), Automatic continuity of derivations of operator algebras, *J. London Math. Soc.*, **5**, 432–8.
153. D.W. Robinson (1977), The approximation of flow, *J. Funct. Anal.*, **24**, 280–90.
154. D.W. Robinson (1987), Smooth derivations on abelian C*-dynamical systems, *J. Australian Math. Soc.*, **A42**, 247–64.
155. D.W. Robinson (1986), Smooth core of Lipshitz flows, *Res. Inst. Math. Sci.*, **22**, 659–69.
156. D.W. Robinson, E. Størmer and M. Takesaki (1985), Derivations of simple C*-algebras tangential to compact automorphism groups, *J. Operator Theory*, **13**, 189–200.
157. N. Rosenblum (1958), On a theorem of Fuglede and Putnam, *J. London Math Soc.*, **33**, 376–77.
158. D. Ruelle (1966), State of physical systems, *Commun. Math. Phys.*, **3**, 116.

159. D. Ruelle (1969), *Statistical Mechanics, Rigorous Results*, W.A. Benjamin, New York.
160. D. Ruelle (1970), Integral representations of states on a C*-algebra, *J. Funct. Anal.*, **6**, 116–51.
161. S. Sakai (1960), On a conjecture of Kaplansky, *Tôhoku Math. J.*, **12**, 31–3.
162. S. Sakai (1966), Derivations of W*-algebra, *Ann. Math.*, **83**, 273–8.
163. S. Sakai (1968), Derivations of simple C*-algebras, *J. Funct. Anal.*, **2**, 202–6.
164. S. Sakai (1971), Derivations of simple C*-algebras, II, *Bull. Soc. Math., France*, **99**, 259–63.
165. S. Sakai (1971), *C*-algebras and W*-algebras*, Ergebnisse series, Vol. 60, Springer, Berlin, Heiderberg, New York.
166. S. Sakai (1976), On one-parameter subgroups of *-automorphisms on operator algebras and the corresponding unbounded derivations, *Amer. J. Math.*, **98**, 427–40.
167. S. Sakai (1975), On commutative normal *-derivations, *Commun. Math. Phys.*, **43**, 39–40.
168. S. Sakai (1976), On commutative normal *-derivations, II, *J. Funct, Anal.*, **21**, 203–8.
169. S. Sakai (1976), On commutative normal *-derivations, III, *Tôkohu Math. J.*, **28**, 583–90.
170. S. Sakai (1977), The theory of unbounded derivations in C*-algebras, Lecture notes, Univ. of Copenhagen and Newcastle upon Tyne, unpublished.
171. S. Sakai (1977), Developments in the theory of derivations in C*-algebras, Proceedings of the International Conference on Operator Algebras, Ideals and Their Applications in Theoretical Physics, Leibzig, unpublished.
172. S. Sakai (1980), Recent developments in the theory of unbounded derivations in C*-algebras, *Proceedings of the International Congress of Mathematicians, Helsinki*, 1978, Academia Scientianim Fennica.
173. S. Sakai (1963), Derivations in operator algebras, *Studies in Applied Math. and Advances in Mathematics*, **8** (Suppl.) (1983), 155–63.
174. S. Sakai (1985), Bounded perturbations and KMS states in C*-algebras, *Proc. Inst. Nat. Sci., Coll. Humanities and Sciences, Nihon University*, **20**, 15–30.
175. S. Sakai (1982), Developments in the theory of unbounded derivations in C*-algebras, *Proceedings of Symposia in Pure Mathematics*, Vol. 38, part 2, pp. 309–31, American Mathematical Society, Rhode Island.
176. S. Sakai (1984), Unbounded derivations in C*-algebras and statistical mechanics, *Contemporary Math.*, **32**, 223–34.
177. S. Sakai (1985), Perturbations and KMS states in C*-dynamical systems, *Contemporary Math.*, **62**, 187–217.
178. S. Sakai (1985), Dynamics in continuous quantum systems, *Contemp. Math.*, **62**, 413–26.
179. K.S. Sibirski (1975), *Introduction to Topological Dynamics*, Noordhoff, London.
180. G.E. Šilov (1947), On a property of rings of functions, *Dokl. Akad. Nauk. SSSR*, **58**, 985–8.
181. I. Singer and J. Wermer (1955), *Math. Ann.*, **129**, 260–4.
182. F.V. Sirokov (1956), Proof of a conjecture of Kaplansky, *Usephi. Math. Nauk.*, **11**, 167–8.

183. H. Takai (1981), On a problem of Sakai in unbounded derivations, *J. Funct. Anal.*, **43**, 202–8.
184. M. Takesaki (1970), *Tomita's theory of Modular Hilbert algebras and its applications, Springer Lecture Notes in Mathematics*, Vol. 128, Springer, Berlin, Heidelberg, New York.
185. K. Thomsen (1985), A note to the previous paper by Bratteli, Goodman, Jørgensen, *J. Funct. Anal.*, **56**, 290–4.
186. J. Tomiyama (1983), The theory of closed derivations in the algebra of continuous functions on the unit interval, Lecture notes, Tsing Hua University, unpublished.
187. J. Tomiyama (1986), On the closed derivations on the unit interval, *J. Ramanujan Math.*, **1**, 71–80.
188. K. Yosida (1968), *Functional Analysis*, 2nd edn, Springer, Berlin, Heidelberg, New York.
189. A.J. Wasserman (1981), Automorphic actions of compact groups on operator algebras, Thesis, University of Pennsylvania.
190. H. Wielandt (1949), Über der unbeschränktheit der operatoren der Quantum mechanik, *Math. Ann.*, **121**, 21.
191. A. Wintner (1947), The unboundedness of quantum mechanical matrices, *Phys. Rev.*, **71** (2), 737–9.
192. H. Araki and G.L. Sewell (1977), KMS conditions and local thermodynamics stability of quantum lattice systems, *Commun. Math. Phys.*, **52**, 103–9.
193. J. Dixmier (1957), *Sur les anneaux d'operateurs dans l'espace Hilbertien*, Gauthier-Villars, Paris, 2nd edn.
194. J. Dixmier (1964), *Les C*-algèbres et leurs représentations*, Gauthier-Villars, Paris.
195. R. Kadison and J. Ringrose (1983), *Fundamentals of the Theory of Operator Algebras*, Vol. I, Academic Press, New York.
196. R. Kadison and J. Ringrose (1986), *Fundamentals of the Theory of Operator Algebras*, Vol. II, Academic Press, New York.
197. I. Gelfand (1941), Normierte Ringe, *Mat. Sb.*, **9**, 3–24.
198. R. Kubo (1957), *J. Phys. Soc. Japan*, **12**, 570–86.
199. C. Martin and J. Schwinger (1959), Theory of Many-Particle Systems I, *Phys. Rev.*, **115**, 1349–73.
200. G. Choquet (1956), Unicité des representations integrals au moyens des points extremaux dans les convexes reticulès, *C. R. Acad. Sci., Paris*, **243**, 555–7.
201. E. Alfsen (1971), *Compact Convex Sets and Boundary Integrals*, Springer, Berlin, Heidelberg, New York.
202. W. Rudin (1974), *Real and Complex Analysis*, 2nd edn, McGraw Hill, New York, 1974.
203. H. Araki (1978), On KMS states of a C*-dynamical system, in H. Araki and R. Kadison (eds.) *Procedings of the second Japan–US Seminar on C*-algebras and Applications to Physics, Los Angeles*, 1977, Springer, Berlin, Heidelberg, New York, 1978.
204. O. Lanford (1970), The KMS states of a quantum spin system, in L. Michel and D. Ruelle (eds.), *Systèmes à un nombre infinies de degrès de Liberte*, CNRS, Paris.
205. D. Ruelle (1970), Symmetry break down in statistical mechanics, in

D. Kastler (ed.), *Cargése Lectures in Physics*, Vol. 4, pp. 169–94, Gordon and Breach, New York, London, Paris.
206. G. Roepstorff (1977), Correlation inequalities in quantum statistical mechanics and their application in the Kondo condition, *Commun. Math. Phys.*, **53**, 143–50.
207. J. Glimm (1960), On a certain class of operator algebras, *Trans. Amer. Math. Soc.*, **95**, 318–40.
208. J. Ringrose (1972), *J. London Math. Soc.*, **5**, 432–8.
209. R. Kadison and J. Ringrose (1971), *Acta Math.*, 227–43.
210. B. Johnson, Memoir of A.M.S.
211. B. Johnson (1967), The uniqueness of the complete norm topology, *Bull. Amer. Math. Soc.*, 537–9.
212. B. Johnson and J. Ringrose (1969), Derivations of operator algebras and discrete group algebras, *Bull. London Math. Soc.*, **1**, 70–4.
213. P. Halmos (1954), Commutators of operators II, *Amer. J. Math.*, **76**, 191–8.
214. N. Jacobson (1935), Rational methods in the theory of Lie algebras, *Ann. Math.*, **36**, 875–81.
215. A. Jaffe and J. Glimm (1969), Singular perturbation of selfadjoint operators, *Pure Appl. Math.*, **XXII** 401–14.

Index